Understanding Landscapes

Understanding Landscapes

Edited by **Alex Vedder**

SYRAWOOD
PUBLISHING HOUSE
New York

Published by Syrawood Publishing House,
750 Third Avenue, 9th Floor,
New York, NY 10017, USA
www.syrawoodpublishinghouse.com

Understanding Landscapes
Edited by Alex Vedder

International Standard Book Number: 978-1-68286-317-6 (Hardback)

Printed in China.

Contents

Preface

Landscapes are an important part of our ecosystem. They vary from tropical rain forests to deserts, from glaciers to wetlands, etc. This book emphasizes upon the scientific and educational aspects in understanding landscapes around the globe. It focuses upon the consequences of human activities and their impacts on landscapes. It explains various models, techniques and approaches in detail to analyse the human involvement, their perceptions and interactions with various landscapes. The book primarily aims to elucidate and suggest novel measures for improvements of landscapes and their surrounding ecosystems. Students, academicians and environmentalists will find this book full of innovative insights that would bring forth new topics for further discussion and research.

This book is the end result of constructive efforts and intensive research done by experts in this field. The aim of this book is to enlighten the readers with recent information in this area of research. The information provided in this profound book would serve as a valuable reference to students and researchers in this field.

At the end, I would like to thank all the authors for devoting their precious time and providing their valuable contributions to this book. I would also like to express my gratitude to my fellow colleagues who encouraged me throughout the process.

Editor

1

Scenario Methodology for Modelling of Future Landscape Developments as Basis for Assessing Ecosystem Services

Matthias Rosenberg[1], Ralf-Uwe Syrbe[1]*, Juliane Vowinckel[1], Ulrich Walz[1]

[1] Leibniz Institute of Ecological Urban and Regional Development, Weberplatz 1, 01217 Dresden, Germany

Abstract

The ecosystems of our intensively used European landscapes produce a variety of natural goods and services for the benefit of humankind, and secure the basics and quality of life. Because these ecosystems are still undergoing fundamental changes, the interest of the society is to know more about future developments and their ecological impacts. To describe and analyze these changes, scenarios can be developed and an assessment of the ecological changes can be carried out subsequently. In the project „Landscape Saxony 2050"; a methodology for the construction of exploratory scenarios was worked out. The presented methodology provides a possibility to identify the driving forces (socio-cultural, economic and ecological conditions) of the landscape development. It allows to indicate possible future paths which lead to a change of structures and processes in the landscape and can influence the capability to provide ecosystem services. One essential component of the applied technique is that an approach for the assessment of the effects of the landscape changes on ecosystem services is integrated into the developed scenario methodology. Another is, that the methodology is strong designed as participatory, i.e. stakeholders are integrated actively. The method is a seven phase model which provides the option for the integration of the stakeholders' participation at all levels of scenario development. The scenario framework was applied to the district of Görlitz, an area of 2100 sq km located at the eastern border of Germany. The region is affected by strong demographic as well as economic changes. The core issue focused on the examination of landscape change in terms of biodiversity. Together with stakeholders, a trend scenario and two alternative scenarios were developed. The changes of the landscape structure are represented in story lines, maps and tables.
On basis of the driving forces of the issue areas „cultural / social values" and „political control", three scenarios were developed up to the time horizons in 2030 and 2050. They are titled „Trend", „Tradition and Ecology" and „Technology and Energy". These scenarios differ markedly in the degree of the future lignite exploitation, in the use of renewable energy and in the environmental compatibility of the agricultural production.
In total, the investigation shows that the integration of the ecosystem services approach into the scenario technology has brought new aspects. However, the procedure became more complex.
For the development of the scenarios a precise definition of the driving forces turned out to be essential. The experiences of the project further show that only two or at most three key driving forces (KDF) can be distinguished really sensibly or can be looked at in their interactions.
It could be shown that from these results itself concrete measures can be derived which support desirable developments or counteract against undesirable effects. By the integration of stakeholders in different working steps, the scenarios can contribute to the sensitization and better perception of future problems and chances of a region.

Keywords:
actors, landscape change, participation, future research, scenarios, drivers, ecosystem services

*Corresponding author. Email: r.syrbe@ioer.de

1 Introduction

1.1 Motivation

Ecosystems provide a large number of services for the benefit of humankind (Costanza & Daly 1992, De Groot et al. 2002, Liu et al. 2009). Such "ecosystem services" (ESS) secure our foundations of life and ensure its quality, e.g. by providing food, the basis for tourism and culture, or protection against soil erosion (MA 2005). However our landscapes are being transformed ever more rapidly (Antrop 2005, Haase et al. 2007). The reasons for this include increased use of renewable energy sources, increased intensification of agriculture, demographic changes, and the still unhampered expansion of areas used for residential and transport purposes. On the base of several scenarios, it is possible to identify what these developments will be like for certain ecosystem services, and how humankind can intervene in a directive manner (TEEB 2009, Carpenter et al. 2006).

The drafting of scenarios is one of a number of possible approaches to investigating future developments; other methods for future research include e.g. Delphi studies (Dörr 2005), prognoses (Jessel 2000), trend projections (Bork & Müller 2002), role playing (Armstrong 2002), neuronal networks (Pijanowski et al. 2002), the analysis of binding planning documents and policy goal statements, and landscape experiments (Oppermann 2008). The scenario technique could function as a bridge concept for interdisciplinary work in research of the human-environment relationship (Santelmann et al. 2004).

The scenario technique is considered an approach to addressing the question of sustainability (Walz et al. 2007), since the assessment of intergenerational justice requires a plausible view into the future, involving, among other things, an investigation of long-term developments.

Scenarios are defined as "plausible and often simplified descriptions of how the future will develop, based on a coherent and plausible set of assumptions on key driving forces (KDFs) and relationships" (MA 2005). Alcamo (2008) states: "A scenario is a description of what the future will look like on the basis of if-then statements, and is typically based on a representation of the initial situation and the description of key driving forces and changes, which will lead toward a certain future condition." Or, to put it more simply: "Scenarios are hypothetical results of events which are designed to highlight the consequences of certain decisions" (Rotmans et al. 2000).

There are two basic forms of methodological approaches to the scenario formulation: first, scenarios may be developed as a kind of narrative storyline, an option which we have preferred here. Second, there are quantitative approaches under which model-based simulations are used. Another manner of distinction is a breakdown by normative scenarios, in which desired versions of the future are depicted and contrasted to projected scenarios describing causal projections caused by driving forces (Nassauer & Corry 2004); see Figure 1. Geoinformation systems (GIS) are often used for the modelling of landscape changes (Steinitz 2003, Zebisch 2004).

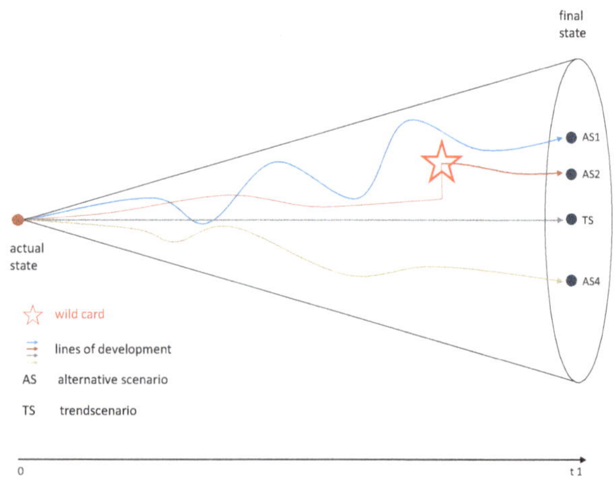

Figure 1: Depiction of a scenario cone
Own Design

However, the distinction is also made between participatory and expert scenarios, with the latter usually structured quantitatively, and the former, qualitatively (Rotmans et al. 2000).

There is a trend toward combined scenario methods (Walz et al. 2007), which both contain qualitative/ participatory elements, and are underpinned by quantitative models. The difficulty is the combination of these two techniques, because the results of participatory processes cannot be directly integrated into quantitative models, and mathematically abstract modeling results are often not understood by actors.

The scenario technique familiar today has been developed in the context of future research and as a method for a prognosis of the effects of nonlinear processes (Kahn & Wiener 1984). The scenario technique achieved great public attention as a result of the study 'The Limits of Growth' by the Club of Rome (Meadows et al. 1972). A scenario technique designed to meet the above stated demands would have to combine methodological elements from environmental research and from corporate planning, but would also have to adapt them to the particular needs of landscape development. As a basis for our methodology, we have in particular used the work of Reibnitz (1991) and of Gausemeier et al. (1996, 2009) in the area of business administration, and also of Alcamo (2008) in the area of environmental science.

The greatest challenge to date facing the further development of scenario methodology has been that of visual preparation and conveyance in a manner appropriate to the recipients of the multifaceted scenario results, which are usually difficult to comprehend. The core paradigm is the basic methodological structure of Integrated Assessment (IA), which has been used in the Visions Project (Rotmans et al. 2000) and elsewhere. Examples of integrated human-environment research projects with the aid of scenarios include the Millennium Ecosystem Assessments (MA 2005), and the Global Environment Outlooks of the UNEP, the fourth generation of which is now available (UNEP 2007); the fifth generation is currently under discussion (UNEP 2011).

At the same time, a growing number of publications are addressing environmental scenarios and their evaluation by means of landscape functions, or through the concept of ecosystem services. These include Fidalgo & Pinto (2005), Nassauer et al. (2002), Dunlop et al. (2002) and Seppelt & Holzkämper (2007). The Fourth Assessment Report (Pachauri & Reisinger 2008) of the Intergovernmental Panel on Climate Change (IPCC) is an investigation which addresses the effects of climate and socio-economic change at the global level, and examines a wide range of ecosystem services. At national level, the project Natural Capital Germany – TEEB DE (2013) deals with changing ecosystem services as well as the development of possible action options.

1.2 Goals of the study

The project upon which this study is based, Landscape Saxony 2050, is oriented towards an investigation area of medium size. In addition to the practical testing with a total of three different scenario runs, the major goal was the development of a consistent and easily applicable methodology, which would particularly take into account the participation of a broad segment of the public and of decision-makers. For this reason, this methodology of scenario development is presented in Chapter 2 in the form of a manual-like set of directions, and the results are only touched on briefly in Chapter 3. In this form, the methodology has the purpose of establishing exploratory scenarios which are to enable the effects of future landscape changes to be assessed on the basis of the Ecosystem Services approach.

The Ecosystem Services (ESS) approach and the combined landscape scenarios should be integrated to the extent that as little additional effort as possible is generated, and the overall methodology remains manageable and comprehensible to the participants. Even in the first work steps of scenario development, the ESS approach should be used in order to filter out the interesting issues, to identify relevant drivers, and to describe the initial situation of the scenarios. During the course of scenario development, the statements on the future are spatialized by GIS in the form of a cartographic foundation for later ESS-referenced evaluation.

The purpose of the step-by-step construction of the scenarios is to make that process an interactive form between the relevant interest groups. These interactions provide the possibility to sharpen the perception for regionally specific future-relevant tasks and issues.

The area under investigation is the district of Görlitz, an area of approx. 2100 sq km characterized by significant population shrinkage and insecure economic perspectives. It is facing major challenges, but also has considerable potentials in the form of its cultural peculiarities and its natural treasures – raw materials, protected areas, large un-fragmented spaces, historic parks, and touristic attractions. These potentials provide possibilities for developments which can be displayed in storylines, maps and tables. On this basis, measures should be discussed and balanced, in order to promote desirable developments and to counter undesirable effects.

2 Methodology

In the scenario technique presented here, the explorative forecast approach is used (Alcamo 2008), a relatively open-ended approach. That means that, in contrast to the formulation of normative scenarios with concrete goal definitions, the range of acceptable developments is not limited. A combined use of quantitative and qualitative procedures is possible.

The methodology developed by the authors consists of seven work phases, which however do not all need to be run through in each case of application or specification. Figure 2 gives an overview of the step-by-step structure of the procedure.

Phases 1 through 6 are run through sequentially, since the results of each phase are needed for the following one. Phase 7 includes first of all the transfer of scenarios, which largely takes place at the end. Secondly, throughout the entire scenario development, the use of various participation methods for the inclusion of a broad segment of the public is provided (Phases 1-7). Based on the assessment and the discussion of the results in Phase 6, a revision or the drafting of further scenarios may become necessary as indicated by the back loop arrow.

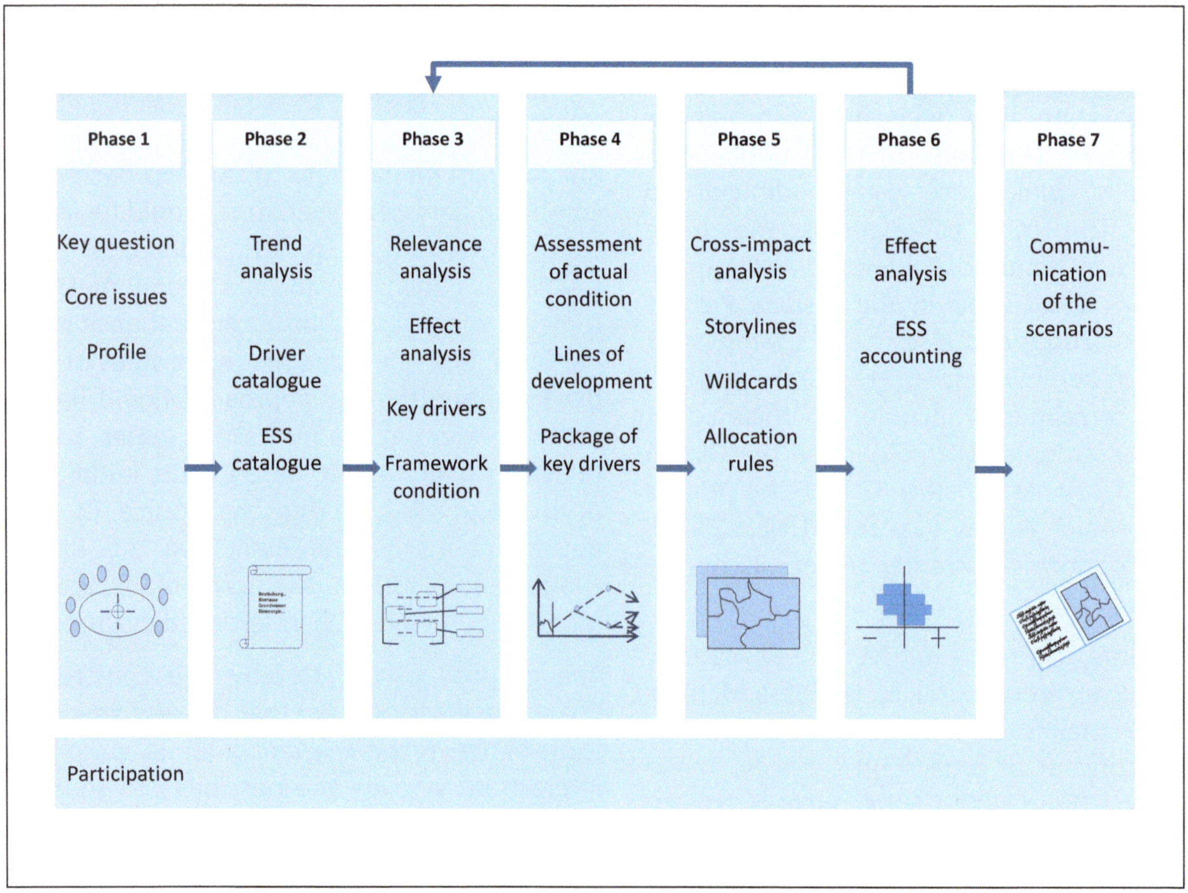

Figure 2: Work phases of scenario development
Own Design

Figure 3: Phase 1 of scenario development
Own Design

2.1 Phase 1

In Phase 1, the scenario process is prepared organizationally, and the object of the investigation is defined through the formulation of the key question by the scenario team, which controls scenario development, and which can consist of experts and/or various actors; if necessary the object of the investigation can be specified more precisely by means of special core issues. The key question determines the overarching goal definition. That includes the essential ancillary conditions, such as the timeframe of the scenarios, and the limitation of the area to be investigated. Once the key question has been defined precisely enough, this phase is completed. However, if the range of issues defined is very extensive, the object of the investigation could be defined more precisely by means of the determination of core issues; see Figure 3. Each core issue thus focuses on one area within the range of problems formulated by the key question. It is useful to record all important information for each core issue in a separate scenario profile, which is established step-by-step in the following work phases. In this phase, the possibilities for participation in the following phases by various actors or by the public are organized and established.

2.2 Phase 2

Phase 2 includes the establishment of those effect factors which are taken into account as the drivers of landscape development in the scenarios, as well as the selection of the ESSs to be processed. It is thus important to determine which factors are significant for establishing the key question and the core issue, and which especially have an effect on the services to be examined. The selection of those services must therefore also be carried out during this stage of work, since ESSs and drivers are mutually conditional. For the ESSs, there are catalogues such as TEEB (2009) or Carpenter et al. (2006), which can facilitate the selection. However, the exact definition and the appropriate selection of effect factors or "drivers" are decisive for the success of the entire project. An exact analysis permits the identification of more than 100 different effect factors, so that the compilation and selection can be very time-consuming. The important thing is that at the end of this work stage, a limited number of drivers – if possible, not more than ten – remain (cf. Syrbe et al. 2013), and that each one of them be clearly defined as a measurable indicator with reference and measurement units, sources and current actual values. For the selection of effect factors, the literature provides a large number of methodological variants, including not only brainstorming, expert discussions and criteria-based ranking, but also effect and preference analyses with the aid of matrices.

The substantive foundation is formed by analyses of the situation or of trends, the results of which are established in the profile. The usual methods for that purpose include spatial analyses of changes to date on the basis of land-use data (CORINE, ATKIS, biotope mapping, long-distance investigative data, historical maps, or statistical data), leading to transfer matrices. Qualitative methods used include reviews of the literature, surveys, and interviews with experts, which may be supplemented for issues relevant for the landscape or for planning by the evaluation of planning documents. The essential result is a specific driver catalogue referenced to the core issue, with initial trend statements.

2.3 Phase 3

In Phase 3, it is necessary to determine which of the drivers ascertained are applicable for all future drafts, and which flow into the scenario in a differentiated manner. The purpose of a scenario development is to investigate various – possibly opposite – developments by assuming different processes for one or more of these drivers. However, such a distinction is only meaningfully feasible for very few drivers. For all other factors, a very definite, uniform development is assumed, characterized as exactly as possible (preferably quantitatively as well), and thus established for all later scenarios. Optimally, prognoses will already be available for these factors, or projections can be calculated by means of trend extrapolation. We call these uniform established quanta "fixed effect factors" or "framework conditions". By establishing framework conditions, the leeway for different possible developments will be greatly reduced, and the development of further scenarios considerably facilitated.

The variable drivers, on the other hand, broaden the scope of possibilities of various future outcomes, and are therefore identified as "key drivers". Determining them is one of the most important milestones in scenario development, and should accordingly be coordinated well with all participants. Since the selection of a large quantity is difficult, the approach should be the other way around: one important key driver should be negotiated, and then supplemented by a second one, and, if absolutely necessary, by a third one. Central to the selection process in terms of substance should be "control instruments", with the aid of which the development can in fact be influenced in order to be able to arrive at substantively significant policy conclusions. The key drivers selected will be described in greater detail. For this purpose, additional research will be necessary, and external experts could be involved.

2.4 Phase 4

In Phase 4, the important issue is to undertake an exact characterization of the initial condition. All selected ESSs are evaluated, for which purpose the EPPS concept (Bastian et al. 2011), which is integrated into the methodology, is particularly well suited. The important thing is to show which service potential a landscape has, and can render sustainably. On this basis, assumptions on future developments of key drivers can be drafted, and their so-called "progression types" defined. The classification shown in Figure 4 is suitable for a schematic description.

For these progression types, decisions must be made as to whether the extrapolation of current trends can be assumed as one of a number of developments for all or for some drivers; unlike for

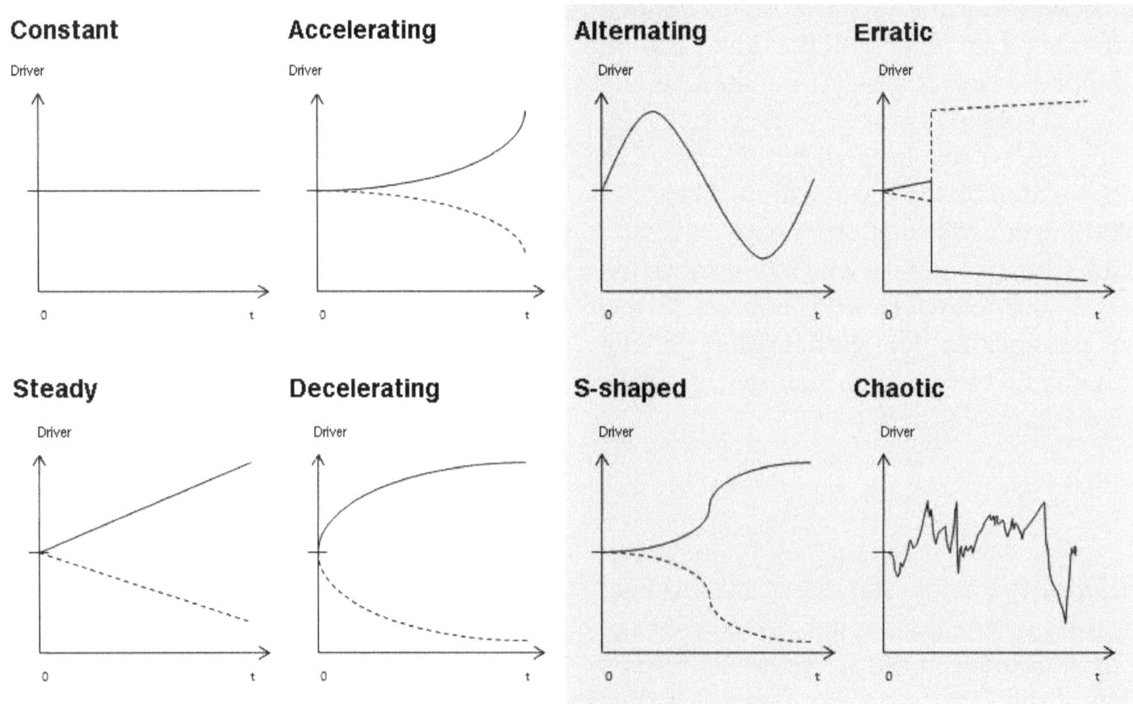

Figure 4: Basic progression types (left) and possible combinations for the key drivers of scenario development
Own Design

the fixed factors, however, non-linear progressions can also be assumed, i.e., if a trend weakens, reverses, or alternates. If the progression types for certain drivers are entered into overview tables as pictograms (Figure 4), it becomes relatively easy to show how changes of various factors are to be considered, and how they can be combined. Not all these combinations will later be analyzed as scenarios, for some of the progression types could be mutually exclusive, or "inconsistent". Only where there are no logical conflicts between certain progression types for all key drivers considered, and are consistent "packages" to be found. The scenario processors will then have to select which items from amongst these packages will in fact be examined. The consistency analysis shows the quality of the work up to that point: for if the drivers are too numerous, too imprecise or poorly selected, one may get an impossibly large number of selections, an unsatisfactory array of them, or none at all. In such cases it will be necessary to change the selection of drivers in Phase 2, possibly replacing some of them, establishing some of them as framework conditions, or eliminating some of them from the selection altogether.

If none of the packages are consistent, that package with the greatest possible degree of consistency is selected, and the inconsistencies are adapted so that non-contradictory packages emerge. The consistency analysis is thus an aid for considering the relationship between certain key drivers.

2.5 Phase 5

Phase 5 represents the core of scenario development. On the basis of driver packages, so-called scenario "trajectories" are defined. The package with the trend extrapolation is, if appropriate, designated as the "trend scenario", or as "business-as-usual" (BAU). Each of the other trajectories receives a simple abbreviation designation, which, while it may not contain all assumed settings, at least identifies one important characteristic of that version of the future; these are called "archetypes".

The decisive step for the implementation of the KDF packages in scenarios is to recognize and to take into account – at least qualitatively – all essential effect quanta in landscape development, which, for reasons of operating practice, is carried out in pairs. This so-called "cross-impact analysis" (Kosow & Gaßner 2008) is often processed in the form of matrices, in which each combination of effect quanta to be taken into account can be noted. Models may depict complex contexts. As a result, a textual description will emerge, the so-called storyline, which should also include justifications. Tables or graphs are suitable for the preparation and as overviews. For later cartographic implementation, and for the assessment, a quantitative evaluation of essential indicators is useful. Care should be taken to determine whether preliminary determinations for the future can be derived from certain developments. These so-called "path dependencies" may result from the exhaustion of certain development, spatial, resource or other potentials which might prevent a later development of the same type from occurring. Often, one type among a number of competing technology lines will end up predominating, and then become the exclusive choice, while others are abandoned.

Rare but consequential single events which considerably confuse a draft scenario for the future can be very interesting. These so-called "wildcards" (e.g. an international bank crash) should be discussed separately from the rest of the scenario development, since they may be important for purposes of precaution, but are often rejected by participants due to their minimal probability. That means that scenario development should first of all be run through to the end without breaks, and wildcards could, if appropriate, be applied afterwards. The wildcard can be "played" at a defined point in time in the scenario run. From this point on, the story line can be rewritten, based on this event, or the model parameter established anew; see Figure 1. The use of wildcards very clearly forces thinking to be oriented towards various alternatives.

A spatialization of the scenarios is indispensable for the ensuing ESS assessment. For this purpose, the potential mapping procedures in Phase 4 are used. This results in the creation of cartographic drafts of future landscape conditions which represent an important test for the quality of the scenarios on the basis of the spatial compilation of particular statements. For this purpose, the Land Use change Modeller (LUMO) has been developed. This GIS-based method permits the localization of certain use changes, or the certification of preferred areas for secondary use and for ESSs, on the basis of stipulations in the scenario storyline. In the LUMO, a distinction is made between the certification of a) development regions, b) area-based changes, c) linear changes, and d) levels of intensity of farming.

In the LUMO, the rules for establishing trend scenarios differ from those for establishing alternative scenarios, due to the different databases and information on landscape development, which is available in such forms as prognoses, or landscape, regional or statewide development plans. This can provide the basis for statements on landscape development based on trend extrapolations, or for the provision of geo-information. Sometimes, however, rules must be set up even at this stage, as is the case for alternative scenarios, if the scenario version exceeds the timeline of the planning process or the prognoses, or if the contents of the scenarios are not covered. The LUMO consists of the following four modules A through D, which, depending on the issue at hand, can be applied singly or in combination:

Module A: Examines the development of residential and transport areas

Module B: Examines the development of agricultural spaces

Module C: Examines the development of spaces of open country with no agricultural use (e.g., nutrient-poor grassland, rocky fields, dwarf-shrub heaths, or ruderal and herbaceous fields)

Module D: Examines the development of regions with regard to their touristically and economically relevant changes.

2.6 Phase 6

In Phase 6, the evaluation of the storylines, tables and maps is carried out on the basis of the ESSs. Depending on how spatially concretely the scenarios have been depicted, they may be evaluated out either spatially differentiated, or non-spatially, on the basis of expert descriptions and statistics. This involves not primarily maps of certain services; rather, this stage is designed to reveal the interactions of services, their so-called trade-offs (Bastian et al. 2012) and synergies. For this purpose, risk and suitability areas are certified. The main goal of this work stage is to draw conclusions regarding the scenarios. It is not the selection of the best storyline that is central here; rather, it is which conditions – i.e., KDFs and their progressions – will lead to the desired developments, and which measures will be useful for achieving that goal. For this purpose, this work step is the one with the highest potential for participatory work. If suitable control instruments are identified during the discussion, it may be necessary to rethink the scenarios on that basis.

Based on a juxtaposition of the scenarios with the actual condition, there follows an accounting of landscape transformation based on selected indicators by the scenario team (s. Table 3). The changes may be statistically evaluated and visualized on the basis of a landscape barometer (Holfeld et al. 2012). Using the fundamental questions – the key question and the core issues – as the point of departure, recommendations for action should be yielded by the scenarios and their assessments. This step is differentiated by target group, as many international studies have impressively shown, e.g. TEEB (2009), or MA (2005). This demands that there be an exact definition of whom the options for action are to be directed towards. On this basis, the preventive and reactive measures can be derived for those addressed. The framework for this step is presented in Bastian et al. (2013).

2.7 Phase 7

Phase 7 includes all measures for the communication and participation of the scenarios with the actors (or the project principal). In spite of its highest number, this work step starts at the very beginning of the development of a scenario, and continues unbroken throughout the entire methodology. However, there can certainly be some switching back and forth between expert dominated work steps and those with greater public participation (so-called loops; cf. Walz et al. 2007), in order to quantify the opinions, translate expert knowledge into generally understandable forms, and give the scenarios a high degree of acceptance by means of broad public participation. In cooperation with experts and stakeholders, a number of participative approaches was tested. The recommendations are summarised in Syrbe et al. (2013).

The possible forms of results' representation are numerous, ranging from verbal through visual to interactive forms of depiction, and depending on target groups and goals. Particularly effective forms would include e.g. pictures, videos, model landscapes, newspaper articles and sketches.

3 Results

This article presents the results for the phases of scenario development all the way to the formulation of the storyline and the cartographic depiction for a selected core issue. We will only enter briefly into a representation and discussion of the results of Phases 6 and 7, since these are the objects of papers on participation and communications methods by Syrbe & et al. (2013), and the observations of the assessment methodologies for ESSs by Holfeld & Rosenberg (2012).

3.1 Phase 1: The key question and core issues

After the establishment of the key question: "How will landscapes in Saxony and their ecosystem services develop through 2050?", the project team carried out a brainstorming process, in which the core issues "biodiversity" and "renewable energies" were identified as particularly relevant for the future development of the landscapes in Görlitz district. In the following, the process of scenario development will be described using the example of the core issue "biodiversity" in the district.

3.2 Phase 2: Drivers and ESSs

Two catalogues have been developed as the basis for the selection of effect quanta and ESSs. The general driver catalogue is broken down into four categories: social, political/legal, economic, and ecological. In this catalogue, the relevant drivers for landscape change are compiled and briefly characterized, especially with regard to indicators, effective mechanisms, control possibilities and relevant actors. The catalogue of ecosystem services contains the possible indicators and assessment methods. Both catalogues must be modified and expanded in accordance with the key question and the respective core issue. There can be no such thing however, as an overall comprehensive catalogue which always characterizes all current drivers and processes from the global to the local levels.

Twelve drivers from the general driver catalogue have been identified as relevant for the core issue biodiversity, and underpinned in a specified catalogue with definitions, indicators and trends (Table 1). For this purpose, the project team has analyzed the development on the basis of statistical key values such as e.g. agriculture and forestry statistics (Statistisches Landesamt des Freistaates Sachsen 2011), farm subsidies by the Common Agricultural Policy (CAP) (Europäische Kommission 2012), Renewable Energy Law subsidies (EEG 2008), and forest increase planning (Upper Lusatian/ Lower Silesian Regional Planning Association 2010). Moreover, a change analysis based on biotope type and land use mapping (BTLNK) processes of 1992-93 and 2005 (SMUL 1992, 2005) has been carried out. The comparison between the two land-use mapping processes shows a growth in area of 73% for "long-term fallows", 60% for "woods and shrubs", 40% for "bodies of water", and 33% for "rocky fields and dwarf shrub heaths". On the other hand, there are major area losses of 78% for "farm maintenance land", 61% for "fallow farmland", 43% for "landfill, waste and excavation areas", and 27% for "transport and infrastructural areas".

3.3 Phase 3: KDFs and framework conditions

The drivers selected in Phase 2 were incorporated into the relevance analysis in which those KDFs were ascertained which are to lead to the differentiation of the scenarios for the core issue (Table 2).

Table 1: Characterization of drivers in the specified driver catalogue (excerpt)

Issue area	Driver	Brief description
Cultural and societal value	Lifestyles	The social dominance of value concepts or behaviour patterns with impact on consumer behaviour, political involvement and land-use, such as consumer associations, organic/slow-food consumption, individual mobility (car-sharing, public transport, bicycles/e-bikes), and regional/local networks
Political control	Support for renewable energies	Renewable Energies Law (EEG), Biofuel Quota Law (BioKraftQuG), Biomass Power and Biofuel Sustainability ordinance
	Support for agriculture and landscape care	Support of agriculture: 1st Pillar – Market order, direct payments; 2nd Pillar: development of rural areas, including agro-environmental measures, support programmes for landscape care, gentle use, restructuring and specific measures for species protection
	Forest expansion programmes	Efforts to increase the forest proportion, supported by promotion of reforestation of previously agriculturally used areas.

Table 2: Presentation of the key drivers and their issue areas

Issue area	Key driver
1. Cultural and societal values	• Lifestyle
2. Political control	• Support for renewable energies (currently via the EEG) • Support for agriculture (CAP, especially ecological priority areas) • Forest expansion programme

The other eight drivers were established as framework conditions for the core issue "biodiversity"; they were not to be differentiated for various scenarios. They were: climate change, demographic change, development of natural site factors (water balance and soil properties), proportion of protected areas, intensity of land use, and demand and exploit ability of raw materials (especially brown coal and copper ore).

3.4 Phase 4: Description of KDFs and their types of progression

For each Key Driving Force (KDF), three development lines were designed and characterized more precisely in a profile. The results flowed into the analysis of the consistence of the type of progression, in order to ascertain a consistent KDF package. From one of these packages, the traditional eco-scenario was drafted and also a contrasting scenario which we designated the "technology-energy scenario" (Figure 5).

3.5 Phase 5: Storyline of the trend scenario

The results of the trend scenario are shown here as an example for the formulation of the storyline and the cartographic localization of use change. It was developed in a workshop together with various professional and volunteer actors from the areas of planning, nature conservation and education in Görlitz district, and has been supplemented by additional research (trend analysis).

Here, excerpts from statements on the first KDF, "lifestyle", are shown. Quantifiable developments which can be localized on a map are described comprehensively on the bases of several essential indicators for the three scenarios in Table 3.

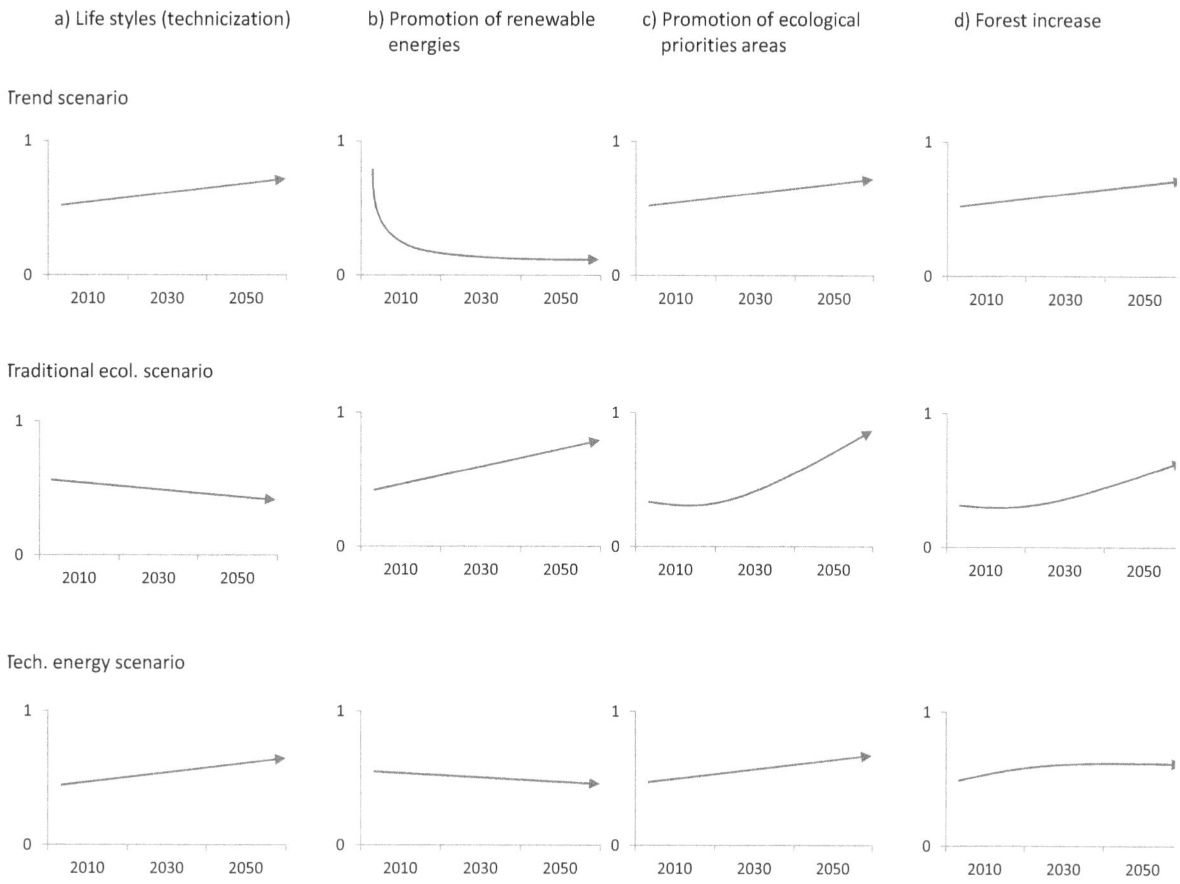

Figure 5: Representation of the progression types and KDF for the trend and alternative scenarios

Storyline cultural and societal values: Lifestyles (traditions, regional identity)

"The demographic and economic development is leading to a further segregation of the population with regard to employment, income and social conditions. Especially well-educated, female young people are continuing to emigrate from the district. During the period through 2050, this will lead, in all age groups, to the loss of competent and committed inhabitants who could identify with the region and thus strengthen regional development. Increasingly, it is older people who live in the district. Nonetheless, dependence on technology and energy is increasing, for the faith in the ability to manage everyday matters by means of technology is becoming ever more dominant, with the change of generations. This is occurring at a time of a global shortage of certain resources, such as copper, and hence is leading to a drastic increase in market prices. Ownership of mobile telephones and computers is becoming normal for all generations. The effect of the media, the conscious filtration of information from the flood of daily events, is leading to the increased manipulation of uncritical groups of the population. The connection to nature and knowledge of various natural processes is declining, since these contexts are being reflected ever less. Ever more tourists are visiting the district, which is increasing in attractiveness as a result of the diversity of its natural spaces and its cultural attractions. Well educated senior citizens are moving to the city of Görlitz, since rents are cheaper there and age-appropriate infrastructure is being developed. These inhabitants are involving themselves in the social life of the city, and strengthening it through their volunteer activities. Nonetheless, local government is ever less financially capable of supporting cultural and social infrastructures.

Young families occasionally move into the district and young people who had previously emigrated are moving back, since real estate is favourably priced here and also because of their ties to their old homeland. In doing so, many consciously risk insecure income situations; however, their interest in life in this region is the dominant factor. In some rural areas, this is leading to the preservation of agriculture for subsidiary income, which provides a contribution to the care and preservation of the cultural landscape.

The growing shortage of natural resources and the increasing environmental problems, such as drought, flooding, soil degradations or epidemics, is leading to stronger policy regimentation as to the type and intensity of certain land uses. The consciousness for organically grown food and the demand for transparency in the food business have grown as a result of a number of food scandals, and have led to a transformation of agricultural production. A strong orientation for the preservation of ecological functions has led to a reduction in nutrient immission into the water, and to an improvement of soil protection.

Continued mining of coal and copper is still accepted by the population of the region, in spite of its environmental effects, since there are no employment alternatives. With the expansion of the opencast mine at Nochten, the lifespan of the Boxberg Power Plant has been secured through 2050. There will be no more opening of further opencast mines, since there is no support in the population for that."

Certification of use change in concrete areas or in certain regions was carried out with the aid of the LUMO regulations, and using the results of the cross-impact analysis. Table 4 shows those use categories for which changes are expected under the scenarios. The result of the modelling is represented for the trend scenario in Figure 6. An animation of all scenarios is presented in Figure 7.

Table 3: Representation of the scenarios and of the indicators derived from the KDFs

KEY DRIVING FORCE	TREND SCENARIO		ALTERNATIVE SCENARIOS			
			Traditional ecol.		Techn. energy	
	2030	2050	2030	2050	2030	2050
Lifestyles (traditions, regional Identity)						
Demand for organic food	↗	↗	↗	↗	→	→
Technologization of everyday life	↗	↗	↘	↘	↑	↑
Agriculture						
Organic agriculture	6%	10%	50%	100%	5%	5%
Ecological priority areas	7%	10%	7%	10%	4%	4%
Energy crops	↗	↗	→	→	↑	↑
Forestry						
Proportion of forested area	35.7%	36.5%	36.5%	38%	35.5%	35.5%
Mining						
Amount of brown coal mined	→	↘	↘	↘	↗	↑
Amount of copper mined	↗	↘	0	0	↑	↑
Renewable energies						
RE share of total power	58%	71%	65%	100%	40%	50%
Economic development						
SME start-ups	↘	↘	→	→	↗	↗

RE - Renewable energies; SME - small and medium enterprise

Table 4: Presentation of the spatial, localizable use changes in the scenarios

Transformation of (row) to (column)*	Arable land	Grass-land	Forest	Eco-priority area	Water	Open-cast mine	Wind-power plants	Fallows
Arable land		+	+	+	-	+	+	+
Grassland	-		+	+	+	+	+	-
Forest	-	-		-	-	+	-	-
Eco-priority area	-	+	+		-	+	-	+
Water	-	-	-	+*		-	-	-
Open-cast mine	+	+	+	+	+		+	+
Wind-power plants	+	-	-	-	-	-		+
Fallows	+	+	+	+	-	+		

*Explanation: E.g., conversions are expected from arable land in grassland in connection with Greening measures to GAP 2014; the reverse dynamism, namely ploughing up of grassland to arable land is prohibited according to EU right and federal state law.

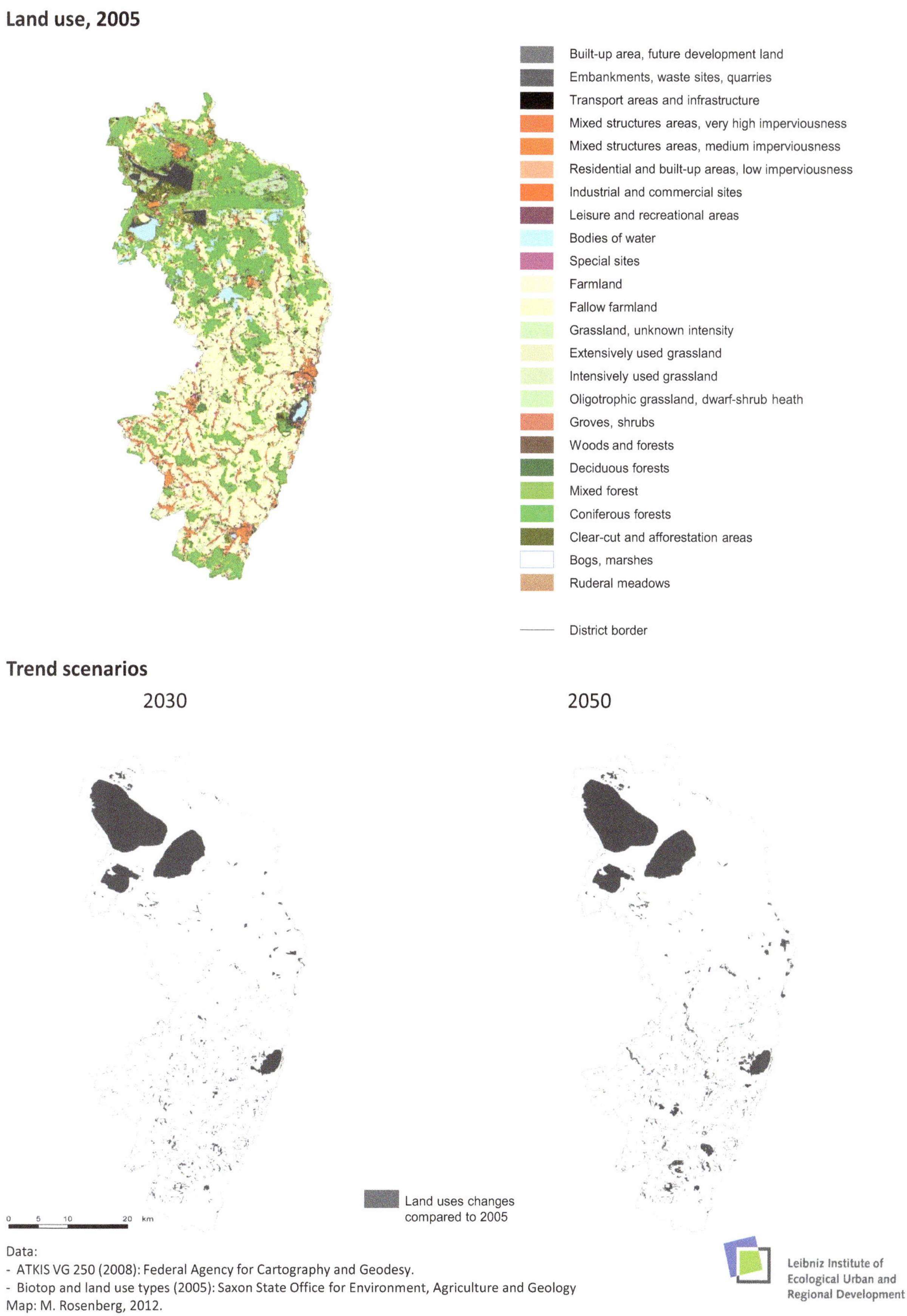

Figure 6: Actual condition of land-use, and trend scenarios, 2030 and 2050

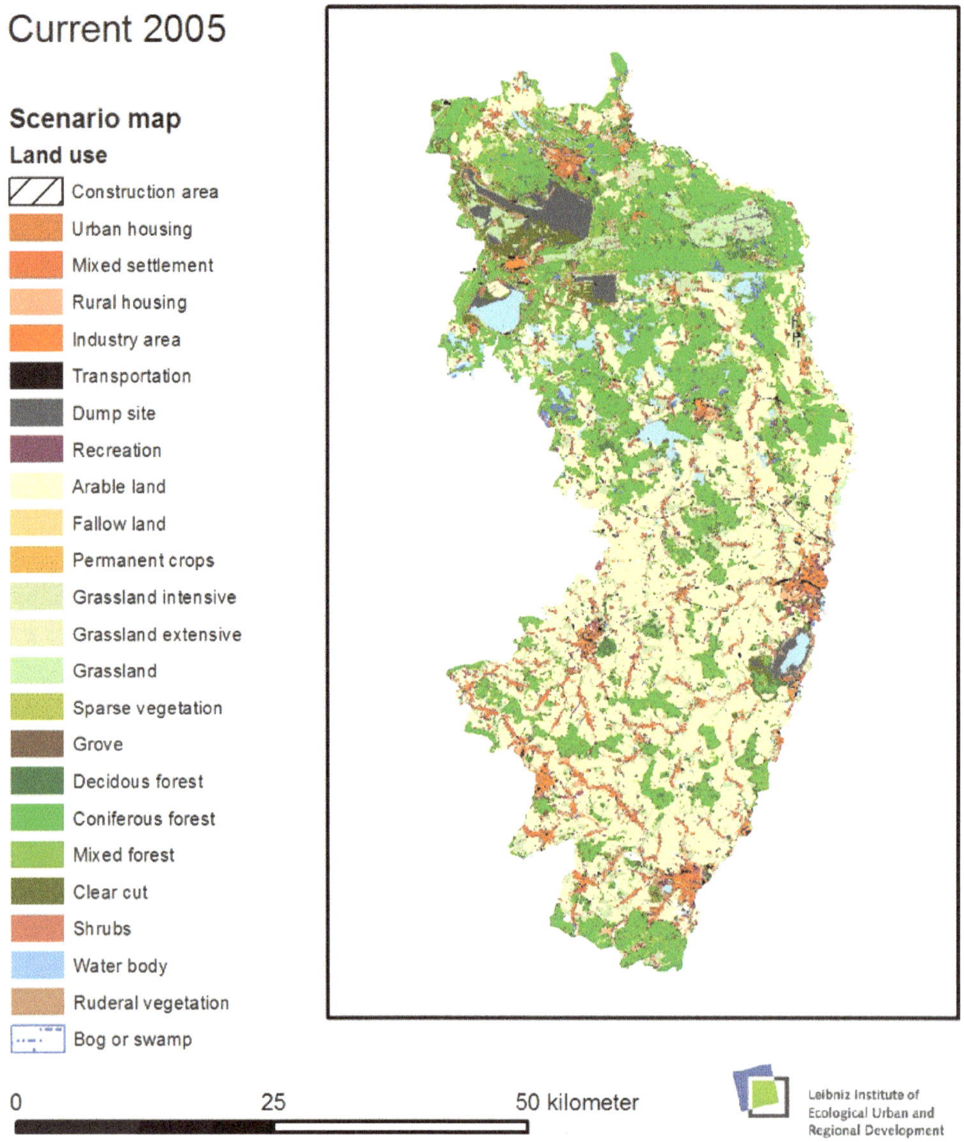

Figure 7: Representation of the actual land use in 2005 and the three scenarios
Basic data: Biotop and land use types (2005): Saxon State Office for Environment, Agriculture and Geology , Map: M. Rosenberg, 2012.

3.6 Phase 6: Evaluation of the scenarios

Based on the qualitative and quantitative changes in Görlitz district as formulated in the scenarios, an assessment of landscape change according to the EES as selected in Phase 2 was carried out. This permitted a comparison of values of the scenarios with the initial condition, and with one another (Holfeld et al. 2012). In the following, only the changes in use are shown, using the example of the trend scenarios. In the mining areas, re-cultivation measures are compensated for by the continuation of opencast mining. An increase in forest areas by 2% through 2050 involves primarily the steep slopes strongly in danger of erosion. As a result of CAP subsidies, the share of ecological priority areas will increase by 7% between 2014 and 2030, and by 10% through 2050 where farming is to be abandoned along the depression lines, on steep slopes and also in selected core areas and corridors of the Saxon biotope networking plan. These areas will then be used as permanent grassland, for the planting of hedges and groves, or will remain fallow.

Renewable energies will provide 58% of power production by 2030, and 71% by 2050 (Prognos AG 2011). This will be made possible primarily by the expansion of wind power and biogas facilities. Overall, the number of wind power plants in Görlitz district is to increase from 123 in 2010 (rated output: 180 MW) to approx. 112 in 2030 (rated power: 260 MW), and then drop to approx. 80 in 2050 (rated output: 350 MW), with the increased output being due to repowering. That will mean that approx. 500 GWh per year of wind power will be produced in 2030, and approx. 720 GWh in 2050. The number of biogas plants will also increase, although the increase will slow down with increased competition for space, and with a slow exhaustion of commercial fertilizer and wastes. In addition, there will also be mostly small cogeneration plants of up to 700 kW_{el}. It is assumed that by 2030, approx. 60 biogas plants will be online. Moreover, an additional 3 to 5 large bio-methane feed-in plants using stock-based substrates with residue utilization are to be built, with an output of 2 to 5 MW.

In addition to the inevitable loss of biodiversity through the mining of coal and through intensified farming, there will also be an overall improvement of ecological functions due to measures for structural enrichment and removal of land from agriculture.

3.7 Phase 7: Communications and participation

Using various participation methods (Syrbe et al. 2013), important actors in the district have been developing the trend scenario in several workshops together with experts from the project team. In selection of the participants, it was seen as important to have a broad thematic background appropriate to the selected core issue "biodiversity". The alternative scenarios and the GIS-based localization of utilization change were developed by the project team. As the scenario exercises showed, the methodology is suitable for the development of the landscape scenarios with an integrated assessment of the ESSs involving a smaller circle of experts and a larger number of stakeholders. In addition, it is open for the consideration of various core issues, drivers of landscape development and ESS assessments.

4 Discussion

When addressing landscape change, it is necessary due to the large numbers of direct and indirect driver, to deal with a multi causal network of them, in order to ascertain those factors which are of the greatest significance for the selected core issue (Höchtl et al. 2006, Klijn 2004). Nassauer et al. (2002) actually began the Internet-based identification and coordination of drivers prior to the beginning of the actual workshops. Such pre-coordination proved to be useful in the present study as well, although we worked with questionnaires which were sent out, rather than with list servers.

In the application example, for the assessment of the effects of landscape scenarios on biodiversity, not only such direct drivers as climate change and use intensity, but also such indirect drivers as demographic change and cultural, economic and socio-political factors were taken into account, which are, according to MA (2005) the responsible impulses with regard to change in biodiversity and ecosystems (cf. also BMU 2007). Santelmann et al. (2004) derive their biodiversity scenario primarily from land use and forestry policy. In Nelson et al. (2009), demographic development was added. The GEO scenarios (UNEP 2007) also incorporated societal value change and scientific-technological innovations. If too many factors are incorporated into the development of scenarios as KDFs, not only is the effort required greatly increased, but the overview and the communicability of the scenario is also hampered. For this reason, the authors of the presented application example have limited themselves to the four most important KDFs for the selected core issue.

Albert (2009) is among those who demand that one proceed participatorily from the outset in scenario development, i.e., that one integrate stakeholders in all scenario stages, and that the groups of actors thus be distinguished. One result is that the number of KDFs must be kept to a minimum. Moreover, most participatory approaches address a lower number of KDFs – usually two or three (Rotmans et al. 2000)

– than do the expert scenarios based on models, where the number is often five or six. However, it is possible to move away from the concept of KDFs by e.g. predetermining the direction of the scenarios through the selection of the actors, as was done by Hulse et al. (2004).

The development of scenarios with the aid of the participation of various stakeholders with varying expert backgrounds and experience with scenario exercises is a time-consuming and high-effort process for the team which is to prepare and accompany the scenario exercise. However, the scenario development process also gains more information through improved regional and local knowledge on the part of the participants, as other investigations, too, have shown (Reed 2008). The participation of the authors in other scenario exercises (Demuth et al. 2010, Priess & Hauck 2011) outside the project have shown how difficult it often is for the participants to develop scenarios freely and independently of the trend, or to carry out an exchange regarding wildcards. This also results from the particular professional backgrounds of the participants, and their willingness to engage in discussions. For the development of scenarios, it is therefore necessary to undertake a training session of participants with regard to the development of scenarios. Since there was no time for several day workshops in the application example, only the trend scenario was developed together with the stakeholders. The development of alternative scenarios was carried out in the project team.

What has hardly been explicitly handled to date is the question of the spatial concretization of scenarios. For this purpose, Verburg et al. (2010) selected a regional clustering of European districts, involving the stakeholders in participatory projects (e.g. Hulse et al. 2004). Fritsch (2002) used a set of rules for the purpose of spatialization. Our own set of rules for deriving future land-use structures (LUMO) was first of all applied for this trend scenario, and later, too, for the alternative scenarios. Here, it was shown that especially the cross-impact analysis was of importance, since a number of areas often showed potential for differing follow-up uses (e.g. forest increase, or ecological priority areas). This in turn had an effect on particular ecosystem services, since the type and structure of land-use decisively affected the development of the ecosystems (Spangenberg 2007) and their species diversity (Michel & Walz 2012). Here it is important, to discuss the priorities of follow-up use in expert teams, and to establish selection criteria. Often, planning is carried out with the reference to particular areas of expertise, so that measures and areas are selected from the perspective of particular use categories, such as the development of residential areas, from the perspective of village or urban development, and without equally taking into account the priorities of other use types, such as the site potentials for agriculture. However, the approach presented here combines the three sectors under consideration of competing use potentials, in order on the one hand to depict all spatially relevant land-use changes, and also the multiplicity of use claims.

For the evaluation of ESSs, it is necessary to find quantifiable indicators, and for that purpose, data in sufficient temporal and spatial resolution so that a basis for the allocation of use change and modelling of various ESSs is possible. These data can however not always be made available in sufficient quantities (Holfeld et al. 2012). For this reason, only those ESSs and/or assessment methods such as InVest (Integrated Valuation of Ecosystem Services and Tradeoffs) (Holfeld & Rosenberg 2012) have been considered in the present application example which can be reliably parameterized on the basis of the available data.

5 Conclusions

The methodology presented here was developed and tested in the project Landscape Saxony 2050, and has now also been used in other projects with scenario tasks. After an intensive introductory explanation of the scenario methodology, the participants engaged in discussions at workshops in a very goal oriented manner, so that even difficult questions, such as those involving the spatial impacts could be handled very well. In the example area, scenario changes brought actors from a variety of areas of expertise together, so that interdisciplinary concepts for landscape development and unconventional strategies could be exchanged.

For the development of the scenarios, a precise definition of drivers was essential. On the other hand, viewing entire issue complexes, e.g. the development of the energy industry or of the mining industry, then precluded any further precise processing. The experience of the project shows that more than four key drivers can under no circumstances be recommended, and that ultimately, only two or at most three distinct drivers really make sense, or can be observed in their interactions. Generally, the time required for the drafting of scenarios is reduced with the number of drivers, the involved stakeholders, and the substantive scope of the core issue.

The integration of ESSs has brought new aspects into the scenario technique, but has on the other hand increased the complexity of the procedure. Simplification steps must therefore also be addressed; they may be provided by other application tests in other regions and on alternative core issues. Currently, a visualization possibility, the landscape barometer, is being tested. It is to be able to depict the assessment of the ESSs of particular scenarios and make them comprehensible to a broad public. An operationalization of the set of rules for the purpose of spatialization of use change in the form of a GIS extension is being considered.

6 Acknowledgements

We thank the Saxon State Ministry for Science and Art for funding the project "Landschaft Sachsen 2050" and thus enabling the recent publication. Thanks also to the other members of the project team who have supported this publication with information and ideas.

Literature

Albert, C. 2009. Scenarios for Sustainable Landscape Development – A Comparative Analysis of Six Case Studies. In 7th International Science Conference on the Human Dimensions of Global Environmental Change.

Alcamo, J. (ed.) 2008. Environmental Futures: The Practice of environmental scenario analysis. Elsevier, Amsterdam-Boston.

Antrop, M. 2005. Why landscapes of the past are important for the future. Landscape and Urban Planning, 70(1-2), 21–34.

Armstrong, J.S. (ed.) 2002. Principles of Forecasting: A Handbook for Researchers and Practitioners. Springer, Netherlands.

Bastian, O., Grunewald, K. & R.-U. Syrbe 2012. Space and time aspects of ecosystem services, using the example of the EU Water Framework Directive. International Journal of Biodiversity Science, Ecosystem Services & Management 8 (1-2), 5-16.

Bastian, O., Haase, D. & K. Grunewald 2011. Ecosystem properties, potentials and services – The EPPS conceptual framework and an urban application example. Ecological Indicators 21, 7-16.

Bastian, O., Syrbe, R.-U., Rosenberg, M., Rahe, D. & K. Grunewald 2013: The five pillar EPPS framework for quantifying, mapping and managing ecosystem services. Ecosystem Services 4, 15-24.

BMU - Bundesministerium für Umwelt, Naturschutz und Reaktorsicherheit 2007. Nationale Strategie zur biologischen Vielfalt, Berlin.

Bork, H.-R. & K. Müller 2002. Landschaftswandel von 500 bis 2500 n. Chr.. In: Offenhaltung der Landschaft. Hohenheimer Umwelttagung Stuttgart, Günter Heimbach Verlag 11–26.

Carpenter, S.R., Bennett, E.M. & G.D. Peterson 2006. Editorial: Special Feature on Scenarios for Ecosystem Services. Ecology and Society 11 (2) (32).

Costanza, R. & H.E. Daly, 1992. Natural Capital and Sustainable Development. Conservation Biology 6 (1), 37-46.

De Groot, R.S., Wilson, M.A. & R.M.J. Boumans 2002. A typology for the classification, description and valuation of ecosystem functions, goods and services. Ecological Economics 41 (3), 393-408.

Demuth, B., Heiland, S., Wojtkiewicz, W., Wiersbinski, N. & P. Finck 2010. Landschaften in Deutschland 2030 - Der große Wandel, BfN-Skripte 284.

Dörr, H., 2005. Die Zukunft der Landschaft in itteleuropa: Verantwortung für die Kulturlandschaft im 21. Jahrhundert; Delphi-Umfrage 2002, Arp-Planning. Consulting Research, Wien.

Dunlop, M., Turner, G., Foran, B. & F. Poldy 2002. Decision points for land and water futures. - Resource Futures Program Working Document 2002/08. Canberra, Australia.

EEG - Erneuerbare-Energien-Gesetz 2008. Gesetz vom 25. Oktober 2008 (BGBl. I S. 2074), das zuletzt durch Artikel 2 Absatz 69 des Gesetzes vom 22. Dezember 2011 (BGBl. I S. 3044) geändert worden ist.

Europäische Kommission, 2012. Die Gemeinsame Agrarpolitik nach 2013.

Fidalgo, B. & L. Pinto 2005. Linking landscape functions and preferences in forest landscapes - A tool for scenario building and evaluation. In: Lange, E. & D. Miller (eds.): Proceedings of Our Shared Landscape Integrating Ecological, Socio-Economic and Aesthetic Aspects in Landscape Planning and Management. Ascona, Switzerland.

Fritsch, U. 2002. Entwicklung von Landnutzungsszenarien für landschaftsökologische Fragestellungen. Thesis, Potsdam.

Gausemeier, J., Fink, A. & O. Schlake 1996. Szenario-Management: Planen und Führen nach Szenarien. 2. bearb. Aufl., Carl Hanser Verlag, München, Wien.

Gausemeier, J., Plass, C. & C. Wenzelmann 2009. Zukunftsorientierte Unternehmergestaltung. Strategien, Geschäftsprozesse und IT-Systeme für die Produktion von morgen. Carl Hanser Verlag München.

Haase D., Walz, U., Neubert, M. & M. Rosenberg 2007: Changes to Central European landscapes - Analysing historical maps to approach current environmental issues, examples from Saxony, Central Germany. Land Use Policy 24 (1), 248-263

Holfeld, M., Stein, C., Rosenberg, M., Syrbe, R.-U. & U. Walz 2012. Entwicklung eines Landschaftsbarometers zur Visualisierung von Ökoystemdienstleistungen. AGIT Tagungsband 2012.

Holfeld, M. & M. Rosenberg 2012. Anwendung von Modulen des ÖSD-Modells InVEST. In: Grunewald,K. & O. Bastian (eds.) Ökosystemdienstleistungen - Konzept, Methoden und Fallbeispiele. Springer Akademie Verlag.

Höchtl, F., Bieling, C. & W. Konold 2006. Waldzunahmen versus Offenhaltung der Landschaft - Raum-Zeit-Prozesse, ökologische Auswirkungen, politische Lösungsansätze. Zwischenbericht über das Jahr 2006.

Hulse, D.W., Branscomb, A. & S.G. Payne 2004. Envisioning alternatives: Using citizen guidance to map future land and water use. Ecological Applications, 14(2), 325–341.

Jessel, B. 2000. Von der „Vorhersage“ zum Erkenntnisgewinn. Aufgaben und Leistungsfähigkeit von Prognosen in der Umweltplanung. Naturschutz und Landschaftsplanung, 32(7), 197–203.

Kahn, H. & A.J. Wiener 1984. Ihr werdet es erleben. Voraussagen der Wissenschaft bis zum Jahre 2000., Rowohlt TB-V.

Klijn, J.A. 2004. Driving forces behind landscape transformation in Europe, from a conceptual approach to policy options. In: The New Dimensions of the European Landscapes. p.267.

Kosow, H. & R. Gaßner (2008). Methoden der Zukunfts- und Szenarioanalyse - Überblick, Bewertung und Auswahlkriterien. WerkstattBericht Nr. 103, IZT - Institut für Zukunftsstudien und Technologiebewertung , Berlin.

Liu, J., Zhou, H., Qin, P., Zhou, J. & G. Wang 2009. Comparisons of ecosystem services among three conversion systems in Yancheng National Nature Reserve. Ecological Engineering, 35(5), 609–629.

MA 2005. Ecosystems and human well-being: Synthesis.

Meadows, D.L., Meadows, D.H. & E. Zahn 1972. Die Grenzen des Wachstums. Bericht des Club of Rome zur Lage der Menschheit, Deutsche Verlags-Anstalt.

Michel, E. & Walz, U. 2012. Landschaftsstruktur und Artenvielfalt - art- und lebensraumspezifische Untersuchungen am Fallbeispiel der Bodenbrüter. - In: Strobl, J., Blaschke, T. & G. Griesebner (eds.): Angewandte Geoinformatik 2012. Beiträge zum 24. AGIT-Symposium Salzburg, Wichmann-Verlag, 770-779.

Nassauer, J.I. & R.C. Corry 2004. Using normative scenarios in landscape ecology. Landscape Ecology 19(4), 343–356.

Nassauer, J.I., Corry, R.C. & R.M. Cruse 2002. The Landscape in 2025: Alternative future landscape scenarios: A means to consider agricultural policy. Journal of Soil and Water Conservation 57(2), 44A–53A.

Nelson, E., Mendoza, G., Regetz, J., Polasky, S., Tallis, H., Cameron, R., Chan, K., Daily, G., Goldstein, J., Kareiva, P., Lonsdorf, E., Naidoo, R., Ricketts, T. & R. Shaw 2009. Modelling multiple ecosystem services, biodiversity conservation, commodity production, and tradeoffs at landscape scales. Frontiers in Ecology and the Environment 7(1), 4–11.

Oppermann, B. 2008. Zur Kunst der Landschaftsvorhersage. Gedanken anlässlich des FLL-Fachforums zum Thema Zukunfts-landschaften. Stadt+Grün, (8), 35–38.

Pachauri, R.K. & A. Reisinger 2008. Contribution of Working Groups I, II and III to the Fourth Assessment Report of the Intergovernmental Panel on Climate Change. IPCC, Geneva.

Pijanowski, B.C., Brown, D. G., Shellito, B. A. & G.A. Manik 2002. Using neural networks and GIS to forecast land use changes: a Land Transformation Model. Computers, Environment and Urban Systems, 26(6), 553–575.

Priess, J. & J. Hauck 2011. Szenarien für nachhaltige Landnutzung in Mitteldeutschland. Helmholtz Centre for Environmental Research (UFZ), Leipzig.

Prognos AG 2011. Bedeutung der Braunkohle in Ostdeutschland. Kurzfassung der Studie, Berlin.

Reed, M.S. 2008. Stakeholder participation for environmental management: A literature review. Biological Conservation 141(10), 2417–2431.

von Reibnitz, U. 1991. Szenario-Technik. Instrumente für die unternehmerische und persönliche Erfolgsplanung. Gabler Verlag, Wiesbaden.

Regionaler Planungsverband Oberlausitz-Niederschlesien 2010. Regionalplan Region Oberlausitz-Niederschlesien - Erste Gesamtfortschreibung gemäß § 6 Absatz 5 SächsLPlG, Bautzen.

Rotmans, J., van Asselt, M., Anastasi,C., Greeuw, S., Mellors, J., Peters, S., Rothman, D. & N. Rijkens 2000. Visions for a Sustainable Europe. Futures, 32(9-10), 809–831.

Santelmann, M.V., White, D., Freemark, K., Nassauer, J.I., Eilers, J.M., Vaché, K.B., Danielson, B.J., Corry, R.C., Clark, M.E., Polasky, S., Cruse, R.M., Sifneos, J., Rustigian, H., Coiner, C., Wu, J. & D. Debinski 2004. Assessing alternative futures for agriculture in Iowa, U.S.A. Landscape Ecology, 19(4), S.357–374.

Seppelt, R. & A. Holzkämper 2007. Multifunctional use of landscape services. Applications and results of optimization techniques of land use scenario development. In Proceedings 7. IALE World Congress. Wageningen.

SMUL - Sächsisches Staatsministerium für Umwelt und Landwirtschaft, 1992. Biotoptypen- und Landnutzungskartierung 1992/1993.

SMUL - Sächsisches Staatsministerium für Umwelt und Landwirtschaft, 2005. Biotoptypen- und Landnutzungskartierung 2005.

Spangenberg, J.H., 2007. Biodiversity pressure and the driving forces behind. Ecological Economics, 61(1), 146–158.

Statistisches Landesamt des Freistaates Sachsen 2011. Flächenerhebung nach Art der tatsächlichen Nutzung.

Steinitz, C., Bassett, H., Flaxman M., Goode, T., Maddock, T., Mouat, D., Peiser, R. & A. Shearer. 2003. Alternative Futures for Changing Landscapes, Island Press.

Syrbe R.-U., Rosenberg, M. & J. Vowinckel 2013. Szenarioentwicklung und partizipative Verfahren. In: Grunewald, K. & O. Bastian (eds.) Ökosystemdienstleistungen - Konzept, Methoden und Fallbeispiele. Springer Akademie Verlag, 110-119.

TEEB 2009. TEEB For Policymakers. The Economics of Ecosystems & Biodiversity. TEEB DE (2013). http://www.teebweb.org/wp content/uploads/Study%20and%20Reports/Reports/Synthesis%20 report/Synthesis_German.pdf

UNEP 2007. GEO-4 - Global Environment Outlook, United Nations Environment Programme (UNEP).

Verburg, P., van Berkel, D., van Doorn, A., van Eupen, M. & H. van den Heiligenberg 2010. Trajectories of land use change in Europe: a model-based exploration of rural futures. Landscape Ecology, 25(2), 217–232.

Walz, A., Lardelli, C., Behrendt, H., Grêt-Regamie, A., Lundström, C., Kytzia, S. & P. Bebi 2007. Participatory scenario analysis for integrated regional modelling. Landscape and Urban Planning, 81(1-2), 114–131.

Zebisch, M. 2004. Modellierung der Auswirkungen von Landnutzungsänderungen auf Landschaftsmuster und Biodiversität. Thesis, TU-Berlin.

2

Detecting Land Use And Land Cover Changes In Northern German Agricultural Landscapes To Assess Ecosystem Service Dynamics

Marion Kandziora*[1], Katja Dörnhöfer[2], Natascha Oppelt[2], Felix Müller[1]

[1] Kiel University, Institute for Natural Resource Conservation, Department of Ecosystem Management, Olshausenstr. 40; 24098 Kiel; mkandziora@ecology.uni-kiel.de

[2] Kiel University, Department of Geography, Remote Sensing & Environmental Modelling, Olshausenstr. 40; 24098 Kiel

Abstract

Land use and land cover (LULC) and their changes in share and number of classes can be documented by remote sensing techniques. Information on LULC is needed for the assessment of ecosystem services and is used as input data for mapping and modelling. This information is important for decision-making and management of ecosystems and landscapes. In this study, LULC were analysed in two agricultural areas in Northern Germany by means of a pixel-based maximum likelihood classification approach of 11 Landsat TM 5 scenes between 1987 and 2011 followed by a post-classification refinement using the tool IRSeL. In this time period, grassland declined by about 50 % in both case study areas. This loss in grassland area can be associated with changes in provisioning ecosystem services as the supply of fodder and crops and the number of livestock declined from 1987 to 2007. Furthermore, an on-going increase in maize cultivation area, which is nowadays more and more used as biomass for biogas production, documents the addition of another provisioning service, i.e., biomass for energy. Combining remote sensing and research on ecosystem services supports the assessment and monitoring of ecosystem services on different temporal, spatial, and semantic scales.

Keywords:
provisioning services, quantification, crop rotation, remote sensing, agricultural management practice

*Corresponding author. Email: mkandziora@ecology.uni-kiel.de

1 Introduction

The detection and monitoring of changes in land use / land cover (LULC) and thus in the supply of ecosystem services at various spatial scales are important research tasks. Ecosystem services contribute to human well-being in various ways, and can be subdivided into provisioning, regulating and cultural services (Kandziora et al. 2013a). The ecosystem services concept needs to be incorporated in management and decision-making, even though there are still challenges in the spatially explicit quantification, mapping and modelling of ecosystem service supply and demand (Crossman et al. 2012), and in the classification and valuation (de Groot et al. 2010).

In recent times, mapping and modelling of ecosystem services has gained increasing relevance and several approaches have been developed to expedite these methods. The assessment of ecosystem services based on LULC has been documented in several case studies (e.g., Koschke et al. 2013; Kroll et al. 2012). As detailed LULC maps are lacking for several countries or regions, remote sensing data is considered a valuable source for LULC data (Kennedy et al. 2009). Some studies have combined remote sensing techniques to examine e.g., potential impacts of land use changes to urban ecosystem services (Estoque and Murayama 2012). The use of models such as InVEST (Polasky et al. 2011) requires LULC maps as input data to model and map multiple ecosystem services (Leh et al. 2013). Many of these studies emphasize the need to include the gained knowledge of LULC change and ecosystem service change in decision-making and management (Estoque and Murayama 2012) as well as to understand past changes (Lautenbach et al. 2011) and predict future impacts on ecosystem services and human well-being (Lorencová et al. 2013; Sexton et al. 2013).

Deforestation and growing urbanisation have been often found to be the most severe LULC changes in the past decades (MEA 2005). However, changes in agricultural areas due to intensification and changes in crop rotation are present as well. Many studies focus on the assessment of ecosystem services of agricultural areas with little consideration of crop rotation and management practices (Koschke et al. 2013). Agricultural ecosystems generate an ample range of ecosystem services and interactions between the single ecosystem services, such as crop production or erosion and nutrient regulation, which in turn influence crop production (Dale and Polasky 2007; Lorencová et al. 2013). As agricultural land use accounts for a large area globally, the focus of research questions should also be placed on gaining detailed information on crop rotation, management practices and shares of arable land and grassland (Koschke et al. 2013; Lorenz et al. 2013).

Thus, this study was conducted in two case study areas (Bornhöved Lakes District and Kielstau Basin) in Northern Germany that are not influenced by deforestation or urban sprawl but are important supply areas of multiple ecosystem services due to their large extent of agroecosystems. The two case study areas are considered representative landscapes for Northern Germany (Fränzle et al. 2008; Fohrer and Schmalz 2012) and are good examples for the development in agriculture and the supply and changes in ecosystem services over the past decades.

Remote sensing techniques (here: classification of Landsat images) were used to derive land cover information in both case study areas for several time steps. In combination with crop rotation information, land cover data provides information on land use. Here, land cover and land use are used as synonyms in the term land use / land cover (LULC).

Therefore, the objectives of the study are:

(i) identifying the LULC changes in the available Landsat time series;

(ii) using this information to derive crop rotation practices and

(iii) identifying the influence of LULC changes on ecosystem service supply and the changes in selected provisioning services over two decades in both case study areas.

After the methodological explanations and the presentation of the results, the approach is discussed regarding the advantages and disadvantages of combining remote sensing techniques and ecosystem service assessments.

2 Study areas

Both case study areas are situated in the eastern part of the northern German state Schleswig-Holstein and feature similar climatic conditions (8.2°C average temperature and 919 mm/a precipitation in the Kielstau Basin (Fohrer & Schmalz 2012) and 8.1°C average temperature and 697 mm/a precipitation in the Bornhöved Lakes District (Fränzle et al. 2008)). LULC are predominantly agroecosystems with small shares of settlements and forest areas.

The Bornhöved Lakes District is located 30 km south of the capital Kiel (Figure 1). The study area was delimited to a size of 60 km^2 and lies partly within ten municipalities in the two districts of Plön and Segeberg. Located on the outskirts of the Weichselian glaciation, the northern part of the Bornhöved Lakes Districts belongs to the moraine area of the "Ostholsteinisches Hügelland" with its diversified relief. The southern part, the so-called "Trappenkamper Sander" contains mostly fluvioglacial deposits. Six glacially formed lakes (between 0.27-1.4 km^2) are predominate features, which are surrounded by forest areas (Fränzle et al. 2008). The lakes have been landscape protection areas since 1962 and partly conservation areas since 1983 (LLUR 2012; Figure 1). Predominant soils are luvisols, cambic arenosols, and histosols (Fränzle et al. 2008). The Bornhöved Lakes District was the focus of an interdisciplinary ecosystem research project, which has been conducted from 1988 to 2001 (Müller et al. 2006).

The Kielstau Basin is located further north, close to the Danish border (Figure 1) in the district of Schleswig-Flensburg and lies partly within eight municipalities. The Kielstau Basin (50 km^2) is a subbasin of the River Treene and a typical lowland basin with high river groundwater levels and low gradients. The river Kielstau is 17 km long and has several tributaries (Fohrer and Schmalz 2012). Haplic luvisols are the predominant soil type in the eastern part of the basin, while stagnic luvisols occur in the west. Sapric histosols are found along the Kielstau and its tributaries (Kiesel et al. 2009). The main river passes Lake Winderatt, which has been protected since 1972 (LLUR 2012; Figure 1). Since 1989 the former agricultural areas around the lake have been protected; nowadays they are used as extensive pastures in combination with a recreational area for hiking, nature protection, and environmental education. Since 2005 the basin is monitored and modelled intensively regarding water quality, sediment yield, water level, and hydrobiology (Fohrer and Schmalz 2012).

Based on the characteristics and peculiarities of both case study areas and the available data from past and recent monitoring and research programmes, the case study areas are considered suitable for the assessment and quantification of changes in LULC and ecosystem service dynamics.

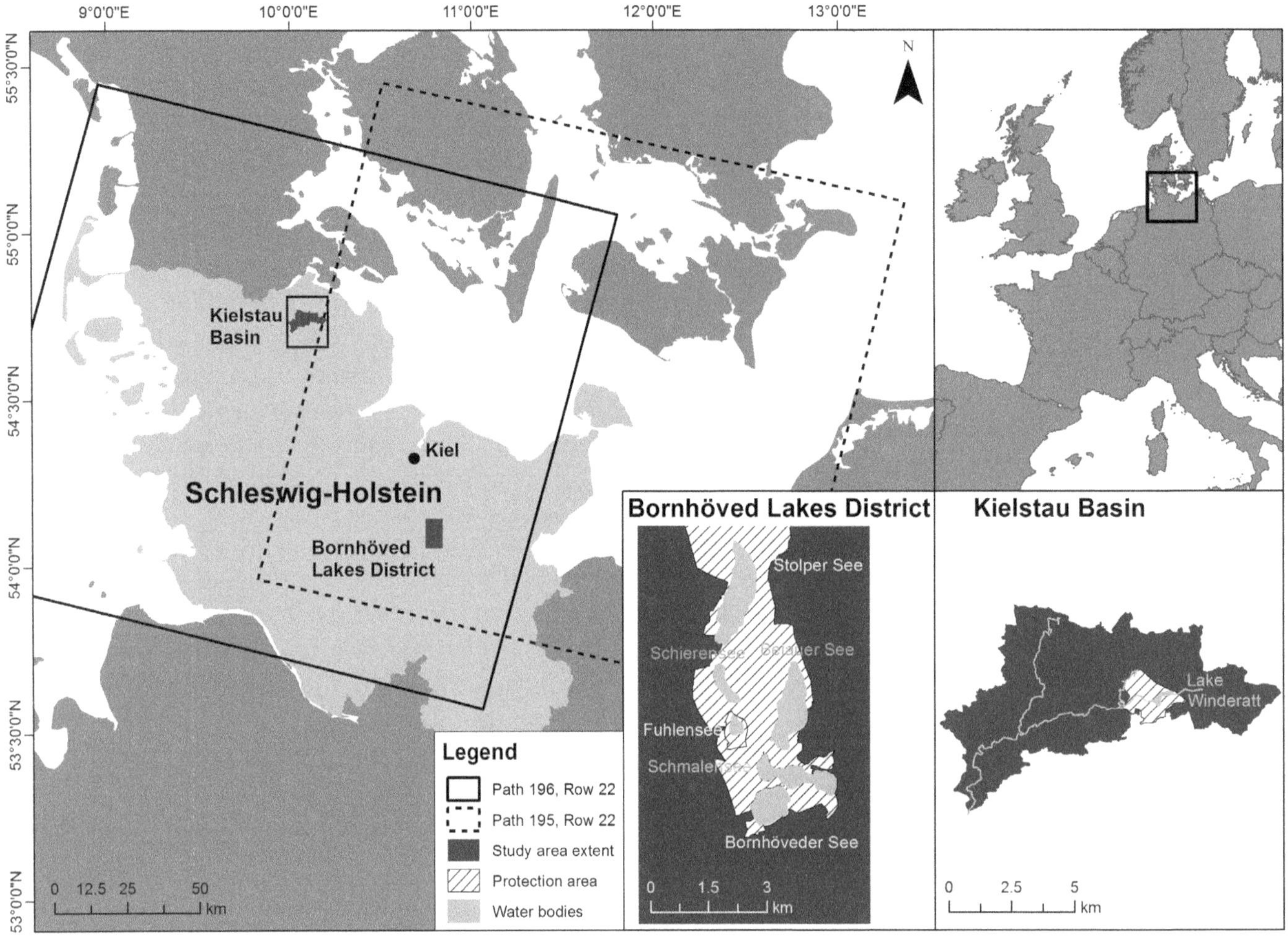

Figure 1: The Kielstau Basin and the Bornhöved Lakes District study areas and the location of the Landsat scenes

3 Methods

Figure 2 schematically illustrates the workflow to retrieve LULC maps, LULC changes, and crop rotations for assessing ecosystem services and their changes. The first step was the classification of available remote sensing data (Landsat TM). Based on the derived LULC classes and available data for quantification, assessments for selected ecosystem services were conducted. The focus was on provisioning services (crops, fodder, biomass for energy, livestock, and mineral resources) (Kandziora et al. 2013a) because they are the group of ecosystem services that can be directly deduced from LULC data, whereas regulating and cultural services strongly depend on the interaction of several ecosystem services and other factors like geological material, climate, topography, management measures and human perceptions. In this study regulating services and cultural services are not considered because respective modelling results for the quantification of regulating services and further quantification approaches on cultural services will be included in the next methodological assessment.

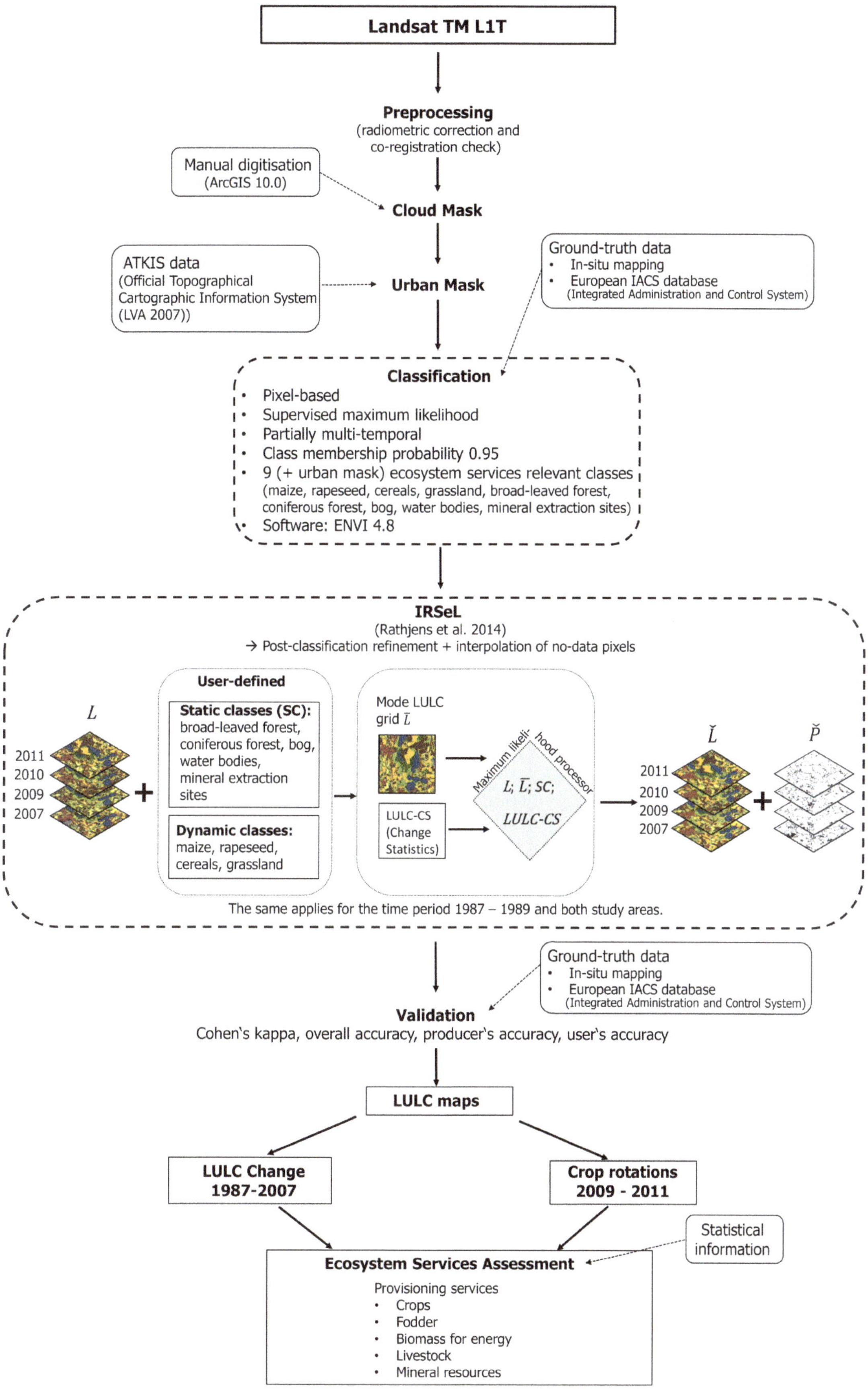

Figure 2: Methodological procedure for the assessments in the case study areas

3.1 Remote sensing preprocessing and land cover classification

For both study areas, eleven Landsat TM images were available to derive LULC maps. Table 1 provides a detailed list of the acquisition dates. Landsat path 196, row 22 entirely covers both study areas. Additionally, the Bornhöved Lakes District is located within Landsat path 195, row 22 (Figure 1); scene subsets were used to classify the study areas (Kielstau Basin: 702 x 502 pixels; Bornhöved Lakes District: 597 x 524 pixels). The United States Geological Survey Earth Resources Observation and Science Center provided L1T data (level-one-terrain-corrected). For both study areas high cloud coverage was the main limiting factor for image availability. Images were geo-referenced on Transverse Mercator UTM zone 32N (WGS84) and showed highly accurate co-registration with Rout-mean-squared errors (RMSE) < 1 pixel (Table 1). Landsat TM has a spatial resolution of 30 m in the optical wavebands and 120 m in the thermal waveband (USGS 2013), whereas only the reflective bands with a pixel resolution of 30 m were used in this study. Conducted radiometric correction encompassed the conversion of the original grey values [*Digital Number*] into radiances [*W/(m² μm)*] according to the equations suggested by Chander et al. (2009). Referring to the discussion of Song et al. (2001) atmospheric correction was considered unnecessary. The green (band 2), red (band 3), near-infrared (band 4) and mid-infrared (band 5 and 7) wavelengths formed the spectral input for the classification; the blue wavelengths (band 1) were not included due to their sensitivity to atmospheric influences. In the case where more than one suitable data set per year was available all scenes were merged and used as one dataset for a multi-temporal classification of the respective year (Table 1). Thus, class separability benefited from phenological or seasonal characteristics of land cover types (Pax-Lenney and Woodcock 1997). Clouded areas were manually digitised and masked

Table 1: Landsat TM acquired for the period 1987 – 2011 (path 195, 196, row 22), co-registration accuracy and available ground truth data

Year	Path 195	Path 196	Kielstau Basin		Bornhöved Lakes District		Co-Registration Accuracy		Ground truth data
			Usage	Cloud coverage [%]	Usage	Cloud coverage [%]	RMSE-Y	RMSE-X	
1987		18th April	x	0	x	0	0.7017	0.6851	Field mapping Bornhöved Lakes District
	27th April			-	x	0	0.6022	0.6425	
		7th July	x	0	x	0	0.7154	0.6574	
1989		25th May	x	0	x	0	0.8124	0.7094	Field mapping Bornhöved Lakes District
	5th July			-	x	0	0.6719	0.6875	
2007		25th April	x	1.8	x	0	0.6853	0.6578	ATKIS*, IACS**
2009		3rd July	x	0	x	0	0.7018	0.6295	ATKIS, IACS, field mapping
		20th August	x	0		0	0.8063	0.9315	
2010		4th June	x	0	x	9.4	0.7350	0.6581	ATKIS, IACS, field mapping
		7th August	x	2.2	x	0	0.6243	0.5223	
2011		6th May	x	0	x	0	0.6527	0.5176	ATKIS, IACS, field mapping

*Official Topographical Cartographic Information System, ** Integrated Administration and Control System prior to the classification. Pixels with sealed land cover often show a wide spectral variability (mixed pixel problem) resulting in an overestimation of sealed areas (Yuan et al. 2005). Therefore, a sealed area mask retrieved from the ATKIS 2007 database (Official Topographical Cartographic Information System; LVA 2007) was used to extract sealed areas from all data sets preventing erroneous extents. Actual land cover, as well as their relevance for ecosystem service supply, determined the definition of nine spectral classes, i.e., maize, rapeseed, cereals, grassland, broad-leaved forest, coniferous forest, bog, water bodies, and mineral extraction sites (the last class only in the Bornhöved Lakes District). Supervised image classification requires ground truth data to both train and validate the classifier (Campbell and Wynne 2011). High quality ground truth data originated from different sources (Table 1) including detailed field mappings, ATKIS database (LVA 2007), and annual agricultural data on field block level of the European IACS (Integrated Administration and Control System) database (MELUR 2010). The ground truth dataset was randomly divided into two datasets: one intended to train the maximum likelihood classifier and the second one for accuracy assessment. After selecting training pixels, their class histograms were checked for normality before applying a maximum likelihood classification implemented in the image processing software ENVI 4.8 (ITTVIS 2010). Using the statistics of the training pixels, the maximum likelihood classifier calculates probability density functions for each class. The pixels were then allocated to the class that showed the highest probability considering all spectral bands. A pixel remained unclassified, i.e., was not assigned to any class if all probabilities were below a user-defined threshold which was constantly set to 0.95 (Campbell and Wynne 2011).

3.2. *Post-classification refinement with IRSeL and validation*

Post-classification refinements performed with IRSeL (Improvement of Remotely Sensed Land Cover) (Rathjens et al. 2014) interpolated "no data pixels" (unclassified or cloud masked) and reduced potential classification errors. IRSeL is a tool that derives crop rotation or land use change statistics from remotely sensed LULC data sets L (Figure 2). Furthermore, IRSeL generates a map of average LULC $\bar{L}$ retrieved from the classification data set. The user defines LULC classes that are expected to be temporally static. In this study, the classes broad-leaved forest, coniferous forest, water bodies, bog, and mineral extraction sites were separately considered static for the 1980es and 2000es periods. IRSeL then integrated the user-defined static classes, crop rotation statistics, and spatial neighbourhoods of a pixel to allocate a LULC class to "no data pixels" or to re-allocate pixels with potential classification errors, i.e., pixels assigned to a static class according to the average LULC map received the respective static LULC class even if they were classified differently. Pixels that were correctly classified as a dynamic, i.e. agricultural class, retained the original LULC class. IRSeL also interpolated no data, i.e. unclassified or cloud masked pixels, according to the LULC of a pixel's spatial neighbourhood and LULC in the previous/following year (crop rotation statistics). Pixels that were classified as a static class, however, which did not match the static classes in the average map, were revised. The resulting maps $\hat{L}$ showed a land cover without "no data pixels" (Rathjens et al. 2014).

Accuracy assessment following Congalton (1991) was carried out on the IRSeL modified LULC maps. The second data set of the ground truth data served as input to generate error matrices as cross-tabulations of ground truth versus classified pixels. The validation was carried out for each data set, consisting of about 2600 (Bornhöved Lakes District) and 1900 (Kielstau Basin) validation pixels. In 1987 and 1989 accuracy measures of the Kielstau Basin were adopted from the Bornhöved Lakes District for maize, rapeseed, cereals, and grassland since the Kielstau Basin lacked field mapping data. Overall accuracy (OA) and kappa coefficient are accuracy measures addressing the entire classification. Overall accuracy is the share of correctly classified pixels (diagonal elements of error matrix) to all reference pixels. The kappa coefficient includes the off-diagonal elements, i.e., erroneous classification, and allows for agreement by chance (Congalton 1991). Values range between 0 and 1. Values close to 1 indicate a high accuracy; values close to 0 point towards poor agreement between classification and ground truth pixels (Foody 2008).

Producer's (PA) and user's (UA) accuracy represent per-class accuracy measures. The user's accuracy is the share of correctly allocated pixels of a class and the sum of all pixels classified as that class. The percentage of all correctly classified pixels to all reference pixels of the class is the producer's accuracy (Campbell and Wynne 2011). A high value close to 1 means higher accuracies and lower errors of commission (1-UA) and errors of omission (1-PA), respectively. A widely reported, but highly discussed, accuracy target for remotely sensed land cover is 85 % correct allocation (Foody 2008).

3.3 Producing land cover maps, land use / land cover change, and crop rotation maps

The validated data sets represented the final LULC maps. LULC change was assessed by calculating the percentage change between 1987 and 2007 with respect to the area of 1987 (Eq. 1).

$$x_i = \frac{A_2 - A_1}{A_1} * 100, \qquad (1)$$

Where represents the rate of change [%] while A_1 and A_2 denote the area of LULC class i at time t_1 (1987) resp. t_2 (2007). Generating a difference map of the 1987 and 2007 LULC information emphasised spatial changes. Changes between agricultural classes (maize, rapeseed, and cereals) were considered as differences due to crop rotations. Combining field mapping data from 1987 and the IACS (Integrated Administration and Control System) data from 2007 enabled accuracy assessment of the LULC change map as previously described. In the Kielstau Basin the lack of ground truth data prevented validation of the LULC changes for the 1980s. Combining the classification results from 2009-2011 revealed 3-years of crop rotation pattern for both study areas.

3.4 Quantification of provisioning services by statistical data

The focus of this study was on provisioning services, which were subdivided into crops, fodder, biomass for energy, livestock, and mineral resources. Table 2 provides definitions for the single ecosystem services, their indicators, and data sources for quantification. All ecosystem services were quantified for the years 1987 and 2007 with official statistical data sets to show changes within two decades.

Table 2: Assessed provisioning services with definition, indicator description, units, spatial scale, and data sources

Ecosystem service and definition	Indicator/unit	Spatial scale	Data source(s)
Crops: harvested biomass used as food for human nutrition	Rapeseed yield (dt/ha/a)	Average data for whole district	Statistikamt Nord (2012)
	Winter wheat yield (dt/ha/a)	Average data for whole district	Statistikamt Nord (2012)
Fodder: harvested biomass used for livestock production	Maize yield (dt/ha/a)	Average data for whole district	Statistikamt Nord (2012)
	Rapeseed yield (dt/ha/a)	Average data for whole district	Statistikamt Nord (2012)
	Hay meadow yield (dt/ha/a)	Average data for whole district	Statistikamt Nord (2012)
Biomass for energy: biomass used to generate energy	Maize yield (dt/ha/a)	Average data for whole district	Statistikamt Nord (2012)
Livestock: here only cattle is quantified which is used for human nutrition	Number of cattle in the municipalities	Number applies to the whole area of the municipalities which are part of the study areas	Statistisches Landesamt Schleswig-Holstein (1989) Statistikamt Nord (2012)
Mineral extraction sites: minerals (here: sand and gravel) excavated from under the surface, e.g. for construction purposes	Number of sites (n)	Study area	Expert valuation

Crops were defined as harvest of plants contributing directly to human nutrition. The indicator for quantification is the annual average yield (dt/ha/a) in the administrative district for winter wheat. Winter wheat covers the largest area of cereals (MELUR 2010) and was therefore used as a proxy because no spectral distinction of cereals into barley and rye, e.g., was possible.

Fodder was calculated by the average yield (dt/ha/a) from maize and rapeseed harvest and the harvested grassland (hay meadow yield (dt/ha/a)) (Statistikamt Nord 2012).

Biomass for energy applied to the harvested maize that is used in biogas plants as primary substrate due to its efficiency in generating energy (MELUR 2011).

Agricultural statistical data were used to quantify the provisioning ecosystem services crops and fodder for the years 1987 and 2007 by multiplying the annual yield with the cultivated area.

Livestock was calculated by the number of cattle in the municipalities which the two study areas are located in. Livestock data was available for 1988 and 2007.

Statistical data for quantification were available at different spatial scales: annual average yield data was available at the district level, whereas data on the number of livestock exist at municipality level but only for selected years. Both case study areas are not identical with the borders of the municipalities. They cover up to approx. 2/3 of the area of some of the municipalities; others account for only a very small part of the case study area. Therefore, the whole areas of the related municipalities were used. As a result, the number of livestock applies to a larger area than the respective case study areas.

Mineral extraction is not considered an ecosystem service in most definitions and studies (Haines-Young and Potschin 2010) but can be of importance in environmental management and decision-making. Depending on the size of the extraction sites, other ecosystem services like landscape aesthetics and also biodiversity can be affected. Here it is considered owing to the development of the mineral extraction sites over the analysed time scale. As no information was available on the amount of excavated material, only the number of mineral extraction sites was used as an indicator for LULC change.

The years 1987 and 2007 were compared and it was investigated whether there was an increase in the selected ecosystem service, a decrease, or a new emergence, which obviously is in relation to changes in LULC.

4 Results

4. 1 Land use / land cover classification

Nine LULC classes, grouped as near-natural classes, agricultural, and artificial classes, could be distinguished from the Landsat TM images and sealed areas (from the urban mask). For each year (1987, 1989, 2007, 2009, 2010 and 2011), a LULC map was generated together with a change map. They show the LULC changes between the Landsat acquisitions from 1987 to 2007. Additionally a crop rotation map was created for the years 2009-2011. Figure 3 and 4 show the maps for the Bornhöved Lakes District and the Kielstau Basin, respectively. For both study areas the distinction of agricultural crops performed well and field structures appeared clearly in all classification results. In the Bornhöved Lakes District, mineral extraction sites and bogs were classified with high accuracy. In both areas, the small forested areas showed high accuracy measures. Due to the six lakes in the Bornhöved Lakes District the water class showed a higher number of pixels than in the Kielstau Basin; nevertheless, high accuracy measures were obtained for both study areas.

After post-classification refinements, the LULC maps showed very high correspondence with the ground truth data. Overall accuracy and kappa exceeded 0.9 for both study areas, except for the Kielstau Basin in 1989 (Table 3 and 4). Agricultural classes (maize, rapeseed, and cereals) exhibited tendencies towards accurate performance exceeding producer's and user's accuracies of 0.9 in most years and for both study areas. Grassland performed slightly better in the Bornhöved Lakes District but presented lower accuracies than agricultural classes. Broad-leaved forest, sealed areas and water bodies were on a very high accuracy level. Coniferous forest and bogs were more accurate for the Bornhöved Lakes District while both classes tended to show strong differences between user's and producer's accuracies. The low producer's accuracies resulted from an overestimation of pixels classified as coniferous forest or bog.

Table 3: Accuracy assessment for the Bornhöved Lakes District land use / and cover maps. Upper part of the table presents per class accuracies (PA and UA); the part below the line contains aggregated (kappa and overall) accuracies

Accuracy assessment Bornhöved Lakes District												
	1987		1989		2007		2009		2010		2011	
Class	PA	UA	PA	UA	PA	UA	PA	UA	PA	UA	PA	UA
Maize	0.91	0.99	0.93	1.00	1.00	0.84	0.73	0.96	1.00	1.00	1.00	0.88
Rapeseed	0.92	0.98	0.93	0.86	1.00	1.00	1.00	1.00	1.00	1.00	0.80	0.96
Cereals	0.98	0.85	0.75	0.89	1.00	0.90	1.00	0.86	1.00	0.92	0.97	0.89
Grassland	0.94	0.89	0.83	0.63	0.94	0.94	0.99	0.87	1.00	0.98	0.96	0.94
Broad-leaved forest	1.00	0.94	1.00	0.94	0.98	0.96	1.00	0.96	1.00	0.96	1.00	0.96
Coniferous forest	0.89	0.97	0.87	0.97	0.94	1.00	0.94	1.00	0.96	1.00	0.96	1.00
Bog	0.91	1.00	0.85	1.00	0.71	1.00	0.61	1.00	0.56	1.00	0.70	1.00
Sealed area	0.97	0.99	0.95	0.98	1.00	1.00	1.00	1.00	1.00	1.00	0.91	1.00
Water bodies	1.00	1.00	1.00	0.99	1.00	0.95	1.00	0.95	1.00	0.95	1.00	0.95
Mineral extraction sites	1.00	1.00	0.94	1.00	0.80	1.00	0.89	1.00	0.91	1.00	0.88	1.00
Kappa coefficient	0.9509		0.9199		0.9568		0.9405		0.9735		0.9374	
Overall accuracy	0.9569		0.9309		0.9620		0.9472		0.9765		0.9445	

Table 4: Accuracy assessment for the Kielstau Basin land use / land cover maps. Upper part of the table presents per class accuracies (PA and UA); the part below the line contains aggregated (kappa and overall) accuracies

Accuracy assessment Kielstau Basin												
	1987		1989		2007		2009		2010		2011	
Class	PA	UA	PA	UA	PA	UA	PA	UA	PA	UA	PA	UA
Maize	0.91	0.98	0.93	0.99	1.00	0.94	1.00	0.78	1.00	0.98	1.00	0.98
Rapeseed	0.92	0.98	0.93	0.87	1.00	0.97	1.00	1.00	1.00	1.00	1.00	0.99
Cereals	0.98	0.79	0.75	0.92	0.97	0.93	1.00	1.00	1.00	0.94	0.97	0.97
Grassland	0.94	0.86	0.83	0.66	0.96	0.91	0.78	0.89	0.95	0.85	0.88	0.85
Broad-leaved forest	0.98	0.88	0.99	0.96	0.99	0.88	0.99	0.90	0.99	0.91	0.99	0.86
Coniferous forest	0.46	1.00	0.84	0.95	0.48	0.95	0.76	0.95	0.52	0.98	0.95	0.45
Bog	0.87	1.00	0.76	1.00	0.59	1.00	0.98	0.98	0.57	1.00	0.53	1.00
Sealed area	0.99	0.99	1.00	0.94	0.99	1.00	0.92	1.00	0.99	1.00	0.96	1.00
Water bodies	1.00	1.00	1.00	0.98	1.00	1.00	0.91	1.00	1.00	1.00	0.82	1.00
Kappa coefficient	0.9231		0.8817		0.9293		0.9187		0.9404		0.9168	
Overall accuracy	0.9342		0.8963		0.9389		0.9291		0.9483		0.9282	

4.2 Land use / land cover changes from 1987-2007

Regarding the distribution of LULC classes (Figure 3 and 4) grassland was the dominant land cover in both study areas in the late 1980s. Cereals represented the most often cultivated crop type, followed by rapeseed. Maize made up about 5.6% in the Bornhöved Lakes District and 9.3 % in the Kielstau Basin in 1987. The Kielstau Basin showed small patches of forest area (< 100 ha) where broad-leaved forest prevailed. Larger broad-leaved and coniferous forests were present in the Bornhöved Lakes District. Only small bog sites existed in both study areas. The lakes formed a water area of about 440 ha (7.4 %) in the Bornhöved Lakes District, whereas Lake Winderatt is the only water body in the Kielstau Basin.

Two decades later, forest cover has increased in both study areas, especially the broad-leaved forest of the Bornhöved Lakes District. Slight reductions occurred in the LULC classes bog and water bodies. Mineral extraction sites extended about 60 ha (2007) in the Bornhöved Lakes District and were only marginally present on one site in 1987. One of the three mineral extraction sites which have been visible in the LULC since 2007, started excavating sand and gravel in 1996 (Wandhoff 2013).

With respect to agricultural classes and grassland, strong changes occurred in both case study areas. By 2007 the grassland area decreased by approx. 50 % compared to its extent in 1987 and cereals replaced grassland as the dominant land use. The cultivation of maize increased by 249 % in the Bornhöved Lakes District and increased by 83 % in the Kielstau Basin. The change maps in Figure 3 and 4 emphasize the spatial dimension of changes with an obvious conversion from grassland to agricultural classes. Re-cultivation from agricultural land to grassland partially occurred. Minor changes within near-natural classes were mainly delineated at the lake borders (Bornhöved Lakes District). Figure 5 highlights the retreat of grassland area as well as the shift towards the predominance of crop cultivation. Table 5 shows the share of crops which were cultivated in 2007 on former grassland area. With respect to the accuracy of detected spatial LULC changes (Figure 3 change map) overall accuracy (0.71) and kappa (0.63) were distinctly lower

Table 5: Percentage of agricultural classes in 2007 with respect to former grassland (1987)

	Agricultural classes in 2007		
Grassland in 1987	Maize	Rapeseed	Cereals
Kielstau Basin	26.9	15.8	57.4
Bornhöved Lakes District	44.5	17.0	38.5

Table 6: Overall accuracy (OA), kappa and per class accuracies (PA and UA) for the LULC change map (1987-2007) of Bornhöved Lakes District

Change class	**PA**	**UA**
No change	0.74	0.46
Changes due to crop rotations	0.83	0.84
From grassland to agricultural	0.76	0.67
From agricultural to grassland	0.59	0.92
Changes within near-natural classes	0.35	0.51
From agricultural to mineral extraction sites	0.44	0.97
Kappa coefficient	0.6271	
Overall accuracy	0.7128	

compared to the accuracies of individual LULC maps (Table 3, 4 and 6). Per-class accuracies varied strongly, i.e., changes from agricultural classes to grassland (user's and producer's accuracy > 0.83) performed better than changes in near-natural classes (user's and producer's accuracy < 0.51). According to user 's accuracy (> 0.92) changes from grassland into agricultural classes and from agricultural classes to mineral extraction sites performed best.

Depending on the year, the changes were calculated with regard to the reference year (here 1987), the increase or decrease in the agricultural classes varied. 1987 was chosen due to the higher accuracy compared to 1989. Table 7 shows the changes of all LULC classes for the years 2007, 2009, 2010, and 2011 compared to the reference year. Standard deviation was higher in the Bornhöved Lakes District. The increase in maize cultivation areas in both study areas was largest in 2011 compared to 1987.

4.3. Changes in crop rotation from 2009-2011

Crop rotation could only be assessed for the years 2009-2011 as no continuous time series of Landsat images were available for other years. Both study areas exhibited similar tendencies for both cultivated crops and land cover change (Figure 3 and 4, crop rotation). A mix of cereals and rapeseed represented the most important crop rotation followed by a combination of permanent grassland, maize and cereals because of relatively fertile soils, as shown by Schleuß (1992). Mono-cropping of maize between 2009-2011 was observable but showed a low share of 5-6 %; however, including maize in shorter crop rotations was the second most important practice (mixture of maize and cereals).

Table 7: Change of area [%] of LULC classes in the years 2007, 2009-2011 compared to 1987 and standard deviation (Stdev)

LULC class	Bornhöved Lakes District					Kielstau Basin				
	2007	2009	2010	2011	Stdev	2007	2009	2010	2011	Stdev
Maize	249	89	142	304	85	83	62	57	113	22
Rapeseed	113	110	24	12	47	47	19	41	28	11
Cereals	-24	-11	10	-3	12	28	33	11	26	8
Grassland	-46	-31	-42	-54	8	-51	-44	-32	-52	8
Broad-leaved forest	37	37	37	37	0	14	14	14	14	0
Coniferous forest	-28	-28	-28	-28	0	169	169	169	169	0
Bog	11	11	11	11	0	-75	-75	-75	-75	0
Sealed area	0	0	0	0	0	0	0	0	0	0
Water bodies	-3	-3	-3	-3	0	9	9	9	9	0
Mineral extraction sites	188	188	188	188	0	-	-	-	-	-

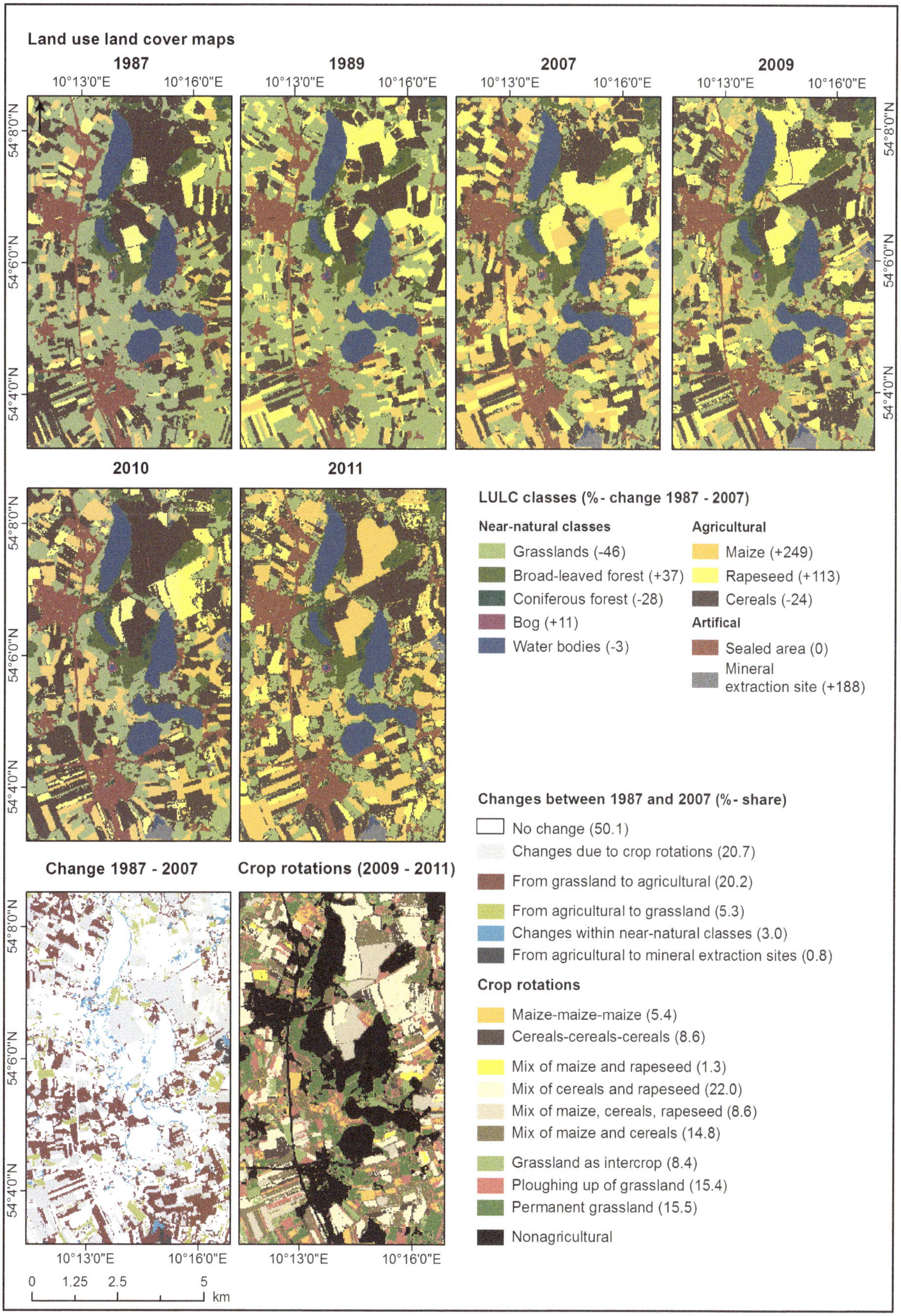

Figure 3: Land use / land cover maps for the Bornhöved Lakes District. Changes [%] in area between 1987 and 2007 are listed in parentheses (see legend) and illustrated spatially in the third row. The third row presents retrieved crop rotations from 2009 – 2011. Share [%] is shown in parentheses (see legend)

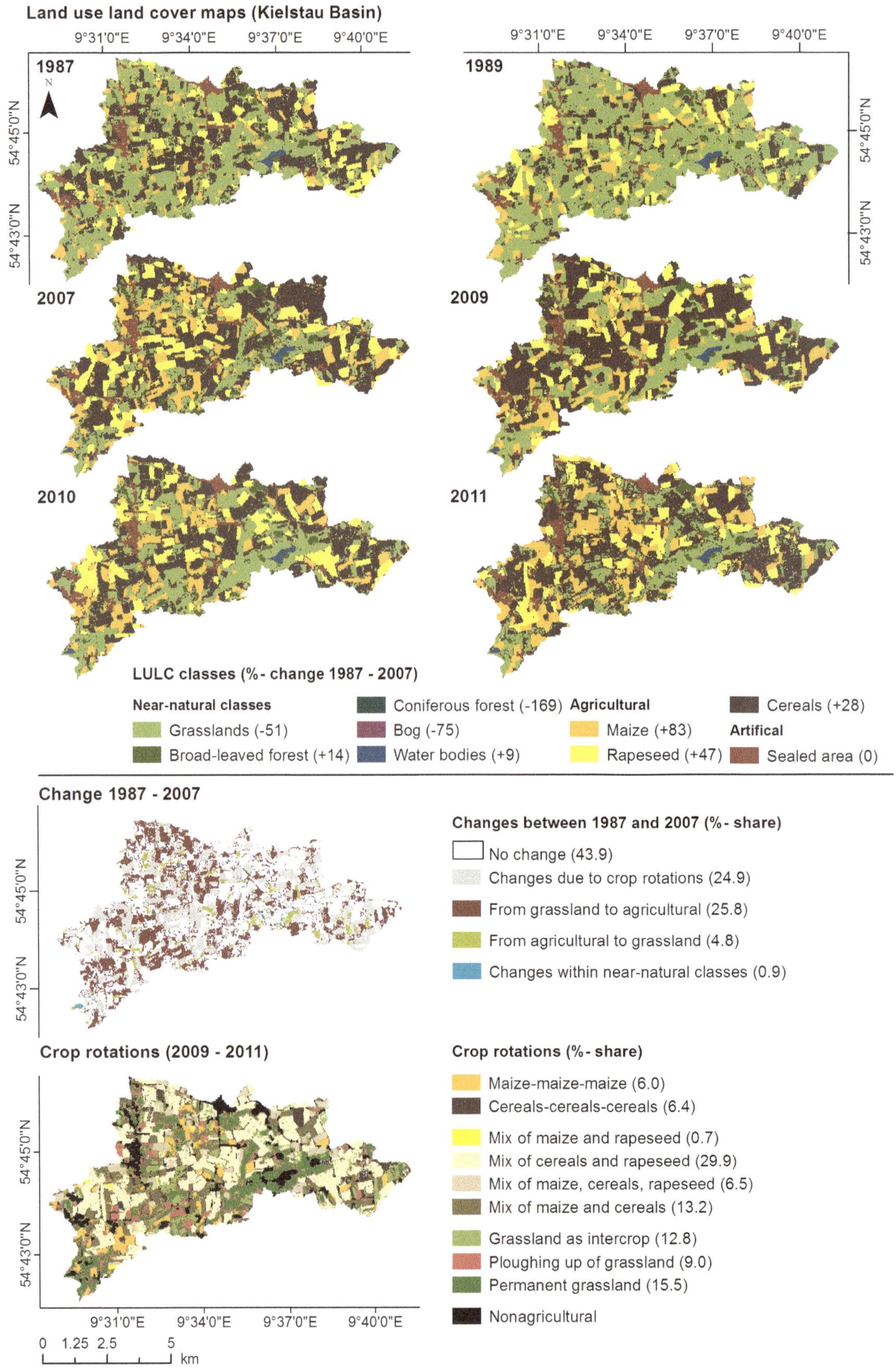

Figure 4: Land use / land cover maps for the Kielstau Basin. Changes [%] in area between 1987 and 2007 are listed in parentheses (see legend) and illustrated spatially in the fourth row. The fifth row presents retrieved crop rotations from 2009 – 2011. Share [%] is shown in parentheses (see legend)

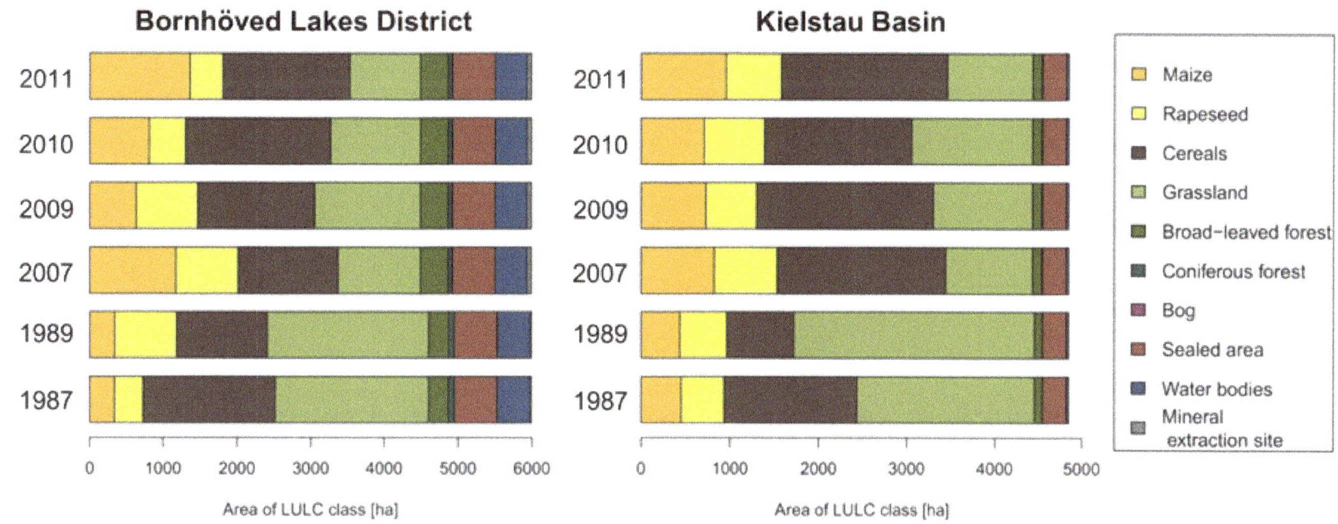

Figure 5: Land use / land cover class areas [ha] of the Bornhöved Lakes District (left) and the Kielstau Basin (right) for the investigated years (1987, 1989, 2007, 2009, 2010, and 2011)

4.4 Changes in ecosystem service supply from 1987-2007

Due to the clear changes in LULC shares for the agricultural areas in both case study areas from 1987 to 2007, several changes in provisioning ecosystem services could be detected. First of all, the decrease in grassland areas was changing the supply of fodder for livestock. This is also connected to the decrease in livestock (Table 8) as the number of cattle has decreased by about one third in both case study areas during 20 years. The harvested fodder from grassland has decreased by about 50 % due to the reduced area as well as lower average yields in 2007 in the districts Plön and Schleswig-Flensburg (Statistikamt Nord 2012).

Second, the increase in maize cultivation is connected to the installation of two biogas plants in each study area (Steen 2013; Hirschberg and Hirschberg n. d.), resulting in a shift in provisioning services from fodder and food supply to the addition of biomass for energy (Table 9). The average harvest for crops differed in 2007 due to a larger share of cereal cultivation in the Kielstau Basin (Table 8).

The LULC maps reveal an addition of two mineral extraction sites in the Bornhöved area due to suitable geological conditions since the beginning of the 1990s.

Table 9 sums up the changes in ecosystem services from 1987 to 2007. The Bornhöved Lakes District shifted the provisioning services supply from fodder, crops, and livestock to biomass for energy, fodder, crops, and livestock production. Mineral extraction was increased by two additional sites in the Bornhöved Lakes District. The LULC changes in the Kielstau Basin resulted in a decrease in the supply of fodder and livestock as well and added biomass for energy supply. Mineral extraction sites are not present in this area.

Table 8: Changes in provisioning services in both study sites from 1987-2007

	Bornhöved Lakes District		Kielstau Basin	
	1987	**2007**	**1987**	**2007**
Maize (dt/a yield)	95860	461108	96826	318345
Rapeseed (dt/a yield)	10602	31438	14480	27145
Cereals (dt/a yield)	133660	98574	105264	144889
Grassland (dt/a yield)	171472	89389	155110	85976
Livestock (number of cattle) *	11624	7692	15264	9468
Mineral extraction sites	1	3	0	0

* data for 1988

Table 9: Changes in selected provisioning ecosystem services from 1987 to 2007. o = ecosystem service was available in this year; / = ecosystem service did not exist in this year; + or – represent an increase / decrease compared to 1987 (based on data from Statistikamt Nord 2012; Statistisches Landesamt SH 1989; expert valuation)

Ecosystem Services	Bornhöved Lakes District		Kielstau Basin	
	1987	**2007**	**1987**	**2007**
Maize (fodder)	o	-	o	-
Maize (biomass for energy)	/	+	/	+
Rapeseed (fodder, crops)	o	+	o	+
Cereals (crops)	o	-	o	+
Grassland (fodder)	o	-	o	-
Livestock (cattle)	o	-	o	-
Mineral extraction sites	o	+	/	/

5 Discussion

Classifying both study areas with a pixel-based maximum likelihood classification approach resulted in highly accurate LULC maps for six years as a basis for delineating LULC changes and assessing ecosystem service dynamics. Nevertheless, the remote sensing approach involves uncertainties which have to be considered for the assessment of ecosystem services together with the uncertainties of the used indicators for quantification of ecosystem services dynamics.

5.1 Remote sensing data set and approach: LULC maps, LULC changes and crop rotation maps

Obtained LULC maps showed very high accuracies that strongly exceeded the 85 % benchmark for thematic mapping in remote sensing applications (Foody 2008). Both study areas were relatively small; detailed, high quality ground truth data sets and partially multi-temporal classification facilitated accurate classification explaining the very high accuracies. Using more than one image of the growing season incorporates phenological characteristics in the classification, which is essential for differentiating grassland and crop types accurately (Pax-Lenney and Woodcock 1997; Franke et al. 2012). Thus, per class accuracies (producer's and user's accuracy) for agricultural classes and grassland exceeded 0.75 throughout. The user's accuracy thus expresses that, e.g., over 86 % (grassland 1987, Kielstau Basin) of the respective LULC class could be located in nature by the end-user. Producer's accuracy means that the analyst correctly classified 94 % (grassland 1987, Kielstau Basin) of the respective LULC class. Clearly visible field structures further underpinned a plausible classification. Moreover, applying a sealed area mask and performing post-classification refinement with the tool IRSeL improved the results, e.g., reduced salt-and-pepper effect of pixel-based classification approaches. With respect to the detection of LULC changes, retrofitting with IRSeL allowed for temporal consistency so that changes resulted rather from LULC changes than classification errors (Rathjens et al. 2014).

Nevertheless, the validation of the Bornhöved Lakes District change map revealed lower accuracies. Post-classification comparison of individual LULC maps to detect changes is subject to spatial inaccuracies and classification errors (Foody 2002). Applying IRSeL on the classification results reduced introduced errors by classification. However, the often disregarded positional accuracy of satellite imagery plays an important role (RMSE in Table 1). The less accurate performance of changes within near-natural LULC classes which depicted small or narrow shapes (Figure 3) underpinned potential errors owing to positional differences. Changes within near-natural classes mainly occurred at lake shores. Mixed spectral signals of water and land in one pixel pose a problem of coarse spatial resolution in classifying the land water interface (Campbell and Wynne 2011). In contrast, changes of LULC in large fields showed higher accuracies; the larger area superimposed positional errors and is not as affected by mixed land covers. Nevertheless, classifying a heterogeneous landscape into 9 or 10 semantic classes on the basis of 30 m spatial resolution always represents a model of reality and involves errors, for instance the explained mixed land covers in one pixel (Boyd and Foody 2011).

Using remote sensing techniques and Landsat imagery, however, enabled the differentiation between agricultural crops, i.e., maize, rapeseed, and cereals which is an important advantage in the assessment and mapping of ecosystem services (Kandziora et al. 2013b). Because they provide different ecosystem services, the distinction between these crop types represents a valuable database for a detailed ecosystem service assessment as different crops account for different provisioning services and influence regulating services, such as erosion control, and cultural services differently (de Vries et al. 2010). Depending on the availability of imagery and ground truth data remote sensing presents a further advantage to obtain LULC data over long time series (Boyd and Foody 2011). Therefore, this study benefited from the long data archive of Landsat TM imagery (Cohen and Goward 2004). Thus, LULC changes could be analysed in both study areas for the time period of two decades. Selecting the year 1987 as change reference was reasonable due to the higher accuracies compared to the 1989 maps. Furthermore, the LULC map of 2007 is the first data set after the commencement (2000) and first amendment (2004) of the German Renewable Energies Act (EEG) (UBA 2011). Since biogas plants were subsidised in the EEG, distinct changes in the agricultural landscape were expected between the two time steps as agricultural statistics showed tendencies towards increased cultivation of maize in Schleswig-Holstein (Statistikamt Nord 2012), which is also noticeable in the further increased area until 2011 (Figure 5). Regarding land cover patterns, both study areas exhibited similar tendencies and were chosen to prove the changes in agricultural landscapes in Schleswig-Holstein. The statistical agricultural data for whole Schleswig-Holstein matched the detected LULC change dynamics: a reduction of grassland area was going along with increased maize and cereal cultivation (Statistikamt Nord 2012; UBA 2011). The strong increase of maize is founded on the more profitable cultivation related to the EEG as has been confirmed by previous studies in nearby locations (Oppelt et al. 2012; MELUR (2011).

Statistical information on cultivated crops revealed that further crops, such as sugar beets and potatoes, were cultivated in Schleswig-Holstein but to a smaller extent (Statistikamt Nord 2013;

MELUR 2010). Deriving separated information from these crops based on Landsat in this region was problematic; depending on image acquisition dates, the spectral behaviour of these crops resolved by Landsat represented a mixture of bare ground and green vegetation. For the sake of similar spectral behaviour and the difficult differentiation from maize in a classification, a separate class was not considered. Therefore, agricultural classes contained a bias, however, constant within the time series.

The LULC grassland included several types of grassland related land use practices, e.g., permanent unused grassland, meadows, or grassland for fodder production. With the data used in this study, conclusions on the practices of the more extended grasslands in the 1980es could be solely drawn by incorporating additional agricultural data. Further information on intensity of use, fertilizer use, and animal distribution must be included when analysing, e.g., regulating ecosystem services and biodiversity. The indicator yield from hay meadows is also only a proxy indicator due to the limited distinction of grassland types. The same applies to the use of winter wheat as a proxy for cereals because other crops, such as rye and barley, could not be distinguished by the applied remote sensing method.

Considering the analysis of the data of three consecutive years, 2009-2011, with regard to crop rotations, information on current grassland use practices was gathered. In both study areas, 15.5 % of the arable land was grassland throughout the three years indicating that this is permanent grassland or meadows. Grassland as intercrop indicated the utilization for fodder. The change class of "ploughed up grassland" were pixels that were grassland in 2009 (and 2010) and a crop fruit in 2010 and/or 2011. The shares of this practice were higher in the Bornhöved Lakes District and a considerable area of grassland conversion within that short period can be deduced. However, the analysis of three consecutive years may lead to a biased view. As explained previously the exact former usage of the grassland areas cannot be derived solely from the LULC maps. Thus, a definite conversion from unused grassland to arable land cannot be deduced. Even though the exact usage is not known, the LULC grassland retreated while the area of arable land increased. With respect to crop rotation practices on arable land, high energy yields per area (Rode 2005) and the self-compatibility of maize lead to tendencies of short crop rotations or even maize mono-cropping (Schultze and Köppel 2007) in parts of Schleswig-Holstein. Due to the discontinued time series of the remote sensing data, no long-term crop rotation practices could be detected. This problem has already been described by Meyer (2000) for the Bornhöved Lakes District. MELUR (2011) recommends, in general, a site-specific crop rotation.

5.2. Ecosystem services assessment: quantification by statistical data

The subgrouping of the provisioning services gives the opportunity to analyse the changes in the supply of the individual ecosystem services. However, the quantification of the individual ecosystem services raises the problem of assigning them indicators. Some data of the agricultural statistics can be assigned to several ecosystem services (e.g., winter wheat for crops and fodder, maize for fodder and biomass for energy). The use of rapeseed for biodiesel production is not considered in this assessment because it applies to another spatial scale since further processing of the biomass is needed, whereas a local relation between the production of maize and the installation of biogas plants can be assumed. Approximately 50 % of the maize yield is used for fodder and 50 % for energy production in biogas plants (MELUR 2011). Other substrates, such as manure and slurry, are used in biogas plants as well. Though the tracking of the substrates to a specific location is difficult (Hirschberg and Hirschberg n. d.; Steen 2013). The correlation between the increase in biogas plants due to energy laws and subsidies and the increase in maize cultivation areas is evident. A further increase in maize area and number of biogas plants in Schleswig-Holstein in the near future is expected (MELUR 2011). For instance, a third biogas plant is planned in the Kielstau Basin (Steen 2013).

The increase in maize cultivation area and the decrease in grassland area are influencing other ecosystem services negatively, especially regulating ecosystem services: e.g., erosion and nutrient loss is increasing because of the low plant cover during most of the year. Negative consequences for the protection of water bodies due to changes in the water and nutrient balance, loss of biodiversity, and abiotic diversity (BfN 2009; MELUR 2012) are expected. The loss of grassland implicates a decrease in the multifunctionality of landscapes. Grasslands contribute notably to landscape aesthetics and landscape beauty, which are cultural ecosystem services. Loss of grassland area in recent years has occurred in other regions in Germany as well (UBA 2011). For that reason, a further analysis of the changes in regulating and cultural services is needed by means of additional data sets and quantification methods (e.g., modelling) compared to the LULC-based quantification method applied here.

Agricultural production area and yields are sensitive to climatic variations (Lorencová et al. 2013), which results in fluctuations of harvest yields and cultivation area. Crop rotation is highly influenced by weather conditions during the sowing periods, such as late sowing due to wet or cold conditions, with a possible shift in future crop rotations. This must be considered when assessing time series. Statistical data have exposed an additional decline in permanent grassland area and an increase in maize cultivation area until 2010 (Statistikamt Nord 2013).

A further increase in the population of wild boars is expected by MELUR (2011), however, in different intensities in the districts in Schleswig-Holstein. The expansion of wild boar is favoured by good food and habitat conditions, which is given by the increase in maize cultivation areas. The main habitats can be found in the southern districts of Schleswig-Holstein. Therefore, the number of hunted animals (which could be used as a quantification indicator for the provisioning service "wild food"; Kandziora et al. 2013a) in 2007 varies largely between the case study areas. The Kielstau Basin is located in a region with a smaller amount of game; especially the number of wild boars is much lower in comparison. An expansion to the north is expected (MELUR 2008; 2012), as well as the further increase in population due to favourable food and habitat conditions.

Mineral resources were analysed in this assessment although they are not a provisioning ecosystem service in its classical meaning. The CICES classification excluded them because they depend more on 'geodiverstiy' than on biodiversity and the prevailing processes operating on the long time scale of ecosystem development (Haines-Young and Potschin 2010). But other authors (Brown et al. 2007; Kandziora et al. 2013a) argue for their inclusion. In the case of the Bornhöved Lakes District they provide an example of how other human activities can influence LULC and indirectly the supply of ecosystem services through the decline of agricultural area. There are several mineral extraction sites to the south and east of the study area because of the geological settings of the Kalübber and Trappenkamper Sander (Fischbock and Venebrügge 1990). Thus, this LULC class plays a major role in this research area. On the other hand, this land use might be of limited temporal scale, as the recent development in the Bornhöved Lakes District shows a partial conversion from one of the mineral extraction sites to fallow/agricultural land and the opening of a new one on former agricultural land in 2012.

In general, there is need for further development of suitable indicators for quantifying all existing ecosystem services in a case study area with spatial and temporal explicitness. Derived LULC maps reveal high potential for the application in LULC-based ecosystem services modelling. This information must be included in decision-making and environmental management.

6 Conclusion

The combined remote sensing and ecosystem services quantification approach enabled the detection of LULC changes and resulting shifts in the supply of ecosystem services in both case study areas in Northern Germany. The outcome of this study can be summarised by clear changes in the shares of agricultural and near-natural LULC classes: a decrease in grassland and a change in crop cultivation from 1987 to 2007 revealed a shift from fodder and crop production to the production of biomass for energy generation and a decrease in the supply of fodder and livestock. The crop rotation maps from 2009-2011 exposed further ploughing of grassland in combination with starting maize mono-cropping. In order to draw detailed conclusions for decision-making and management of agroecosystems a further analysis of the effects of these supply changes in provisioning ecosystem services on regulating and cultural ecosystem services is necessary.

The study demonstrated the potential of LULC classifications for the assessment of ecosystem services and their dynamics in more detail. This fact is often neglected. Remote sensing techniques can provide further methodological opportunities for integrated approaches in the assessment of ecosystem services at various spatial and temporal resolutions (e.g., Blaschke 2010; Kennedy et al. 2009). Combining remote sensing and ecosystem services research in interdisciplinary studies may enhance the assessment and monitoring of ecosystem services at different temporal, spatial, and semantic scales.

Acknowledgments

The authors would like to thank the Schleswig-Holstein state offices LVA, LLUR and MLUR/MELUR for providing ground truth data sets as well as data to quantify ecosystem services. The Statistikamt Nord kindly provided further data for quantification of ecosystem services and Britta Schmalz kindly supplied additional information for the Kielstau Basin. Furthermore, the USGS and NASA are thanked for the provision of Landsat TM data.

References

BfN (Bundesamt für Naturschutz) 2009. Where have all the flowers gone? Grünland im Umbruch. Hintergrundpapier und Empfehlungen des BfN. http://www.bfn.de/fileadmin/MDB/documents/themen/landwirtschaft/Gruenlandumbruch_end.pdf (Date: 18.08.2013).

Blaschke, T. 2010. Object based image analysis for remote sensing. ISPRS Journal of Photogrammetry and Remote Sensing 65 (1), 2–16.

Boyd, D. & G. M. Foody 2011. An overview of recent remote sensing and GIS based research in ecological informatics. Ecological Informatics 6 (1), 25–36.

Brown, T. C.; Bergstrom J. C. & J. B. Loomis 2007. Defining, valuing, and providing ecosystem goods and services. Natural Resources Journal 47 (2), 329-376.

Campbell, J.B. & R.H. Wynne 2011. Introduction to remote sensing. Guilford Press, New York.

Chander, G.; Markham, B. & D. Helder 2009. Summary of current radiometric calibration coefficients for Landsat MSS, TM, ETM+, and EO-1 5ALI6 sensors. Remote Sensing of Environment 113 (5), 893–903.

Cohen, W. B. & S.N. Goward 2004. Landsat's Role in Ecological Applications of Remote Sensing. BioScience 54 (6), 535–545.

Congalton, R.G. 1991. A review of assessing the accuracy of classifications of remotely sensed data. Remote Sensing of Environment 37(1), 35–46.

Crossman, N. D.; Burkhard, B. & S. Nedkov 2012. Quantifying and mapping ecosystem services. International Journal of Biodiversity Science, Ecosystem Services & Management 8 (1-2), 1-4.

Dale, V. H. & S. Polasky 2007. Measures of the effects of agricultural practices on ecosystem services. Ecological Economics 64, 286-296.

de Groot, R. S.; Alkemade, R..; Braat, L.; Hein, L. & L. Willemen 2010. Challenges in integrating the concept of ecosystem services and values in landscape planning, management and decision making. Ecological Complexity 7, 260–272.

de Vries, S. C.; van de Ven, G. W. J.; van Ittersum, M. K. & K. E. Giller 2010. Resource use efficiency and environmental performance of nine major biofuel crops, processed by first-generation conversion techniques. Biomass and Bioenergy 34, 588-601.

ESRI 2010. ArcGIS 10.0.

Estoque, R. C. & Y. Murayama 2012. Examining the potential impact of land use/cover changes on the ecosystem services of Baguio city, the Philippines: A scenario-based analysis. Applied Geography 35 (1-2), 316–326.

Fischbock, A. & G. Venebrügge 1990. Bevölkerung, Erwerbsstruktur und Wandel der Flächennutzung im Hauptforschungsraum Bornhöveder Seenkette. Interne Mittelungen, Heft 4, 233-235.

Fohrer, N. & B. Schmalz 2012. Das UNESCO Ökohydrologie-Referenzprojekt Kielstau-Einzugsgebiet - nachhaltiges Wasserressourcen-management und Ausbildung im ländlichen Raum. Hydrologie und Wasserbewirtschaftung 56 (4), 160-168.

Foody, G. M. 2008. Harshness in image classification accuracy assessment. International Journal of Remote Sensing 29 (11), 3137–3158.

Foody, G. M. 2002. Status of land cover classification accuracy assessment. Remote Sensing of Environment 80 (1), 185–201.

Franke, J.; Keuck, V. & F. Siegert 2012. Assessment of grassland use intensity by remote sensing to support conservation schemes. Journal for Nature Conservation 20 (3), 125–134.

Fränzle, O.; Kappen, L.; Blume, H.-P. & K. Dierssen (eds.) (2008). Ecosystem Organization of a Complex Landscape - Long-Term Research in the Bornhöved Lake District, Germany. Springer, Berlin Heidelberg. Ecological Studies 202.

Haines-Young, R. & M. Potschin 2010. Proposal for a common international classification of ecosystem goods and services (CICES) for integrated environmental and economic accounting. Background document. European Environmental Agency. http://unstats.un.org/unsd/envaccounting/ceea/meetings/UNCEEA-5-7-Bk1.pdf (Date: 10.08.2013)

Hirschberg, K. & L. Hirschberg n.d. Gut Perdöl. http://www.perdoel.de/gut_perdoel.html (Date: 10.08.2013).

ITTVIS 2010. Envi. Version 4.8.

Kandziora, M.; Burkhard, B. & F. Müller 2013a. Interactions of ecosystem properties, ecosystem integrity and ecosystem service indicators - A theoretical matrix exercise. Ecological Indicators 28, 54-78.

Kandziora, M.; Burkhard, B. & F. Müller 2013b. Mapping provisioning ecosystem services at the local scale using data of varying spatial and temporal resolution. Ecosystem Services 4, 47-59.

Kennedy, R. E.; Townsend, P. A.; Gross, J. E.; Cohen, W. B.; Bolstad, P.; Wang, Y. Q. & P. Adams 2009. Remote sensing change detection tools for natural resource managers: Understanding concepts and tradeoffs in the design of landscape monitoring projects. Monitoring Protected Areas 113 (7), 1382–1396.

Kiesel, J.; Schmalz, B. & N. Fohrer 2009. SEPAL – a simple GIS-based tool to estimate sediment pathways in lowland catchments. Adv. Geosci 21, 25–32.

Koschke, L.; Fürst, C.; Lorenz, M.; Witt, A.; Frank, S. & F. Makeschin 2013. The integration of crop rotation and tillage practices in the assessment of ecosystem services provision at the regional scale. Ecological Indicators 32, 157-171.

Kroll, F.; Müller, F.; Haase, D. & N. Fohrer 2012. Rural–urban gradient analysis of ecosystem services supply and demand dynamics. Land Use Policy 29, 521–535.

Lautenbach, S.; Kugel, C.; Lausch, A. & R. Seppelt 2011. Analysis of historic changes in regional ecosystem service provisoning using land use data. Ecological Indicators 11, 676-687.

Leh, M.D.K.; Matlock, M. D.; Cummings, E. C. & L.L. Nalley 2013. Quantifying and mapping multiple ecosystem services change in West Africa. Agriculture, Ecosystems and Environment 165, 6– 18.

LLUR (Landesamt für Landwirtschaft, Umwelt und ländliche Räume) 2012. Data on protected areas in Schleswig-Holstein. Personal communication.

Lorencová, E.; Frélichová, J.; Nelson, E. & D. Vačkář 2013. Past and future impacts of land use and climate change on agricultural ecosystem services in the Czech Republic. Land Use Policy 33, 183–194.

Lorenz, M.; Fürst, C. & E. Thiel 2013. A methodological approach for deriving regional crop rotations as basis for the assessment of the impact of agricultural strategies using soil erosion as example. Journal of Environmental Management 127, S37-S47.

LVA (Landesamt für Vermessung und Geoinformation Schleswig-Holstein) 2007. ATKIS (Official Topographical Cartographic Information System). Landesamt für Vermessung und Geoinformation Schleswig-Holstein. Personal communication.

MEA (Millennium Ecosystem Assessment) 2005. Ecosystems and human well-being: synthesis; a report of the Millennium Ecosystem Assessment. Island Press, Washington DC. 137 pp.

MELUR (Ministerium für Energiewende, Landwirtschaft, Umwelt und ländliche Räume) 2012. Jahresbericht 2012. Jagd und Artenschutz. http://www.schleswig-holstein.de/MELUR/DE/Service/Broschueren/Umwelt/pdf/Jagd_und_Artenschutz_2012__blob=publicationFile.pdf (Date: 01.08.2013).

MELUR (Ministerium für Energiewende, Landwirtschaft, Umwelt und ländliche Räume) 2011. Empfehlungen zur Optimierung des Maisanbaus in Schleswig-Holstein. http://www.schleswigholstein.de/UmweltLandwirtschaft/DE/LandFischRaum/05_Pflanzenproduktion/001_Maisanbau/Maisanbau_node.html (Date: 01.08.2013).

MELUR (Ministerium für Energiewende, Landwirtschaft, Umwelt und ländliche Räume) 2010. IACS (Integrated Administration and Control System) land use data. Ministerium für Energiewende, Landwirtschaft, Umwelt und ländliche Räume des Landes Schleswig-Holstein - Ministry of Energy, Agriculture, the Environment and Rural Areas Schleswig-Holstein. Personal communication.

MELUR (Ministerium für Energiewende, Landwirtschaft, Umwelt und ländliche Räume) 2008. Jahresbericht 2008. Jagd und Artenschutz. http://www.schleswig-holstein.de/MELUR/DE/Service/Broschueren/Umwelt/pdf/Jagd_und_Artenschutz_08__blob=publicationFile.pdf (Date: 15.08.2013).

Meyer, M. 2000. Entwicklung und Modellierung von Planungsszenarien für die Landnutzung im Gebiet der Bornhöveder Seenkette. Dissertation Christian-Albrechts-Universität zu Kiel.

Müller, F.; Schrautzer, J.; Reiche, E.-W. & A. Rinker 2006. Ecosystem based indicators in retrogressive successions of an agricultural landscape. Ecological Indicators 6, 63–82.

Oppelt, N.; Rathjens, H. & K. Dörnhöfer 2012. Integration of land cover data into the open source model SWAT. Proceedings of the First Sentinel-2 Preparatory Symposium held on 23-27 April 2012 in ESA-ESRIN, Frascati.

Pax-Lenney, M. & C. E. Woodcock 1997. Monitoring agricultural lands in Egypt with multitemporal Landsat TM imagery: How many images are needed? Remote Sensing of Environment 59 (3), 522–529.

Polasky, S.; Nelson. E.; Pennington, D. & K.A. Johnson 2011. The Impact of Land-Use Change on Ecosystem Services, Biodiversity and Returns to Landowners: A Case Study in the State of Minnesota. Environ Resource Econ 48, 219–242.

Rathjens, H.; Dörnhöfer, K. & N. Oppelt, N. 2014. IRSeL – An approach to enhance continuity and accuracy of remotely sensed sensed land cover data. International Journal of Applied Earth Observation and Geoinformation 31, 1-12.

Rode, M. 2005. Energetische Nutzung von Biomasse und der Naturschutz. Natur und Landschaft 80, 403–412.

Schleuß, U. 1992. Böden und Bodenschaften einer norddeutschen Moränenlandschatft: ökologische Eigenschaften, Vergesellschaftung und Funktion der Böden im Bereich der Bornhöveder Seenkette. EcoSys 2.

Schultze, C. & J. Köppe 2007. Gebietskulissen für den Energiepflanzenanbau? Steuerungsmöglichkeiten der Planung. Naturschutz und Landschaftsplanung 39, 269–272.

Sexton, J. O.; Urban, D. L.; Donohue, M. J. & C. Song 2013. Long-term land cover dynamics by multi-temporal classification across the Landsat-5 record. Remote Sensing of Environment 128, 246–258.

Song, C.; Woodcock, C. E.; Seto, K. C.; Lenney, M. P. & S.A. Macomber 2001. Classification and Change Detection Using Landsat TM Data: When and How to Correct Atmospheric Effects? Remote Sensing of Environment 75 (2), 230–244.

Statistikamt Nord 2013. Statistisches Jahrbuch Schleswig-Holstein 2012/2013. http://www.statistik-nord.de/uploads/tx_standocuments/JB12SH_gesamt.pdf (Date: 15.08.2013).

Statistikamt Nord 2012. Data on livestock, harvest yields, cultivation area 1987-2011. Personal communication.

Statistisches Landesamt Schleswig-Holstein 1989. Statistische Berichte des Statistischen Landesamt Schleswig-Holstein. Gemeindeergebnisse der allgemeinen Viehzählung 1988.

Steen, H. 2013. Daten zur Biogasanlagen im Kielstau-Gebiete. Personal communication.

UBA (Umweltbundesamt) 2011. Data on the environment 2011 edition - Environment and agriculture. http://www.umweltdaten.de/publikationen/fpdf-l/4129.pdf (date: 15.08.2013).

USGS (U.S. Geological Survey) 2013. FAQ about Landsat Missions: What are the band designations for the Landsat satellites? http://landsat.usgs.gov/band_designations_landsat_satellites.php (Date: 27.08.2013).

Wandhoff, V. 2013. Kiesgrube Vierhusen. Personal communication.

Yuan, F.; Sawaya, K. E.; Loeffelholz, B. C. & M. E. Bauer 2005. Land cover classification and change analysis of the Twin Cities (Minnesota) Metropolitan Area by multitemporal Landsat remote sensing. Remote Sensing of Environment 98 (2–3), 317–328.

Competing Wind Energy Discourses, Contested Landscapes

Markus Leibenath[1]*, Antje Otto[2]

1 Leibniz Institute of Ecological Urban and Regional Development, Weberplatz 1; 01217 Dresden; Germany
M.Leibenath@ioer.de

2 University of Potsdam, Institute of Earth and Environmental Science, Karl-Liebknecht-Str. 24-25, 14476 Potsdam-Golm Germany; anotto@uni-potsdam.de

---

Abstract

The impairment of landscapes is a concern constantly raised against wind energy developments in Germany as in other countries. Often, landscapes or landscape types are treated in the literature as essentialist or at least as uncontested categories. We analyse two examples of local controversies about wind energy, in which "landscape" is employed by supporters and opponents alike, from a poststructuralist and discourse theoretical angle. The aim is to identify and compare landscape constructs produced in the micro discourses of wind energy objectors and proponents at local level (a) within each case, (b) between the two cases and (c) with landscape constructs that were previously found in macro discourses. One major finding is that several different landscapes can exist at one and the same place. Furthermore there seems to be a relatively stable set of competing landscape concepts which is reproduced in specific controversies. The paper concludes by highlighting practical consequences and by identifying promising avenues of further research.

Keywords:
Social construction of landscapes, local energy politics, poststructuralist discourse theory

---

*Corresponding author. Email: M.Leibenath@ioer.de

1 Introduction

Many authors point out that the impairment of landscapes is a constant concern raised against renewable energy technologies in general and wind energy developments in particular (e.g., Nadaï & Van der Horst 2010; Wolsink 2007). According to Warren et al. (2005: 857), "opposition to the visual despoliation of valued landscapes" is the "key motivation for anti-wind farm campaigners." Wolsink & Breukers (2010) underscore the salience of landscape values for local protests against wind power. Likewise, Wüstenhagen et al. (2007: 2684) attest to the "crucial significance of landscape issues in the attitude towards wind power schemes."

But what do people mean when they talk about "landscape" or "landscape issues"? "Landscape" covers a wide variety of meanings. They oscillate, for instance, between a physical phenomenon and its image or visual impression. "Landscape is both the phenomenon itself and our perception of it" (Wylie 2007: 7). Hence there are authors who stress the material and functional properties of landscapes as do many landscape ecologists, while others treat landscapes as social constructions or try to integrate both views (for an in-depth discussion of possible meanings of "landscape," cf. e.g. Cosgrove 2003; Gailing & Leibenath 2012; Henderson 2003; Olwig 2001).

In this paper, we adopt a social constructionist view of landscape (for a comparison between constructionist and other approaches to landscape research, cf. Gailing & Leibenath 2010). More specifically, we look on the social construction of landscapes from a poststructuralist, discourse theoretical angle (for a more in-depth discussion of different types of constructionist landscape research, cf. Gailing & Leibenath 2013; Kühne 2009; Leibenath 2013). In this perspective, to put it simply, landscapes are conceptualized as relational systems of meaning (cf. Wylie 2007: 94)—or discourses—which consist of linkages between material objects, actions, words and subjects. This is a non-essentialist understanding of "landscape" because, as discourse analysts, we are describing second-order observations of what people call "landscape" or—in a broader sense—of how people relate themselves to material objects and to others through words and actions. It is self-evident that such landscapes can be assessed neither mechanically, for instance by means of landscape metrics in combination with geographic information systems, nor "objectively" without consciously reflecting upon the perspectives of the observed and the observers, as well as upon the contexts in which observations are made (these and other tenets of post-positivist research are discussed in greater detail in, e.g., Fischer 1998). What is needed instead is an interpretive approach such as discourse analysis.

A number of authors have employed discourse analysis to study wind energy politics. Szarka (2004) examined wind energy policy discourses and related discourse coalitions in three European countries and at the EU level. Concepts of landscape, space or place play a marginal role therein and are mentioned only as part of anti-wind protest discourses (" 'industrialisation' of 'sensitive' landscapes by massive wind turbines": 325). Similarly, Jessup (2010) looked at macro discourses without reference to specific projects in Australia and the United Kingdom and established storylines in the discourses of objectors and supporters of wind energy. According to Jessup, landscape again appears exclusively in the objectors' storyline "valuing landscape and place" (26).

Gee (2010) analysed local "views of the landscape" (189) which emerged in discussions about wind power developments off Germany's North Sea coast and "whether the meanings given to the sea and the coastal landscape influence acceptance of offshore wind farms" (186). She speaks of landscapes exclusively in the singular. This is in line with her finding that even the supporters of the disputed wind farms basically adhere to the idea of protecting "nature or the seascape," although they "are willing to trade this for forms of energy generation that are safe and climate-friendly" (194). This means that all respondents basically share a common understanding of the North Sea and the coastal landscape.

Other authors also address landscape in wind energy conflicts. Wolsink (2007), for instance, stresses that investors and key stakeholders such as the members of an environmental organisation evaluated a number of predefined landscape types such as "agricultural areas," "mound landscape," and "recreational areas" (2697) differently with regard to their suitability as sites for the construction of a wind farm. He thus states "that what matters is the type of landscape that is selected for the siting of wind turbines" (2696). And: "Hence, the planning and decision-making process has to focus on the type of landscape" (2702). Wolsink treats landscapes (or landscape types) as essentialist categories. There are given landscape types and there are the evaluations or assessments of these landscape types expressed by members of a major stakeholder organisation. In Wolsink's study, only the latter seem to be an issue of deliberation. Thus, in the case he describes, what landscape type a certain stretch of land represents and whether it has to be regarded as "nature area" or "industry, harbour area" (Wolsink 2007: 2697) are not contested.

There is accordingly a strand of literature (Gee 2010; Jessup 2010; Szarka 2004; Wolsink 2007) in which "landscape" appears to be a relatively homogenous, uncontested notion in wind energy conflicts. This is contrasted by the findings of Leibenath and Otto (2012), who scrutinized landscape constructs in German wind energy politics and basically discern two concepts which they label "landscape as beautiful, valuable area" and "landscape as an area shaped by humans" (126). The first of these two macro-discourses can actually be traced back to Germany's romantic landscape discourse, which has been described by historians such as Blackbourn (2007). A third landscape concept, which Leibenath and Otto (2012: 126) term "landscape as something which is subjectively perceived," occurred only marginally.

Against this background, the article pursues three research questions with an eye to local controversies about onshore wind energy projects in Germany. The first is whether there are local debates on wind turbines in which competing landscapes are constructed. The second question is what similarities and differences exist between landscape constructs from different local wind energy discourses. And, finally, the extent to which landscape constructs in discourses at the local level overlap with landscape constructs that have been identified in wind energy discourses at the macro level. To answer these questions, we present a case study on a dispute about a single wind turbine in the municipality of Ingersheim in Baden-Wuerttemberg, and contrast it with another case study on a planned wind farm in the town of Wolfhagen in Hesse (Leibenath & Otto 2013b; Otto & Leibenath 2014).

The findings are not intended to be generalised and to contribute to a general theory of landscape discourses related to wind energy developments which would allow predictions about other cases. Instead, our overall aim is to show whether and, if so, how landscapes become part of political conflicts about wind energy, to help better understand such conflicts and to examine the social construction of landscape from the viewpoint of politics and power.

The paper is organised in three parts. First, we outline our poststructuralist understanding of discourse, space and the discursive construction of landscapes. This sets the frame for the second part of the paper in which the methods, contexts and results of our empirical analyses are presented and discussed. Finally we draw conclusions on practical implications of our findings and further research needs.

2 Discourse, Space, and the Discursive Construction of Landscapes

Like most other scholars dealing with discourse (cf. the overview given by Feindt & Oels 2005), Laclau & Mouffe pay attention to the constitutive role of language in constructing social facts. However, they reject the distinction between discourse as something merely linguistic and an extra-discursive, non-linguistic reality. They take discourses to be structured totalities which relate linguistic elements such as words and utterances to objects and practices in a contingent manner (Laclau 1993; Laclau & Mouffe 1985).

From this perspective it is a characteristic of discourses that they imply the articulation of a political frontiers between the inside and outside, or "between insiders and outsiders, which exclude particular practices and possibilities" (Griggs & Howarth 2012: 309). A logic of equivalence to some extent dissolves the differences between a number of elements by relating them positively to one element on the inside and negatively to elements on the discursive outside. Elements to which many others are related are called nodal points or, in the extreme, empty signifiers tending to be free of any particular meaning precisely because they are related to so many other elements (Howarth & Stavrakakis, 2000; Laclau & Mouffe, 1985).

It is in principle undecidable what a discursive element "really" means. Therefore it becomes a matter of power to institute a certain discourse, to establish an antagonistic frontier between inside and outside (cf. Howarth 2010: 309 f.) and to define—by articulating relations of equivalence or difference—what, e.g., a landscape is and what it is not. All these decisions are contingent in the sense that they are made on an "undecidable terrain. The conclusion is . . . that as undecidability operates at the very ground of the social, objectivity and power become indistinguishable" (Laclau 1993: 435). Every discourse privileges certain types of knowledge and devaluates others. It has its regime of truth which determines what can be said and what not. Hence, power and knowledge are inextricably interwoven (cf. Dryzek 1997: 9; Foucault 2005 [1976]; Waage & Benediktsson 2010: 5). This type of power is both repressive and productive. Moreover, power relations of this kind can be studied at any level or scale. For instance, Foucault (1983 [1976]) recommended to study the workings of power at the micro level of everyday practices and how they relate to more global strategies.

Poststructuralist discourse theory also offers a specific approach to landscape or space. Commonly, space is treated by many social scientists as a phenomenon with physical and social dimensions (cf. Bourdieu 1991; Lefebvre 1991 [1974]). By contrast, Laclau & Mouffe (1985) assert that no physical object can gain social presence without being mediated by discourses and that therefore all social facts are discursive facts (cf. also Laclau 1990; Natter & Jones III 1993). If one adheres to their anti-essentialist, post-foundational perspective, then space has to be regarded as a system of differential relations between physical objects, practices and linguistic elements—i.e., as discourse or as a part thereof. In any case, it is impossible to draw a line between space and its representations, because physical space (understood as ensembles of material objects), practices, and linguistic elements such as place names are ineluctably intertwined. Another consequence is that spaces are not seen as something absolute, but firstly as contingent effects of power and secondly as contested or at least contestable phenomena (Jones III & Natter 1999). The same applies to landscapes.

As discourse researchers, we seek to deconstruct existing ontologies of "landscape" and analyse the power-laden mechanisms of inclusion and exclusion, as well as the relations between linguistic and non-linguistic elements by which they come to bear. In the following we speak of "landscape concepts," meaning specific ontologies of landscape, which we address as a discursive structure. These can either be supra-local concepts which include the word "landscape," or local, place-specific discourses in which toponymic expressions such as "Saxon Switzerland" or "North Kent Marshes" usually play a greater role.

3 Data and Analytical Methods

The empirical findings are based on case studies in two German municipalities: Ingersheim and Wolfhagen. Departing from a comprehensive survey about local debates on wind energy in Germany (Leibenath & Otto 2013a), the two cases were chosen because they were among the few in which "landscape" figures prominently in the discourses of both objectors and proponents. This is an embedded multiple-case design, as each of the two cases comprises several embedded units of analysis, i.e., several discourses (cf. Yin 2003: 42). The cases were selected according to the logic of the "most different systems design" rather than the logic of the "most similar systems design" (Blatter et al. 2006: 177 f.): They "share certain family resemblances" (Howarth 2005: 334) with regard to the phenomenon of interest (i.e., landscape discourses produced by opponents and supports), but differ in many other ways (e.g., different size of the schemes, different physical environments, different histories, different political conditions at state level etc.).

In both cases we adopted a multi-data approach (see table 1). We relied primarily on non-reactive linguistic data such as websites of the organisations involved, leaflets, local newspaper reports, and letters to the editors of local newspapers. With regard to this type of written material, we tried to establish as comprehensive a corpus as possible.

Furthermore we conducted semi-structured, open-ended interviews with key persons. They were selected on the basis of our empirical knowledge gained from document analysis. Moreover we asked interviewees who else we should talk to in order to get a fuller picture. The interview guidance included questions on the personal background of interviewees, the linguistic structures of the given discourse (e.g., "Why are you [Why is your group] fighting for [against] the wind energy scheme?"), related non-linguistic practices, the historic development of the discourse, context conditions, potential turning points, and the coalitions that the discourse produced.

Table 1: Overview of analysed documents and conducted interviews in the case studies

	Ingersheim	Wolfhagen
Document analysis	Corpus of about 300 documents mainly from 2010-2012. In-depth analysis of ten written documents: ▪ four leaflets (EGIU 2010a; Müller 2010a; 2010b; 2010c), ▪ one website entry (EGIU 2010c), ▪ one open letter (Huber & Huber 2010), ▪ four letters to the editors of local newspapers (Eisenmann 2011; Grimm 2011; Haecker 2010; Schieber 2010), and ▪ two press releases (SHB 2010; SPD 2011).	Corpus of 500 documents dating from 2008-2012 In-depth analysis of eight written documents: ▪ two leaflets (BI 2010a; Götte & Degenhardt-Meister 2009), ▪ a series of leaflets (SVF 2010), ▪ two brochures (BI 2010b; SWG 2008), and ▪ three letters to the editor of the local newspaper (Dux 2009; Kneißl 2009; Wassmuth 2009).
Semi-structural interviews	Three interviews with six persons: ▪ one individual interview with the mayor of Ingersheim, ▪ one multi-person interview with three representatives of the local energy cooperative that runs the wind turbine, and ▪ one multi-person interview with two members of the anti-wind protest initiative.	Eight interviews with eleven persons: ▪ five individual interviews with supporters of the proposed wind park, ▪ one individual interview with an objector, ▪ one individual interview with a local journalist, and ▪ one multi-person interview with four objectors.

All interviews were intended to be individual. However, some interviewees proposed bringing one or more others, which we accepted. As a result, interviews were both individual and collective. They were conducted in July 2012 for Ingersheim and from July to September 2011 for Wolfhagen. They lasted between one and three hours and were all taped and transcribed verbatim.

In applying Laclau & Mouffe's logics of equivalence, we sought to identify the main nodal points and establish the antagonistic frontiers between inside and outside in objectors' and proponents' discourses on the wind power scheme, focusing on discursive constructs of landscapes. The following quote from the Ingersheim case shows how an antagonistic boundary is constructed between the discourse's inside ("the landscape up there") and outside ("perfect idyll"):

"We find the landscape up there . . . the landscape is already impaired by three parallel high voltage power lines and by intensive farming. This is certainly not a perfect idyll" (interview statement).

Nodal points are those elements that appear repeatedly in the texts and which are related to many other elements. For instance, the following quotes from different texts were the basis for us to identify "health risks of wind turbines" as a nodal point in the objectors' discourse in the Ingersheim case:

- "Long-term health risks are looming because of harmful immissions (noise, infrasound and others) from wind turbines, which are not yet sufficiently researched."
- "The wind turbine has significant negative effects, e.g., harmful immissions (infrasound, noise, shadow, electric smog) and resulting health risks for citizens living in the vicinity."
- "The considerable harassment and the health risks of wind energy were played down."
- "There is no doubt that wind turbines produce infrasound. [...] A growing number of scientists points to the health risks of infrasound."

Nodal points structure and organise a discourse. In our analysis, they were the key to deriving discursive structures from individual texts and interviews. For this purpose, we analysed the entire corpus and examined statements for and against the wind energy scheme. Furthermore, we examined a small number of key documents in more detail. These texts were chosen firstly because they included the major nodal points of the discourses in question and secondly because they offered rich chains of equivalential elements in relation to the nodal points.

4 Ingersheim

4.1 Context of the Case

Ingersheim is a municipality with a population of 6000 (Gemeinde Ingersheim n.d.). It lies 25 km to the north of Stuttgart. A nuclear power plant is located in Neckarwestheim, only ten kilometres from Ingersheim, see figure 1. The giant steam cloud emerging from its cooling tower can easily be seen from there.

In 2002 a group of citizens who had previously dealt with photovoltaic projects identified the elevation between Ingersheim and neighbouring Besigheim as a promising site in terms of wind conditions (interview statement). In 2010 this group proposed to establish an energy cooperative and to erect a wind turbine there with a tower height of 138 metres, a rotor diameter of 82 metres and a total height of 179 metres—700 metres from the village and only 450 metres from some detached farmsteads. The generator was supposed to have a capacity of two megawatts. The estimated construction costs amounted to € 3.6 million and were to be recovered by selling shares at € 2500 each (EGIU 2010b).

Shortly after the plans were first made public in March 2010, the citizens' group organized public meetings (or hearings) in Ingersheim and neighbouring communities. The energy cooperative was founded by 11 persons the same month and formally registered in December 2010. Finally, permission to construct the wind turbine was given by the district authority in January 2011. In April 2011 membership of the energy cooperative had reached 350 and sufficient funds were thus available to finance the project. The members of an action group lobbying against the planned wind turbine, which had already

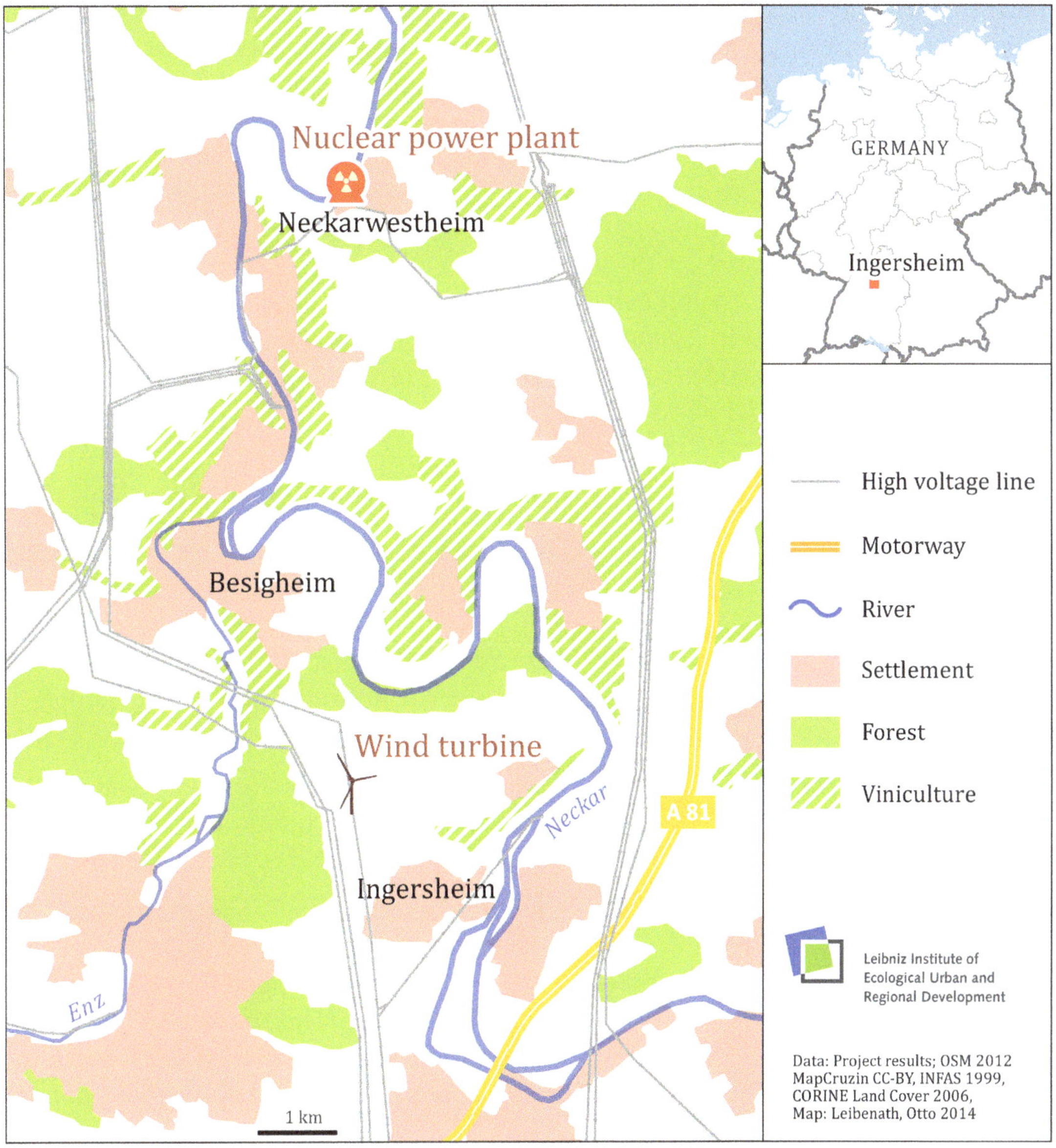

Figure 1: Location of the Ingersheim wind turbine.

formed in February 2010, tried to stop installation by, among other things, legal action. However, their efforts failed. The turf-cutting ceremony for the wind turbine was therefore celebrated in September 2011 and the facility was finalised in March 2012.

4.2 Discursive Constructions of Landscapes in the Proponents' Discourse

Let us take a look at who belonged to the discourse coalition in favour of erecting the wind turbine. First of all there was the energy cooperative with a core group of 10-20 people. It enjoyed the support of all political parties represented in the Ingersheim council, Ingersheim's mayor, the head of the district authority, and representatives of the local parish. Media, such as local newspapers, but also public radio and TV stations advocated building the wind power plant. This was also the case for the newly elected state government, state and district organizations of the environmental non-governmental organization BUND (one of Germany's major environmental NGO's), and the German Wind Energy Association.

The proponents of the project do not use any place names but simply speak of "the landscape," "the landscape here," "the landscape up there" or "up there on the site." They reject the notion that this was an "unimpaired landscape," a "retreat for birds and fauna" or part of a "major flyway of migratory birds." "They [i.e., the opponents] claim we were destroying an unimpaired landscape, a recreational area, and we were expelling pedestrians, walkers and tourists." These elements in the discursive outside are articulated as equivalents of "nature conservation," "conserving everything" and "making believe there was a perfect idyll." Instead they argue that "the landscape is already impaired by three parallel high voltage power lines and by intensive farming" and that "landscapes have to change." According to them, this is a "cleared usable area," a "densely populated suburban area" and a "cultural landscape, characterized by settlement, industry and infrastructure," which "has been modified again and again." Should there be any wildlife, it would surely be compatible with the new facility: in other areas "nests of the threatened skylark have been found between wind turbines." Furthermore, supporters of the wind power scheme point to the mandatory "compensation measures such as planting orchards and transforming intensively used areas into semi-natural fallow and meadow areas." This exemplifies two major antagonistic nodal points in the proponents' discourse, namely a "used and usable landscape which has been modified many times" (inside) and "perfect idyll" (outside).

Concerning the question as to how the landscape is affected by the wind turbine and how this has to be evaluated, three different and slightly contradictory nodal points can be observed in this discourse's inside: "The wind turbine represents a landmark, a symbol of innovation and of the future," "lesser evil" and a "matter of taste."

- The nodal point "landmark" amounts to an outright appraisal of the wind turbine as "aesthetically beautiful." This statement is connected to an equivalential chain of elements such as "the wind turbine integrates well into the landscape," "the colours are fitting," "tourist attraction," "Ingersheim acquires a more positive image because of the wind turbine," and "the wind turbine represents a landmark, a symbol of innovation and the future." It also includes non-linguistic elements, for instance spontaneous barbecues on the building site during the construction phase and it being a popular destination for walkers, a public celebration shortly after the facility was activated, romantic photos of the new wind turbine as well as photos taken from the top of the tower, presentations and public discussions on the ground floor of the power plant, which also serves as an exhibition space—and of course the turbine itself as a material object.

- The nodal point "lesser evil" is articulated as an equivalent of "the view of the wind turbine has to be preferred to the cloud of the nuclear power station Neckarwestheim," "it has to be compared with the nuclear power plant" and "the nuclear power plant can do much more damage to the landscape than a wind turbine."

- Statements related to the third nodal point "matter of taste" stress the importance of individual attitudes and judgements, for instance: "depending on the personal viewpoint, the cultural landscape is upgraded or disturbed by the wind turbine," "how to assess changes in the landscape is a question of personal attitude" and "some people like wind turbines and regard them as symbols of progress."

All three nodal points are positively related to the overarching nodal point of installing a wind turbine while being opposed to "nuclear power."

Indeed "getting rid of nuclear power" is articulated as the central nodal point in the supporter's discourse. "Nuclear power" is "symbolized by the steam cloud of Neckarwestheim always in sight." It also corresponds to "iodine tablets have been distributed here in case an incident happens," "the nuclear disaster of Chernobyl when our children could not play outside," and "Fukushima."

Another pair of contrasting nodal points is "good wind conditions" and "false data." Those in favour of the wind turbine articulate equivalences between the said landscape and elements, such as "professional wind measurement," "wind forecast from a renowned expert," "good wind condition proven through long-term measurement in 2002/2003" and "sound data." By contrast, they accuse the opponents of using "false data"—an element which is related to "insufficient wind conditions for operating a wind turbine economically."

4.3 Discursive Constructions of Landscapes in the Opponents' Discourse

This discourse is produced mainly by inhabitants of Besigheim. The protest is led by a citizens' initiative which has about 100 members and more than 1,400 supporters who participated in a petition against the wind power scheme. According to initiative representatives, most of their supporters are from the centre-right political camp of the Liberal Democratic Party FDP and the Christian-Democratic Union CDU, including the mayor of Besigheim (interview statement). Many of these persons identified themselves publicly with this discourse through actions such as attending the handover of petitions to the head of the district authority in December 2010. The discourse coalition also includes the long-time chairpersons of Ingersheim's BUND chapter and the board of the "Schwäbischer Heimatbund," a traditional non-governmental organisation with more than 5,000 members, which advocates the preservation of Swabian historical landscapes and monuments.

The opponents of the wind power scheme raised a broad range of concerns. However, some issues are brought up repeatedly by many different speakers and can thus be regarded as the nodal point of this discourse. One of them is the site and the surrounding landscape, which is termed a "century-old cultural landscape." Furthermore it is articulated in an equivalential chain of elements such as "poor wind conditions," "unfavourable topography with steep valleys," "open land with high visual sensitivity," "century-old cultural landscape with steep-sloped vineyards, orchards and agriculture," "nearly undeveloped rolling hill," "farming," "regional green belt," "particularly lovely landscape" and finally a "landscape worth protecting." According to the objectors, this landscape is "home" and "our native landscape," and it provides "identity."

The corresponding nodal point in the discursive outside can be labelled "irreparable disfigurement." It is related to "large-scale technical intrusion," "monster plant," "impairing the visual quality of the landscape in an irreparable way," "total blight," "permanent mutilation of the scenery," "loss of valuable farmland," and "the claim that it was possible to compensate for this intervention." An interesting non-linguistic element in this regard was the action of demonstrating the height of the planned wind turbine with the help of a helicopter (Willy 2010).

Another important nodal point in this discourse's interior is "energy transition" in combination with "wind energy" and "climate protection and nuclear phase-out." Almost all objectors argue that "we accept the energy transition . . ., we are basically in favour of renewable energies," that "wind energy is an important element of energy generation" and that "supporting the energy transition is positive." However, this principal backing of wind energy hinges on the condition that "it makes sense economically." Therefore the equivalential chain further extends to elements such as "areas with constantly high wind speed," "other locations with better wind conditions in other parts of Baden-Wuerttemberg," "in eastern and northern Germany or England" and "wind power offshore, in England or Scotland."

This nodal point, too, has a clear-cut counterpart in the discursive outside, namely to erect a wind turbine at this location. Thus, the wind turbine as such and the action of building it together constitute a nodal point that is often related to the slogan "ecological and economic nonsense" (in German: "Ökologie- und Ökonomie-Unfug"). It is further articulated as part of an equivalential chain of elements such as "building wind turbines at sites which neither make sense in economic terms nor in terms of energy policy," "the energy yield does not justify the damage to the landscape" and "doing the energy transition a disservice." Many other economic items are articulated in this context, for instance "inefficiency," "waste of governmental subsidies," "the profitability of the scheme is out of proportion to the environmental intrusion," "devastating cost-benefit ratio at this location," "devaluation of real estate" and "marginal benefits for a few investors [that] does not justify disadvantages at the expense of the entire population of Ingersheim and Besigheim."

The last major nodal point to be discussed here has to do with human health and the quality of life. It is subsumed under the statement "health risks of wind turbines" and belongs to the discursive outside. It is articulated as an equivalent of "health effects of the planned turbine were not assessed," "negative effects of infrasound," "shadow flicker can lead to concentration disorders," and "loss of quality of life." This nodal point with its equivalents corresponds weakly to elements on the discursive inside such as "the distance between the wind turbine and residential buildings has to be enlarged," which cannot be regarded as a nodal point. Instead, "harmful effects on health" is chiefly articulated in an equivalential relation with the other nodal points on the outside, namely "irreparable disfigurement" and "ecological and economic nonsense."

The opponents' discourse displays a number of breaches or contradictions. For instance, while most members of the discourse coalition reject nuclear energy by employing terms such as "nuclear energy plight" and "the devil of nuclear power," others speak of nuclear energy as a "helpful transitory energy." Another example can be seen in statements such as "new wind turbines lead to additional carbon dioxide emissions because backup power plants have to operate" or "photovoltaic panels and wind mills . . . [are] sacred buildings of a new creed" which stand in stark contrast to the nodal point "energy transition."

5 Wolfhagen

5.1 Context of the Case

The town of Wolfhagen has approximately 14,000 inhabitants (SWDM n.d.) and lies 20 kilometres east of Kassel in northern Hesse, see figure 2.

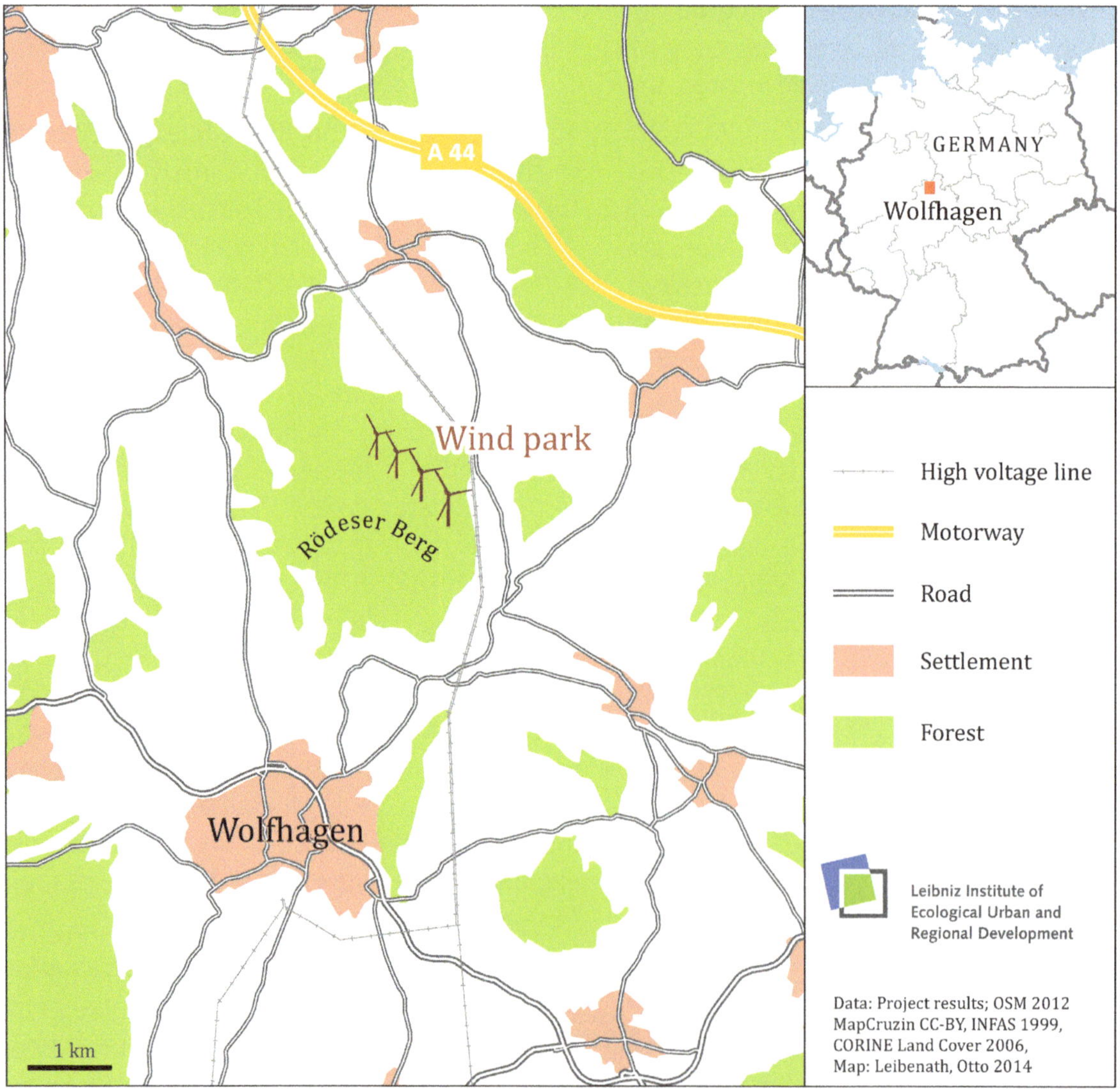

Figure 2: Location of the Wolfhagen wind park.

The argument about wind turbines in Wolfhagen is embedded in a public debate on climate change, climate protection and renewable energies. In 2008, the local council unanimously passed an energy concept, according to which all of Wolfhagen's households were to be provided with locally produced electricity from renewable sources by 2015. To this purpose, a wind power scheme provided for the erection of four turbines with a capacity of three megawatts each and a total height of 186 metres. It was planned for the facility to be built and operated jointly by the municipal utility company and a local energy cooperative. But where were the wind turbines to be constructed? A commission was established to answer this question. It included "representatives of potentially affected municipal districts, of the Climate Offensive Wolfhagen, of the municipality, forest owners, conservation agencies as well as of other organisations" (SWG 2008: 2). After several months of discussing, assessing and weighing up the facts, the commission agreed on the Rödeser Berg—a wooded hill or low mountain not far from the town centre—as the preferred site of the wind park.

After the siting decision was made public in late 2008, several citizen's initiatives and other organisations were founded, which lobbied for or against the planned wind park. The final permission to build the wind farm was issued in January 2014 and the four turbines, which are currently under construction, are to be completed by the end of 2014 (SWG 2014).

5.2 Discursive Constructions of Landscapes in the Proponents' Discourse

The discourse in favour of the wind park was produced by a coalition comprising council members from three parties (Social Democrats, Christian Democrats and the so-called Wolfhagen List), the independent mayor, the initiative "ProWind Wolfhagen—Energy Transition Now" and an informal group of citizens who prepared the foundation of the local energy cooperative.

The proponents' discourse centres on the municipal energy concept and the planned wind farm as its cornerstone. The energy concept is articulated as an equivalent of "security, jobs," "necessary climate protection," "responsibility," "sustainability," "independence," and—regarding the prospect of supplying energy by means of a civic cooperative—"democratization." The wind turbines are depicted as "beacons of progress," "pillars of climate protection" and "symbols of acting responsibly and sustainably at the regional level." The related antagonistic nodal point is "climate change" in combination with "not protecting the climate," "multinational corporations," and—especially since the Fukushima disaster on February 11, 2011—"nuclear power."

This is the background for articulating a very distinct toponymic landscape concept of the Rödeser Berg, the area where the wind park is supposed to be built. According to the supporters, the Rödeser Berg is a place where a significant contribution towards saving the world's climate can be made. This nodal point is intertwined with the municipal energy concept and the issue of climate protection. Consequently, it is opposed to "not protecting the climate" and related elements such as "growing numbers of extreme meteorological events" and "Mediterranean brushwood on the Rödeser Berg instead of the forest we know."

The second nodal point "commercial forest" is positively related to "created and shaped by humans," "ready to be harvested," "devastated by the latest hurricane," and "one of many similar forest complexes." This is underpinned by photos such as an aerial image of the windthrow areas on the Rödeser Berg. The proponents maintain that the Rödeser Berg is neither a "primary forest" nor a habitat of threatened species. The attitude of conservationists who claim that the Rödeser Berg was a "hunting habitat of red kite" or a "resting area of wild cat" is classified as "ignorance," "local egoism" and "romantic transfiguration by certain nature lovers."

Finally the supporters of the wind park describe the Rödeser Berg as the most promising and appropriate site. Such statements belong to an equivalential chain which also includes "effective," "sufficient distance to other buildings," and "least environmental problems." By contrast, any alternative site would imply "building more facilities," having a lower energy yield, and aggravating the conflicts with nature conservation.

5.3 Discursive Constructions of Landscapes in the Opponents' Discourse

The opposing discourse coalition has enlarged gradually. At the end of the analysed period it comprised the council members of the Green Party and their newly founded local chapter, the citizen's initiative "No Wind Power in our Woods" and the "Wolfhagen Civic Alliance," which is also represented on the local council.

The objectors' concept of the Rödeser Berg is even more elaborate than that of the supporters. The Rödeser Berg functions as a kind of master signifier in this discourse. First it is termed an "intact beech and oak forest." Furthermore it is described as "large," "silent," "valuable," "natural," and as a "resting area for animals," as well as the "habitat of many threatened animal and plant species." Cases in point are the red kite whose population density in this area is "three times higher than the Hessian average," black stork, "rare bat species," and the wild cat. The opponents of the proposed wind scheme invested much effort in proving the existence of cats, kites and bats by means of sophisticated techniques. They argue that the wind park is going to cause the "destruction of the forest ecosystem," a "permanent disturbance" and an "impairment or destruction of species diversity." This is also illustrated by photos, e.g. by one of a natural-looking deciduous forest which is juxtaposed to another one showing the massive concrete base of a wind turbine amidst spruce trees.

Another nodal point is "alternatives," although on the discursive inside (and not on the outside as in the supporters' discourse). By this the opponents mean "sites which are already impaired" such as "commercial and industrial areas" and existing wind power sites, which could be upgraded. They reject "massive intrusions in our nature and our landscape scenery." They also doubt the suitability of the Rödeser Berg in terms of wind conditions. For them, installing wind turbines on the Rödeser Berg amounts to a "catastrophe for the landscape." In the objectors' discourse, the Rödeser Berg is "our beautiful landscape," characterized by "uniqueness," "recreation," "tourism" and "home." By contrast, the planned wind turbines are denoted as "monsters," "atrocious," "gigantic," "visual madness," and an "extremely dubious industrial project."

Although the Rödeser Berg plays a key role in the opponents' discourse, there are other nodal points that are differently accentuated by different speakers. One is the planning and decision-making process for the wind farm so far, which they relate to "misleading people," "political pressure," "blocking out critics," and "retaining expert assessments." Instead, they demand a referendum and the acknowledgement of the "technically due assessment."

6 Discussion

Both Ingersheim and Wolfhagen are examples of local controversies about wind energy in which "landscape" is employed by supporters and opponents alike. Our first research question was whether the opposing camps discursively construct the same or different landscapes. We found that different landscape concepts, which are intertwined with opposed political claims, were produced in each of these debates. We are therefore confronted with political landscapes, laden with power. Seemingly neutral statements such as "there is always a strong wind blowing up there" or "the rare bird xy is nesting there" are highly political because by uttering these phrases speakers position themselves on one side or the other of a battle about wind turbines at a specific location.

With regard to the second research question, which concerns the similarities and differences between landscape constructs from different local wind energy discourses, it can be said that in both cases the discursive constructions of landscapes are integral components of the respective discourses. In the Wolfhagen controversy, they are related to the conflict between local conservation and global climate protection that, e.g., Gee (2010) also identified in her study. By contrast, in Ingersheim these issues figure only on the margins of the discourses. Here, the entire debate is literally overshadowed by the cloud of the nearby nuclear power plant. Therefore it is to some extent logical that the opponents of the Ingersheim wind turbine stress the harmful effects on human health that possibly emanate from wind turbines. Notwithstanding the different contexts of the cases, the landscape concepts of objectors and proponents display striking similarities in cross-case comparison. In both cases the opponents of wind schemes construct an idyllic, beautiful, and valuable landscape, which is home to both humans and a rich biodiversity of plants and animals and in need of protection. In the proponents' discourses, by contrast, cultivated and exploited landscapes appear that are constantly undergoing changes.

This leads to the third research question regarding possible overlaps between landscape constructs at the micro and macro levels. Our brief literature review in the introductory section showed that there is one macro discourse about landscape which focuses on the notion of "landscape as beautiful, valuable area" (Leibenath & Otto 2012: 126). The landscape discourses produced by those who oppose wind energy developments in our cases overlap broadly with this macro discourse. The same holds true for the supporters' landscape discourses in relation to the second macro discourse of "landscape as an area shaped by humans" (Leibenath & Otto 2012: 126). Thus there seem to be a relatively stable set of landscape concepts that is reproduced in specific controversies. However, this hypothesis needs to be assessed in further empirical studies.

7 Conclusion

Our analyses of the debates about wind energy schemes in Ingersheim and Wolfhagen demonstrate that landscapes can be a contested issue in such controversies at the local level. If this is the case, it becomes almost impossible to speak objectively or neutrally about the sites in question. This has direct practical consequences. First, if basically all landscapes are political and contingent constructs, virtually any landscape can be contested, even if it not actually is. Therefore almost any argument about visual despoliation and disfigurement of landscapes by wind turbines can be countered by (re-)producing a different discourse of landscapes as dynamic and of wind turbines as enhancing the scenic quality of a given landscape, and vice versa. Secondly, Wolsink's (2007) recommendation to consult key stakeholders in the process of wind power planning and to select the types of landscape with the highest acceptability in the eyes of the stakeholders has to be questioned, because it cannot always be said in the first place who the relevant stakeholders are and what landscape type category a given site falls into. For instance, in the Wolfhagen case, the local council and the municipal utility company set up a site-finding commission which was supposed to represent all relevant stakeholders and to identify a site that would be acceptable to everybody. However, after the decision had been made, a range of new interest groups was established who voiced quite opposite views of the Rödeser Berg landscape.

There are many promising avenues for further research. One is to shed more light on the role of key actors in producing wind energy-related landscape discourses in local conflicts. Although discourse theory assumes that all subjective identities are discursively constructed and that discourse therefore precedes individuality (Howarth 2005: 317 and 320; Howarth 2010: 314), it would be interesting to combine inquiries into discourses with other branches of social scientific research and raise questions such as how the commitment of key stakeholders is linked to their personal biographies and how they became stakeholders at all. In order to obtain a fuller understanding, it would also be revealing to take the relations and interactions between stakeholders into account and to assess the degree to which their involvement is influenced by categories such as trust. This would help close the neglect of individual agency in discourse analytical research often lamented (cf. Leipold 2014: 18). Another possible direction for future inquiry is to conduct longitudinal analyses of the intertwinement of energy and landscape discourses and to see how they develop over time. It is likely that new conceptions of landscapes, of generating and using energy, and of spatial planning and decision-making will emerge and influence each other. The prevalence of certain concepts and discourses among given groups of interest could be mapped repeatedly by means of quantitative tools such as Q-methodology (Ellis et al. 2007) or surveys (Smith & Hay 2008). Finally it would be promising to elucidate why landscapes are only in a few local wind energy debates explicitly contested and why this issue is considered by just one of the conflicting parties or even entirely ignored in most cases.

Funding

This work was supported by the German Research Foundation (Deutsche Forschungsgemeinschaft—DFG) [grant number LE 2255/2-1].

References

BI (= Bürgerinitiative [BI] Wolfhager Land – Keine Windkraft in unseren Wäldern) 2010a. Liebe Wolfhager Bürger und Leser des Stadtanzeigers! http://www.kein-windrad-im-wald.de/uploads/media/Stadtanzeiger_WOH_Mai-2010_01.pdf (Date: 06.03.2012).

BI (= Bürgerinitiative [BI] Wolfhager Land – Keine Windkraft in unseren Wäldern) 2010b. Statt Anzeiger. BI Wolfhager Land ‚Keine Windräder in unseren Wäldern'. http://www.kein-windrad-im-wald.de/uploads/media/BI-StattAnzeiger_2010-11_03.pdf (Date: 06.03.2012).

Blackbourn, D. 2007. The Conquest of Nature: Water, Landscape and the Making of Modern Germany. Norton, New York, London.

Blatter, J.; F. Janning & C. Wagemann 2006. Qualitative Politikanalyse. Eine Einführung in Methoden und Forschungsansätze. Studienbrief für die Fernuniversität Hagen, Hagen.

Bourdieu, P. 1991. Physischer, sozialer und angeeigneter physischer Raum. In Wentz, M. (eds.): Stadt-Räume. Campus, Frankfurt.

Cosgrove, D. 2003. Landscape: Ecology and semiosis. In Palang, H. & G. Fry (eds.): Landscape Interfaces. Cultural Heritage in Changing Landscapes. Kluwer, Dordrecht, Boston, London.

Dryzek, J. S. 1997. The Politics of the Earth: Environmental Discourses. Oxford University Press, New York.

Dux, B. 2009. [letter to the editor]. Hessische/Niedersächsische Allgemeine Zeitung 15.09.2009.

EGIU (= Energiegenossenschaft Ingersheim und Umgebung e. V. i. G.) 2010a. Fakten statt Emotionen. http://www.eg-ingersheim.de/index.php?option=com_phocadownload&view=category&download=26:positionen-und-stellungnahme-seite-1&id=12:positionen&Itemid=40 (Date: 16.04.2013).

EGIU (= Energiegenossenschaft Ingersheim und Umgebung e. V. i. G.) 2010b. Informationen zum Genossenschaftsmodell. http://www.eg-ingersheim.de/index.php?option=com_phocadownload&view=category&download=24:flyer-genossenschaft-seite-1&id=3:flyer&Itemid=24 (Date: 15.04.2013).

EGIU (= Energiegenossenschaft Ingersheim und Umgebung e. V. i. G.) 2010c. Landschaft. http://www.eg-ingersheim.de/index.php?option=com_content&view=article&id=17:landschaft&catid=17:windoekologie&Itemid=39 (Date: 16.04.2013).

Eisenmann, A. 2011. Zum Thema Windkraft in Ingersheim: Landschaftseingriff nicht zu rechtfertigen. Bietigheimer Zeitung 28.07.2011.

Ellis, G.; J. Barry & C. Robinson 2007. Many Ways to Say ‚No', Different Ways to Say ‚Yes': Applying Q-Methodology to Understand Public Acceptance of Wind Farm Proposals. Journal of Environmental Planning and Management 50, 517-551.

Feindt, P. H. & A. Oels 2005. Does discourse matter? Discourse analysis in environmental policy making. Journal of Environmental Policy & Planning 7, 161-173.

Fischer, F. 1998. Beyond empiricism: Policy inquiry in postpositivist perspective. Policy Studies Journal 26, 129-146.

Foucault, M. 1983 [1976]. Sexualität und Wahrheit: Der Wille zum Wissen. Suhrkamp, Frankfurt.

Foucault, M. 2005 [1976]. Vorlesung vom 14. Januar 1976. In Defert, D.; F. Ewald & J. Lagrange (eds.): Michel Foucault. Analytik der Macht. Suhrkamp, Frankfurt.

Gailing, L. & M. Leibenath 2010. Diskurse, Institutionen und Governance: Sozialwissenschaftliche Zugänge zum Untersuchungsgegenstand Kulturlandschaft. Berichte zur deutschen Landeskunde 84, 9-25.

Gailing, L. & M. Leibenath 2012. Von der Schwierigkeit, „Landschaft" oder „Kulturlandschaft" allgemeingültig zu definieren. Raumforschung und Raumordnung 70, 95-106.

Gailing, L. & M. Leibenath 2013. The social construction of landscapes: Two theoretical lenses and their empirical applications. Landscape Research.

Gee, K. 2010. Offshore wind power development as affected by seascape values on the German North Sea coast Land Use Policy 27, 185-194.

Gemeinde Ingersheim n.d. Zahlen, Daten, Fakten. http://www.ingersheim.de/website/de/gemeinde/zahlen_daten_fakten (Date: 21.02.2014).

Götte, A. & I. Degenhardt-Meister 2009. ProWind Wolfhagen – Energiewende jetzt – [leaflet], Wolfhagen.

Griggs, S. & D. Howarth 2012. Poststructuralist policy analysis: Discourse, hegemony and critical explanation. In Fischer, F. & H. Gottweis (eds.): The Argumentative Turn Revisited : Public Policy as Communicative Practice. Duke University Press, Durham, NC.

Grimm, E. 2011. Beelzebub. Ludwigsburger Kreiszeitung 16.07.2011.

Haecker, W.-D. 2010. Taugt nicht als Vorzeigeobjekt für die Energiewende. Bietigheimer Zeitung 08.09.2011.

Henderson, G. 2003. What (else) we talk about when we talk about landscape: For a return to the social imagination. In Wilson, C. & P. Groth (eds.): Everyday America: Cultural Landscape Studies after J. B. Jackson. University of California Press, Berkeley, Los Angeles, London.

Howarth, D. 2005. Applying Discouse Theory: the Method of Articulation. In Howarth, D. & J. Torfing (eds.): Discourse Theory in European Politics. Identity, Policy and Governance, Houndmills, Basingstoke.

Howarth, D. 2010. Power, discourse, and policy: articulating a hegemony approach to critical policy studies. Critical Policy Studies 3, 309-335.

Howarth, D. & Y. Stavrakakis 2000. Introducing discourse theory and political analysis. In Howarth, D.; A. J. Norval & Y. Stavrakakis (eds.): Discourse Theory and Political Analysis: Identities, Hegemonies and Social Change. Manchester University Press, Manchester, New York.

Huber, K. & W. Huber 2010. Offener Brief von Karin und Wolfgang Huber. http://www.gegenwind-husarenhof.de/huber.html (Date: 24.04.2013).

Jessup, B. 2010. Plural and hybrid environmental values: a discourse analysis of the wind energy conflict in Australia and the United Kingdom. Environmental Politics 19, 21-44.

Jones III, J. P. & W. Natter 1999. Space 'and' representation. In Buttimer, A.; S. D. Brunn & U. Wardenga (eds.): Text and Regional Image: Social Construction of Regional Knowledges. Selbstverlag des Instituts für Länderkunde, Leipzig.

Kneißl, R. 2009. [Letter to the editor]. Hessische/ Niedersächsische Allgemeine Zeitung 06.03.2012.

Kühne, O. 2009. Grundzüge einer konstruktivistischen Landschaftstheorie und ihre Konsequenzen für die räumliche Planung. Raumforschung und Raumordnung 67, 395-404.

Laclau, E. 1990. New Reflections on the Revolution of Our Time. Verso Press, London, New York.

Laclau, E. 1993. Discourse. In Goodin, R. E. & P. Pettit (eds.): A Companion to Contemporary Philosophy. Blackwell, Oxford, Cambridge (MA).

Laclau, E. & C. Mouffe 1985. Hegemony & Socialist Strategy. Towards a Radical Democratic Politics. Verso Press, London.

Lefebvre, H. 1991 [1974]. The Production of Space. Blackwell, Oxford.

Leibenath, M. 2013. Konstruktivistische, interpretative Landschaftsforschung: Prämissen und Perspektiven. In Leibenath, M.; S. Heiland; H. Kilper & S. Tzschaschel (eds.): Wie werden Landschaften gemacht? – Sozialwissenschaftliche Perspektiven auf die Konstituierung von Kulturlandschaften. Transcript, Bielefeld.

Leibenath, M. & A. Otto 2012. Diskursive Konstituierung von Kulturlandschaft am Beispiel politischer Windenergiediskurse in Deutschland. Raumforschung und Raumordnung 70, 119-131.

Leibenath, M. & A. Otto 2013a. Local debates about 'landscape' as viewed by German regional planners: Results of a representative survey in a discourse-analytical framework. Land Use Policy 32, 366– 374.

Leibenath, M. & A. Otto 2013b. Windräder in Wolfhagen – eine Fallstudie zur diskursiven Konstituierung von Landschaften. In Leibenath, M.; S. Heiland; H. Kilper & S. Tzschaschel (eds.): Wie werden Landschaften gemacht? – Sozialwissenschaftliche Perspektiven auf die Konstituierung von Kulturlandschaften. Transcript, Bielefeld.

Leipold, S. 2014. Creating forests with words — A review of forest-related discourse studies. Forest Policy and Economics 40, 12-20.

Müller, W. 2010a. Mach mit (1): Husarenhof-Bürgerinitiative zur Verhinderung einer Windkraftanlage. Kein Ökologie- und Ökonomie-Unfug am Standort Husarenhof/Lerchenhof. http://www.gegenwind-husarenhof.de/sonstiges/190510Info-Blatt1Ablehnung.pdf (Date: 23.04.2013).

Müller, W. 2010b. Mach mit (3): BI ‚Gegenwind-Husarenhof' zur Verhinderung einer Windkraftanlage (WKA). Öko- und Ökonomie-Unfug? Nein! http://www.gegenwind-husarenhof.de/sonstiges/210510InfoBlatt3%D6koUnfug.pdf (Date: 25.05.2013).

Müller, W. 2010c. Mach mit (5): ‚Gegenwind Husarenhof' – BI zur Verhinderung einer Windkraftanlage (WKA): Ablehnungsgründe. http://www.gegenwind-husarenhof.de/sonstiges/150510InfoBlatt5Ablehnungsgr%FCnde.pdf (Date: 24.04.2013).

Nadaï, A. & D. Van der Horst 2010. Wind power planning, landscapes and publics. Land Use Policy 27, 181-184.

Natter, W. & J. P. Jones III 1993. Signposts toward a poststructuralist geography. In Jones III, J. P.; W. Natter & T. Schatzki (eds.): Postmodern Contentions: Epochs, Politics, Space. Guilford Press, New York.

Olwig, K. R. 2001. Landscape as a contested topos of place, community and self. In Adams, P.; S. Hoelscher & K. Till (eds.): Textures of Place. Exploring Humanist Geographies. University of Minnesota Press, Minneapolis.

Otto, A. & M. Leibenath 2014. The interrelation between collective identities and place concepts in local wind energy conflicts. Local Environment.

Schieber, M. 2010. Leserbrief zum Thema Windrad in Ingersheim: Windrad und Sonnenkollektoren. Bietigheimer Zeitung 29.03.2010.

SHB (= Schwäbischer Heimatbund) 2010. Pressemitteilung. Windkraftanlage Ingersheim, Kreis Ludwigsburg. ‚180 Meter sind zu hoch!' http://www.gegenwind-husarenhof.de/sonstiges/Pressemitteilung%20SHB%2012.8.2010.pdf (Date: 24.04.2013).

Smith, N. & C. Hay 2008. Mapping the political discourse of globalisation and European integration in the United Kingdom and Ireland empirically. European Journal of Political Research 47, 359-382.

SPD (= SPD im Kreis Ludwigsburg) 2011. SPD im Kreis Ludwigsburg bekräftigt Unterstützung des Ingersheimer Windrades [press release of 07.01.2011].

SVF (= Stadtverordnete der Fraktionen von CDU SPD und Wolfhager Liste/FDP) 2010. Kommunalvertreter-Serie zum Energiekonzept Wolfhagen. http://wolfhagen-energenial.de/download/Kommunalvertreter-Serie%20zum%20Energiekonzept%20Wolfhagen_Artikel%201%20-%2015.pdf (Date: 07.03.2012).

SWDM (= Stadt Wolfhagen – Der Magistrat) n.d. Einwohnerzahlen (Haupt- und Nebenwohnungen) Juni 2013. http://www.wolfhagen.de/de/rathaus/zahlen_fakten/einwohnerzahlen.php?navanchor=1110036 (Date: 21.02.2014).

SWG (= Stadtwerke Wolfhagen GmbH) 2008. Bürger-Windpark Rödeser Berg. Die Bürgerinformation der Stadtwerke Wolfhagen. Nr. 1 November 2008. http://www.buergerforum-qlb.de/uploads/media/2008-12-03_SWW_Windpark_Broschuere.pdf (Date: 07.03.2012).

SWG (= Stadtwerke Wolfhagen GmbH) 2014. Regierungspräsidium Kassel erteilt Genehmigung für Windpark am Rödeser Berg. Windpark sichert Wolfhager Zukunft. https://www.stadtwerke-wolfhagen.de (Date: 21.02.2014).

Szarka, J. 2004. Wind power, discourse coalitions and climate change: breaking the stalemate? European Environment: The Journal of European Environmental Policy (Wiley) 14, 317-330.

Waage, E. R. H. & K. Benediktsson 2010. Performing Expertise: Landscape, Governmentality and Conservation Planning in Iceland. Journal of Environmental Policy & Planning 12, 1-22.

Warren, C. R.; C. Lumsden; S. O'Dowd & R. V. Birnie 2005. ‚Green on Green': Public perceptions of wind power in Scotland and Ireland. Journal of Environmental Planning and Management 48, 853-875.

Wassmuth, K. 2009. Einseitige Ziele: Zum Thema Windkraft auf dem Rödeser Berg. Hessische/Niedersächsische Allgemeine Zeitung 06.03.2012.

Willy, M. 2010. Helikopter-Demo gegen Windrad. Stuttgarter Zeitgung 19.05.2010.

Wolsink, M. 2007. Planning of renewables schemes: Deliberative and fair decision-making on landscape issues instead of reproachful accusations of non-cooperation. Energy Policy, 2692-2704.

Wolsink, M. & S. Breukers 2010. Contrasting the core beliefs regarding the effective implementation of wind power. An international study of stakeholder perspectives. Journal of Environmental Planning and Management 53, 535-558.

Wüstenhagen, R.; M. Wolsink & M. J. Bürer 2007. Social acceptance of renewable energy innovation: An introduction to the concept. Energy Policy 35, 2683-2691.

Wylie, J. 2007. Landscape (Key Ideas in Geography). Routledge, London.

Yin, R. K. 2003. Case Study Research. Design and Methods (Third Edition). Sage, Thousand Oaks, London, New Delhi.

4

Ecosystem Service Potentials, Flows and Demands – Concepts for Spatial Localisation, Indication and Quantification

Benjamin Burkhard[1,2] *, Marion Kandziora[1], Ying Hou[1] & Felix Müller[1]

1 Kiel University, Institute for Natural Resource Conservation, Department of Ecosystem Management, Olshausenenstr. 40; 24098 Kiel; Germany, bburkhard@ecology.uni-kiel.de

2 Leibniz Centre for Agricultural Landscape Research ZALF, Eberswalder Straße 84, 15374 Müncheberg; Germany

Abstract

The high variety of ecosystem service categorisation systems, assessment frameworks, indicators, quantification methods and spatial localisation approaches allows scientists and decision makers to harness experience, data, methods and tools. On the other hand, this variety of concepts and disagreements among scientists hamper an integration of ecosystem services into contemporary environmental management and decision making. In this article, the current state of the art of ecosystem service science regarding spatial localisation, indication and quantification of multiple ecosystem service supply and demand is reviewed and discussed. Concepts and tables for regulating, provisioning and cultural ecosystem service definitions, distinguishing between ecosystem service potential supply (stocks), flows (real supply) and demands as well as related indicators for quantification are provided. Furthermore, spatial concepts of service providing units, benefitting areas, spatial relations, rivalry, spatial and temporal scales are elaborated. Finally, matrices linking CORINE land cover types to ecosystem service potentials, flows, demands and budget estimates are provided. The matrices show that ecosystem service potentials of landscapes differ from flows, especially for provisioning ecosystem services.

Keywords:
ecosystem service matrix, providing units, benefit areas, landscape, mapping

*Corresponding author. Email: bburkhard@ecology.uni-kiel.de

Highlights

- Ecosystem service definitions, indicators for ecosystem service potentials, flows and demands are suggested.
- Spatial ecosystem service relations referring to service providing units, benefitting areas, rivalry and scales are discussed.
- Ecosystem service potentials of landscapes differ from ecosystem service flows, especially for provisioning ecosystem services.
- Differentiation between potentials and flows is difficult for regulating ecosystem services.
- Exemplary matrices linking land cover types to ecosystem service potentials, flows, potential vs. flow comparisons, demands, flow-demand budgets are provided.

Focal characteristics of the described method

- Focal ecosystem service(s): regulating, provisioning and cultural services
- Focal ecosystem type(s): generally applicable; example Europe
- Focal temporal scale(s): generally applicable
- Focal spatial scale(s): generally applicable
- Focal target group of the method(s): scientists, decision makers, students
- Focal purpose of the method: ecosystem service indication, quantification and localisation

1 Introduction

The more popular the ecosystem service concept has got, the higher the demand for appropriate indication, quantification and spatial localisation methods has become (Alkemade et al. 2014; Crossman et al. 2013; Burkhard et al. 2013). Prospects of the ecosystem service idea to become a major tool in environmental management are promising, but at the same time expectations of and pressure from practitioners and scientists are raised (Honey-Rosés & Pendleton 2013; Daily et al. 2009; Kienast et al. 2009). Major challenges in all ecosystem service assessment efforts are the high complexity of the topic itself and the need for universal and rather easy-to-apply approaches (Crossman et al. 2013, Seppelt et al. 2012). Different classification systems (TEEB 2010; Costanza 2008a; MA 2005) and varying understanding of ecosystem service supply-benefit delivery chains among scientists (Fisher et al. 2009; Boyd & Banzhaf 2009) have inhibited broad-scale practical applications so far. Focussing on the different ecosystem service delivery components, Villamagna et al. (2013) have recently discussed several highly relevant questions on how ecosystems produce services, how to consistently quantify ecosystem service flows, how services relate to each other and how landscape changes affect future service delivery.

Most of the currently available spatial ecosystem service studies focus on ecosystem service supply (see Crossman et al. 2013; Martínez-Harms & Balvanera 2012; Egoh et al. 2012 for reviews), whereas the demand side has not been sufficiently considered. The integration of societal needs for goods and services enhances currently applied function-oriented landscape planning approaches and environmental management strategies. This reveals the full application potential of the ecosystem service framework (Chan et al. 2012; de Groot et al. 2010).

Since the first publication of the ecosystem service 'matrix', which links land cover types to ecosystem service supply capacities (Burkhard et al. 2009 in this journal), the method has successfully been applied to quantify and map ecosystem services in several case studies (e.g. Kandziora et al. 2013b; Kaiser et al. 2013; Vihervaara et al. 2010 and 2012; Müller et al. subm.). It has also inspired the development of other ecosystem service mapping studies (e.g. Clerici et al. 2014; Baral et al 2013; Maes et al. 2011). In 2012, an improved version, including demands for ecosystem services and ecosystem service budget estimates using the same method, was published (Burkhard et al. 2012a). This method has also been applied in different case studies (e.g. Kroll et al. 2012; Nedkov and Burkhard 2012) and improved further. In Schröter et al. (2012), valuable comments on "how and where to map supply and demand of ecosystem services for policy-relevant outcomes" related to the matrix method were provided. These ideas were elaborated further in a recently published case study by Schröter el al. (2014). Hou et al. (2013) discussed the uncertainties related to the matrix method applied for landscape analyses and showed further aspects to improve it.

In our opinion, the attractiveness of the matrix approach results from its flexibility concerning detailedness and levels of abstraction from rather simple to highly complex. Its potential to integrate all kinds of data, from expert-scores to statistics, interview data, measurements or high-end model outcomes makes it applicable in data-poor as well as data-rich environments. Last but not least, results based on the flexible 0-5 ranking system and the linkage to geobiophysical spatial units (e.g. land cover, biotope, vegetation or soil types) in ecosystem service maps provide wide application ranges in science and, hopefully, in decision making.

In this article, recently gained insights concerning ecosystem service quantification and localisation in space and time are presented, and the 'matrix' method is developed further. We hope to trigger scientific debate by contributing to the Special Issue on 'Concepts and Methods for Ecosystem Service Assessments' in Landscape Online. We collected experience and data from different case study applications, international workshops and conferences as well as the work in the IALE-D working group on Ecosystem Services[1] and within the three Ecosystem Services Partnership[2] (ESP) Thematic Working Groups on Indicators[3] , Mapping[4] and Modelling[5] Ecosystem Services. Based on this experience we are aware that no final solution for highly complex ecosystem service assessments has been found yet and that related challenges are manifold. Therefore, we find it important to share our most recent findings and to exchange methods. As defined by the ESP working groups mentioned above, the development of tools, guidelines and standards for improving analyses of ecosystem services is one major goal of related studies.

Our aim is to contribute to the further development of the ecosystem service concept with our ideas and approaches, and thereby to increase its application potential for sustainable decision making. In the following chapter, we present and discuss different ecosystem service definitions, categorisations and indicators related to ecosystem service supply, demands, flows and their spatial localisation. Indicators, definitions and spatial characteristics for 11 regulating, 14 provisioning and 6 cultural ecosystem services are presented. Chapter 3 gives more detailed explanations of methods for regulating, provisioning and cultural ecosystem service assessments. Exemplary assessment matrices of ecosystem service potentials, flows and demands as well as budget estimates are provided in chapter 4. The methods and exemplary results are discussed in chapter 5, followed by the conclusions with a special focus on further research needs in chapter 6.

The following key questions have been of special relevance for the development of the concept on spatial localisation, indication and quantification of ecosystem service potentials, flows and demands:

- Is a distinction between ecosystem service potentials, flows and demands practical?
- What are appropriate indicators for ecosystem service potentials, flows and demands?
- Where are demands for ecosystem services localised best and how can they be quantified?
- Are there patterns of ecosystem service supply, demand and budget estimates when displaying them in land cover-based ecosystem service matrices?

[1] http://www.iale.de/home/arbeitsgruppen/oekosystem-dienstleistungen.html
[2] http://www.es-partnership.org/esp
[3] http://www.es-partnership.org/esp/79024/5/0/50
[4] http://www.es-partnership.org/esp/79222/5/0/50
[5] http://www.es-partnership.org/esp/79026/5/0/50

2 Background concepts

2.1 *Ecosystem service definitions, categorisations and indicators*

Many different ecosystem service definitions, classification and categorisation systems have been developed during the last decades and are under discussion for application in decision making (Villamagna et al. 2013). Perhaps ecosystem services are too case-specific for applying a common classification system (Burkhard et al. 2012b; Costanza 2008a). Promising attempts for defining and categorising ecosystem services have been undertaken for example by TEEB (2010), MA (2005) or CICES[6]. In our work, we use an approach based on the most commonly applied three ecosystem service categories (regulating, provisioning, cultural services) adding ecosystem functions (structures and processes relevant for ecosystem self-organisation; see concept of ecological integrity; Müller 2005). Based on the comprehensive list of ecosystem function and service definitions published in Kandziora et al. (2013a), we now provide an updated list of ecosystem service definitions. Additionally we differentiate between supply and demand indicators as well as their spatial characteristics. Ecosystem functions have not been specifically considered in this article because they often do not provide direct benefits to humankind (van Oudenhoven et al. 2012; Bastian et al. 2012). Thus, the intended differentiation in potentials and flows as well as demands (see following Chapters 2.2-2.3) is not applicable for ecosystem functions. Comparable problems are discussed for several regulating ecosystem services such as pollination, water flow and nutrient regulation (Chapter 2.2).

We use the following definition for ecosystem services: "Ecosystem services are the contributions of ecosystem structure and function – in combination with other inputs – to human well-being" (Burkhard et al. 2012b, p. 2; see Box 1). The rather new recognition of 'other inputs' into ecosystem services provides an improved representation of conditions in reality. In many human-environmental systems, nature-based ecosystem service contributions are hardly separable from anthropogenic inputs anymore. This point will be further elaborated in the following paragraphs.

Due to the nature of the ecosystem service approach, related indicators include descriptive aspects as well as evaluative items (Müller & Burkhard 2012). Therefore it is mandatory to have a flexible and consistent indicator selection process keeping multiple types of end-users in mind (van Oudenhoven et al. 2012). Unfortunately, many ecosystem service studies, especially at larger spatial scales, tend to be data-driven exercises (Dick et al. 2014). In fact, indicator-indicandum (the object of interest) relations have to be significant for the particular ecosystem service, the studied problem and the actual purpose of the study.

2.2 *Ecosystem service supply*

According to the definition provided in Box 1, ecosystem services contribute to human well-being. A more detailed look into these contributions reveals difficulties distinguishing between ecosystem structures, functions, stocks, actual flows and beneficiaries (the "ecosystem services cascade" components; Haines-Young & Potschin 2010). Villamagna et al. (2013) ask how to separate ecosystem capacity for service production, actual service production or its use, societal demand and various pressures on ecosystem services. Bastian et al. (2012) distinguish between ecosystem properties, potentials and services. This includes the idea that ecosystems provide a certain potential to supply services based on their functioning (van Oudenhoven et al. 2012). The demand for these potential services from society converts them into real ecosystem services. Thus it is important to distinguish between potential supply and actual flow of ecosystem services. Related measures deliver practical, policy-relevant information on the sustainability of service use (Schröter et al. 2012).

We suggest a framework based on ecosystem functions, ecosystem service supply and demand (Figure 1). Ecosystem functions are strongly influenced by land cover and land use. The supply of ecosystem services is based on specific

[6]http://cices.eu/

Box 1: Definitions used within the approach.

Ecosystem services: contributions of ecosystem structure and function – in combination with other inputs – to human well-being (Burkhard et al. 2012b).

Ecosystem service potential: the hypothetical maximum yield of selected ecosystem services (Burkhard et al. 2012a).

Ecosystem service flows: *de facto* used set (bundles) of ecosystem services and other outputs from natural systems in a particular area within a given time period[7].

Demand for ecosystem services: ecosystem goods and services currently consumed or used in a particular area over a given time period, not considering where ecosystem services actually are provided (Burkhard et al. 2012a).

Additional inputs: non-ecosystem-based anthropogenic contributions to ecosystem services, referring for example to fertiliser, energy, pesticide, technique, labour or knowledge use in human-influenced land use systems. These additional inputs (e.g. agro-, forestry or urban system services) converge with (natural) ecosystem service potentials into e.g. agro-, forestry or urban ecosystem services.

Ecosystem service providing units (SPU): spatial units that are the source of an ecosystem service (Syrbe & Walz 2012). Include the total collection of organisms and their traits required to deliver a given ecosystem service (Vandewalle et al. 2009) as well as abiotic ecosystem components (Syrbe & Walz 2012). Commensurate with ecosystem service supply (Crossman et al. 2013). Hotspots are areas that provide large components of particular services in a comparably small area/spot (García-Nieto et al. 2013; Egoh et al. 2008; Gimona & van der Horst 2007).

Ecosystem service benefiting areas (SBA): the complement to ecosystem service providing units. SBAs may be far distant from relevant SPUs (see next point spatial relations). The structural characteristics of a benefiting area must be such that the area can take advantage of an ecosystem service (Syrbe and Walz 2012). Commensurate with ecosystem service demand (Crossman et al. 2013) but several intermediate steps related to complex production and trade schemes may be included (Burkhard et al. 2012a).

SPU - SBA spatial relations: spatial characteristics describing the relationships between the place of service production and where the benefits are realized (Fisher et al. 2009; Syrbe & Walz 2012). Suggested categories include: i) *in situ* (SPU and SBA are realized in the same location), ii) omni-directional (SPU in one location, SBAs in the surrounding landscape without directional bias), iii) directional (SBA in a specific location to flow direction from the SPU), and iv) decoupled (ecosystem service can be traded over long distances, e.g., many provisioning ecosystem services) (after Fisher et al. 2009).

Ecosystem service rivalry: the degree to which the use of one ecosystem service prevents other beneficiaries from using it (Schröter et al. 2014; Kemkes et al. 2010; Costanza 2008a). Non-rival ecosystem services in return provide benefits to one person that do not reduce the amount of benefits available for others (Burkhard et al. 2012b).

Scale (spatial and temporal): the physical dimensions, in either space or time, of phenomena or observations (Reid et al. 2006). Regarding temporal aspects of ecosystem service supply and demand, hot moments are equally important to be identified as spatially relevant hotspots (Burkhard et al. 2013).

[7] In former publications (e.g. Burkhard et al. 2012a), ecosystem service flows (as defined here) were referred to as ecosystem service supply 'capacities'. However, the distinction between service potentials and capacities did not become sufficiently clear. Therefore, the term 'ecosystem service flows' is preferred instead.

ecosystem service potentials and additional system inputs converging in an ecosystem service flow to societies. The ecosystem service potential is thereby comparable to natural capital stocks, yielding a flow of ecosystem services into the future (Costanza 2008b). The additional inputs are related to the economic concept of social, human, financial and manufactured capital assets (Costanza & Daly 1992). Ecosystem service flows in return relate to *de facto* used ecosystem goods and services. This distinction can relatively easy be made for many of the provisioning ecosystem services, such as timber provision (service flow) from a stock of trees (potential) in a forest. For many regulating as well as cultural ecosystem services, this distinction and respective indicator derivation tend to be more difficult, as will be shown in the following chapters.

Our definitions (see Box 1) and indicators for altogether 31 different ecosystem services (Tables 1-3) distinguish between ecosystem service potentials and flows. The indicators need to be tested in empirical case studies. Suggested indicators for ecosystem service potentials and flows fit well with state (how much of the service is present) and performance indicators (how much can be used/ provided in a sustainable way) proposed by de Groot et al. (2010). Benefits based on ecosystem service flows are the basis for human well-being. The valuation of these benefits that forms the end of the 'ecosystem service cascade' (Haines-Young and Potschin 2010) has not been included in the framework here. The conceptual model is constructed as an ecosystem service supply-demand cycle from environment to human society and back. The framework can be linked to the Driver-Pressure-State-Impact-Response (DPSIR) model of human-environmental systems (Müller and Burkhard 2012).

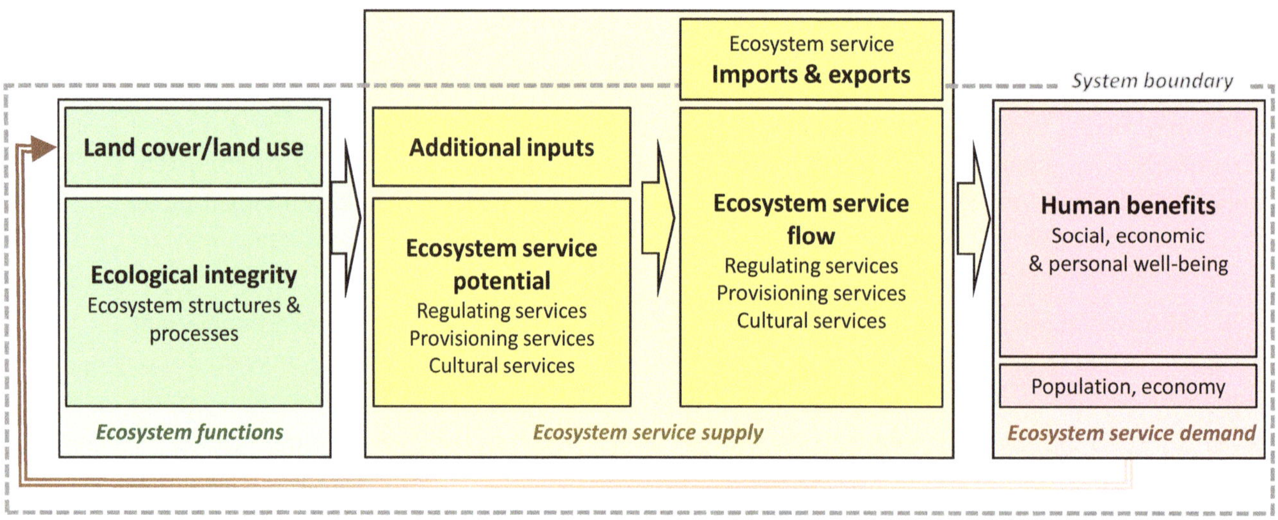

Figure 1: Conceptual model showing relations of ecosystem functions, services and benefits. For definitions, see Box 1.

One important concept regarding ecosystem service supply is the concept of service providing units (SPUs; see Box 1) or areas (SPAs). They include the total collection of organisms, their abundance, phenology, distribution and trait attributes required to deliver certain ecosystem services (Vandewalle et al. 2009) as well as abiotic components (water bodies, soil units) hosting the service supplying ecosystems (Syrbe & Walz 2012). Spatial ecosystem service assessments should preferably refer to these units or to areas affected by related processes (floodplains, catchments) instead of administrative units, which often mark artificial system boundaries. Hotspots (and coldspots as their opposite) of ecosystem service supply are special types of SPUs. They can be either small local point sources or larger sources within larger SPUs. Examples for such hotspots are given in Tables

1-3. Times of particularly high ecosystem service supply, for example due to seasonal variations, can be identified as hot moments. It is highly relevant for landscape management to identify spatial hotspots (García-Nieto et al. 2013; Palomo et al. 2013; Schneiders et al. 2012; Egoh et al. 2008; Gimona & van der Horst 2007) and temporal hot moments of ecosystem service supply and demand.

2.3 Ecosystem service demands and flows

Studies and models on the supply side of ecosystem services on different scales have developed quite far already (see reviews by Crossman et al. 2013; Martínez-Harms & Balvanera 2012; Egoh et al. 2012). But there is still a clear underrepresentation of research on the demand side for ecosystem services (Honey-Rosés & Pendleton 2013; Burkhard et al. 2013). Nevertheless, several studies have included aspects of ecosystem service demands and developed promising assessment methods (Bagstad et al. 2013a; Mubareka et al. 2013; García-Nieto et al. 2013; Kroll et al. 2012; Nedkov & Burkhard 2012; Lautenbach et al. 2011). Ecosystem service flows were assessed for example in Bagstad et al. (2013b), Willemen et al. (2013) and Palomo et al. (2013).

The demand for ecosystem services refers to ecosystem goods and services currently consumed or used in a particular area over a given time period (Burkhard et al. 2012a; Box 1). Demand can change over time and space, independent from actual ecosystem service supply (and *vice versa*; Villamagna et al. 2013). Ecosystem service use will also be driven in the future by demand. For separate demand assessments or regional supply-demand budget calculations (see Kroll et al. 2012; Burkhard et al. 2012), it is not imperative to consider where ecosystem services are actually supplied. However, for integrative large-scale ecosystem service supply-demand assessments, origins and flows of goods and services should be considered. Ecosystem service flows are the spatially explicit routing of an ecosystem service from sources to beneficiaries (Bagstad et al. 2013b). Thus, service benefiting areas (SBAs) are complementary to SPUs, but in contrast, SBAs do not relate primarily to ecosystems or geobiophysical units but to beneficiaries of certain ecosystem services. Therefore, typical locations for SBAs are urban areas or rural settlements and respective assessment units are administrative and/ or planning units (Syrbe & Walz 2012).

Flows of ecosystem goods and services from SPUs to SBAs can take place via service connecting areas (SCAs; Syrbe & Walz 2012) or certain 'carriers' (Bagstad et al. 2013b). SCAs can be of natural origin (natural waterways, gas circulation paths, viewsheds) or human-made/modified (artificial waterways, transport ways, pipelines). For a differentiated analysis of ecosystem service flows, the spatial relations between areas of ecosystem service supply and demand are of special interest. Fisher et al. (2009) and Syrbe & Walz (2012) identified four different types of SPU-SPA spatial relations:

i. *in situ* - where the services are provided and the benefits are realized in the same location,

ii. omni-directional - where the services are provided in one location, but benefit the surrounding landscape without directional bias,

iii. directional - where the service provision benefits a specific location due to the flow direction, and

iv. decoupled - where the ecosystem service can be traded over long distances.

Costanza (2008a) applied a comparable system, based on the ecosystem service classes 'global or local non-proximal', 'directional flow related', 'in situ' and 'user movement related'. Many regulating ecosystem services show in situ, omni-directional or directional (but never decoupled) SPU-SBA relationships. Demand for regulating ecosystem services has to be met locally or regionally (except global climate regulation), whereas provisioning and cultural services can show decoupled supply-demand relationships (Villamagna et al. 2013). This was shown for the example of flood regulating ecosystem services (Nedkov & Burkhard 2012), where SPUs and SBAs have to be physically connected because flood regulation cannot be imported from decoupled remote regions ("local proximal service supply" after Costanza 2008a). In the case of cultural ecosystem services, flows are generally more difficult to grasp, because most of them are intangible assets (see Chapter 3.3).

Rivalry and excludability are further important features when assessing demands for ecosystem services. The degree of rivalry indicates how much the use of one service by an individual or a user group impacts the quality or quantity of that service available to other users (Schröter et al. 2014; Kemkes et al. 2010). Excludability occurs if institutions or technologies exist that prevent other individuals or groups from using the good or service. Costanza (2008a) classified market goods and services (most provisioning services) and open access resources (some provisioning services) as rival. Some recreation services and most regulatory and cultural services were classified as non-rival. Most of the rival goods and services can be made excludable by institutions (Kemkes et al. 2010). Therefore, many ecosystem services are best treated as 'public goods' (Burkhard et al. 2012b).

In Tables 1-3 we provide lists of ecosystem service demand indicators, spatial relationships between SPUs and SBAs and their rivalry for each ecosystem service. Unfortunately, most of the suggested demand indicators do not clearly distinguish between ecosystem service demand (consumption rates) and actual human needs (as defined for example in Ruppert & Schaffer 1969; Maslow 1943). Food demand, for example, can be indicated by average crop consumption in kg or kJ/person per year. Respective numbers may however differ from nutritional needs in the assessed region. Similar problems emerge when indicating demands for many other provisioning but also cultural ecosystem services.

2.4 Spatial ecosystem service assessments

In general, ecosystem service supply, demands and flows are spatially explicit items (Schröter et al. 2012; Burkhard et al. 2012a and 2009; Fisher et al. 2009; Costanza 2008a). Respective models are needed in order to synthesize and quantify our understanding of ecosystem services and in order to understand dynamic, spatially explicit trade-offs as part of larger human-environmental systems (Burkhard et al. 2012b; Raudsepp-Hearne et al. 2010). For improved landscape planning, monitoring and sustainable environmental resource management, a better understanding of where, when and what services are provided by certain pieces of land, landscapes, regions, states, continents and globally has to be developed (Crossman et al. 2012; Swetnam et al. 2010). Spatial visualisations in maps are powerful tools with high potentials (but also risks) for the explanation of complex phenomena (Burkhard et al. 2012a; Wood et al. 2010).

The most recent ecosystem service mapping efforts have been reviewed by Crossman et al. (2013); Martínez-Harms & Balvanera (2012) and Egoh et al. (2012) and related Special Issues have been edited for example by Burkhard et al. (2013) and Crossman et al. (2012). Many mapping studies apply complex ecosystem service models and maps dealing with ecosystem service supply. However, they apply related ecosystem service demands and flows to a smaller degree. Most of the more recent work (e.g. Palomo et al. 2013; Bastian et al. 2013) clearly refers to aspects of practical application and stakeholder involvement and provide clear recommendations on how to improve mapping for application in science, policy and practice. Spatial and temporal scales, both key map attributes, and their appropriate selection are a recurring challenge of ecosystem service science and practical application. Ecosystem service assessment units (SPUs and SBAs) and related indicators, models and maps should match scales of their geobiophysical supply origin, flow and demand units on the one hand. On the other hand, they should match scales of administrative units for better application in decision making (Burkhard et al. 2013). Spatial mismatches can result in misinterpretations or inapplicability of assessment results (Kandziora et al. 2013b).

As ecosystem services are relevant over a broad range of scales in space, time and complexity, various and flexible measurement, model, accounting and assessment tools are needed (Burkhard et al. 2012b). We believe that the concept for ecosystem service assessment presented in this article provides sufficient flexibility to fulfil these demands if applied appropriately. In Tables 1-3 we suggest spatial and temporal scales for related ecosystem service assessments. To keep the approach sufficiently flexible, the scale definitions and transitions between them are rather fuzzy (especially between local and regional scales). Spatial scales include:

- local (e.g. communities, farms, ecosystems),
- regional (e.g. administrative districts, watersheds, landscapes),
- continental (e.g. Europe, Asia) and
- global

Spatial units suggested for quantification of the indicators in Tables 1-3 can refer to 'regions', which can be administrative units (such as states, counties, communities), environmental regions (e.g. biomes) or geobiophysical spatial units (soil associations, watersheds). Several of the regulating ecosystem services relate to specific spatial process units such as catchments. Temporal assessment scales include:

- short-term (e.g. events, peak flows),
- seasonal (e.g. harvest rhythms, tourist seasons, growing seasons),
- annual (e.g. sums, yearly average values),
- medium-term (e.g. decades) and
- long-term (e.g. generations, centuries, millennia) periods.

3 Methodology

In the following, we present and discuss comprehensive tables, providing details on the information discussed above, separately for regulating (Table 1), provisioning (Table 2) and cultural (Table 3) ecosystem services.

3.1 *Regulating ecosystem services*

Regulating ecosystem services are by nature closely related to ecosystem structures, processes and functions. For some regulating ecosystem services, such as nutrient flow, water flow or waste regulation, clear overlaps with ecosystem functions like nutrient or water cycling are obvious in almost all current ecosystem service categorisation systems. This induces a high risk for double-counting when no clear separation between ecosystem functions and services is made and both are jointly valuated (Fisher et al. 2009). Regarding land use impacts, many ecosystem functions and regulating ecosystem services can take place (or would perform even better) without human intervention or demand. Therefore, several ecosystem service definitions, referring to ecosystem services as direct benefits to human societies (e.g. Fisher et al. 2009; MA 2005), are not applicable for all the different regulating services. Many regulating ecosystem services are not perceived as services by the public because they lack clear (direct) benefits to the society, although progress has been made in their evaluation (see Kumar & Wood 2010).

The perhaps most prominent example here is pollination, one of the regulating ecosystem services assessed and mapped frequently (Schulp et al. 2014; Lautenbach et al. 2011). The final good would be the fruit or flower to be consumed or enjoyed, whereas the pollination process itself (the pollen transfer) would be an intermediate service (according to Fisher et al. 2009), or perhaps even better would be treated as ecosystem function (Boyd and Banzhaf 2007). Distinctions between pollination potentials and actual flows are also intricate. Here we used potential habitats for pollinators as well as species numbers and amount of pollinators as indicators for ecosystem service potential (Table 1). The amount of pollinated plants was then used to indicate actual pollination service flow. According to Lautenbach et al. (2011), demand for pollination services is generated by the farmer's decision to plant crops depending on/profit from pollination. Similar to this definition, we use the amount of agricultural, garden or wild plants demanding pollination.

For many regulating ecosystem services it is difficult to distinguish between potential supply and actual flows. For more event-related regulating ecosystem services, such as natural hazard protection, erosion regulation or pest and disease control, we choose service flow indicators to be proportional to amounts of prevented hazards, prevented erosion events or prevented pest and disease outbreaks. Numbers and effects of prevented events are of course difficult to measure in most cases. Furthermore, a clear localisation of demands for regulating ecosystem services is not always possible, in some cases perhaps not even reasonable. Local climate regulation for example continuously takes place over space and time. Thus, there are continuous flows or states, no matter what potential has been indicated for a particular site. Therefore, indicators suggested for local climate regulation potential and flows in Table 1 are relatively similar, except that service flows refer to deviations of local climate components compared to surrounding areas.

Another mostly unsolved problem is the distinction between purely nature-based regulating ecosystem service supply and additional (mostly anthropogenic) inputs, especially in intensive agricultural land use systems (delivering agro[eco]system services; Papendiek et al. 2012). We understand agrosystem services to be additional anthropogenic system inputs such as fertiliser, water, energy, technology, labour or knowledge affecting especially regulating ecosystem service supply (e.g. regulation of nutrients, erosion, natural hazards or water flows). Agrosystem services converge with (nature-based) ecosystem services in agro-ecosystem services (see Box 1 and Figure 1). The concept is related to the idea of Human Appropriation of Net Primary Production (HANPP) (Haberl et al. 2012). However, HANPP compares potential natural vegetation with human-modified systems, whereas agrosystem services refer to human-modified agricultural systems and contrast anthropogenic inputs with natural potentials for actual agro-ecosystem service flows.

For many regulating ecosystem services the spatial localisation, as well as the clear definition of beneficiaries, are problematic, mainly due to the lack of a final good or end-product (Villamagna et al. 2013). For pollination services mentioned above, omni-directional spatial relationships between potential service providing units (suitable habitats for pollinators) and benefitting areas (plants demanding pollination) are likely to occur (Fisher et al. 2009). Similar patterns can be found regarding pest and disease control. Villamagna et al. (2013) suggest using ecological work performed instead of conventionally used environmental quality measures to indicate regulating ecosystem services. For example, elements removed from water should be used to indicate water purification regulating services. This idea, although introducing a lot of uncertainty, was implemented to distinguish potentials from flows for several regulating ecosystem services, for example for water purification and nutrient regulation (see Table 1).

We are aware, that regulating ecosystem service demands and related perceived human benefits may differ considerably. For global climate regulating services for example, human benefits refer to non-desired temperature changes, storm events or coastal hazards. Nevertheless, the indicators suggested in Table 1 relate to regulating processes (e.g. greenhouse gas emissions) relevant for climate regulation. This relation to regulating processes makes more sense in terms of developing, quantifying and localising suitable avoidance strategies (for example the 'polluter pays' principle as well as carbon emission trading).

3.2 Provisioning ecosystem services

In general, the distinction between provisioning ecosystem service potentials and flows can be carried out more easily than the distinction between regulating ecosystem service potentials and flows. Natural capital stocks (e.g. forests, agricultural fields or water bodies) for potential use can be measured and related service flows assessed. Problems emerge when goods and services go through long supply chains from providing units to traders, processing, refinement, finishing, selling and transporting entities before the end product reaches the final consumer (Figure 2). The questions are where to locate the demand (Schröter et al. 2012) and who has to be considered as beneficiary from the large group of actors involved? One exception is direct marketing, where the group of actors involved would be much smaller. Otherwise each involved entity in the production chain needs to supply respective return flows (money or other services), thereby getting a share of the value of the good or service traded. This normally leads to an increasing price of the product. But does it really add up to a higher value of the product?

In such complex supply chains, involving natural, social and manufactured capital, a separation into intermediate and final ecosystem services as suggested by Boyd & Banzhaf (2007) provides relevant aspects but would not sufficiently solve the dilemma. In reality, various ecosystem services in different intermediate stages can be contributing to

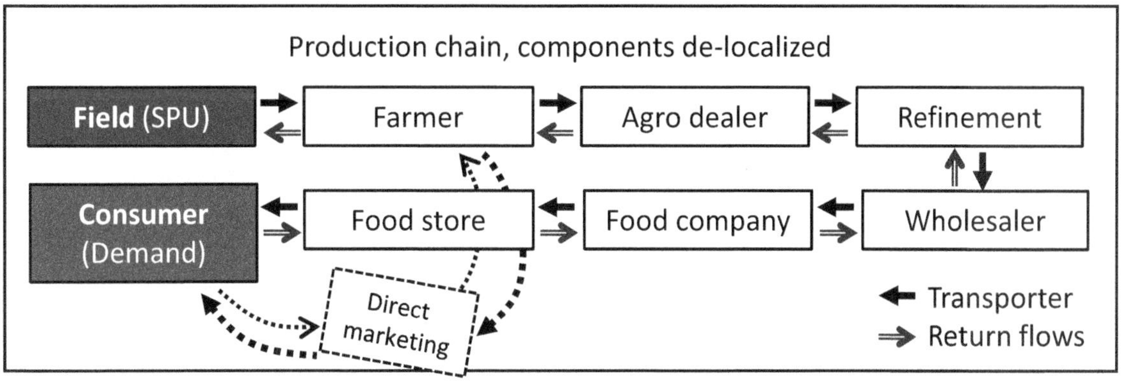

Figure 2: Complex ecosystem service supply chain, example of food provisioning services. Direct marketing could provide a more direct link. Several further 'short-cuts' between supply and demand are possible.

one final product. Schröter et al. (2012) suggested localising the demand for non-spatially confined (thus tradable or actively transportable) ecosystem services at the site of the last contribution of the ecosystem. We prefer to use this definition to locate provisioning ecosystem service flows in combination with the indicators suggested in Table 2. Ecosystem service demand should be located at the site of the final beneficiary, usually the end-consumer. If, however, the demand for an ecosystem service has been quantified at another site, the localisation has to be clearly defined.

Additional inputs modifying ecosystem structures and functionality, as mentioned above, make assessments of provisioning ecosystem services even more complex (Lautenbach et al. 2011). Difficulties emerge especially in highly human-dominated systems such as intensive agricultural areas, greenhouses, aquaculture systems or intensive mass animal farming (Baral et al. 2014; Petz & van Oudenhoven 2012). Other strongly anthropogenic-shaped systems such as cities, other settlements or industrial units have high demands for provisioning goods and services (Burkhard et al. 2012a). Problems with the share of natural ecosystem inputs to the supply of provisioning (agro-)ecosystem services remain, similarly to problems with regulating ecosystem services. Anthropogenic inputs are likely to be higher in assessed provisioning ecosystem service flows than in related potentials for the same service. This effect can be found in many agricultural systems requiring many inputs.

Mineral resources or abiotic energy sources also need a lot of facilities and other human-made inputs and equipment to finally be harvested. In most ecosystem service categorisation systems, mineral resources and abiotic energy sources are not acknowledged as ecosystem services in the strictest sense (see Haines-Young and Potschin 2010). After De Groot et al. (2002), fossil fuels, wind and solar energy are usually non-renewable abiotic resources which can neither be attributed to specific ecosystems nor be called ecosystem services. In the provisional CICES[6] classification, they are (temporally) treated as 'abiotic outputs from natural systems'. However, mineral resources and abiotic energy sources are both highly relevant for policy decisions and land use, as well as resource management strategies. Therefore, their spatial localisation and quantification need to be included in ecosystem service assessments as they can influence the amount and supply of other ecosystem services (Kandziora et al. in review).

The supply of other, more nature-based provisioning ecosystem services, such as freshwater, can be very hard to localise appropriately based on surface data (such as land cover) alone. If the water is not withdrawn from surface reservoirs but from groundwater, the aquifer's location has to be considered as service providing unit. In most cases, the SPU's location will not be related to land cover or land use forms identified on the study area's surface. Wells on the surface are only point sources of freshwater supply, linked to much bigger (underground) SPUs. The same problem emerges when localising SPUs of mineral resource provision based on underground mining. Proxies like groundwater recharge rates were used to calculate freshwater potentials (Kroll et al. 2012). Vigerstol & Aukema (2011), provide a comprehensive review of freshwater ecosystem service modelling tools.

Besides location, temporal aspects are of high relevance when quantifying provisioning ecosystem services. Single events like agricultural harvests, normally taking place only 1-3 times per year (depending on geographical conditions), can definitely be named as hot moments of provisioning ecosystem service supply. During the growing season, the provisioning ecosystem service potential increases constantly, whereas the flows temporarily remain more or less at zero until the final harvest. During harvest, the potential decreases dramatically, whereas service flows (from field to farmer, the first two elements of the supply chain in Figure 2) show a (short-term) peak (Figure 3). Therefore, rural agricultural landscapes are characterized by regular growth and harvest phases, reflected by related changes in their ecosystem service supply (Burkhard et al. 2011). Of course, additional agricultural strategies and crops with various growing and harvesting rhythms exist. Better information on these variations is highly relevant for site-specific landscape management, i.e. to optimise additional inputs. Respective seasonal patterns can be found for regulating (e.g. during storm or rain seasons) and cultural ecosystem service (e.g. tourist season) supply (and demand) as well. Other provisioning ecosystem services, such as timber, show much longer rotation periods taking several decades to grow before being harvested rather suddenly. Therefore, the selection and definition of appropriate temporal assessment scales have to be carried out very carefully.

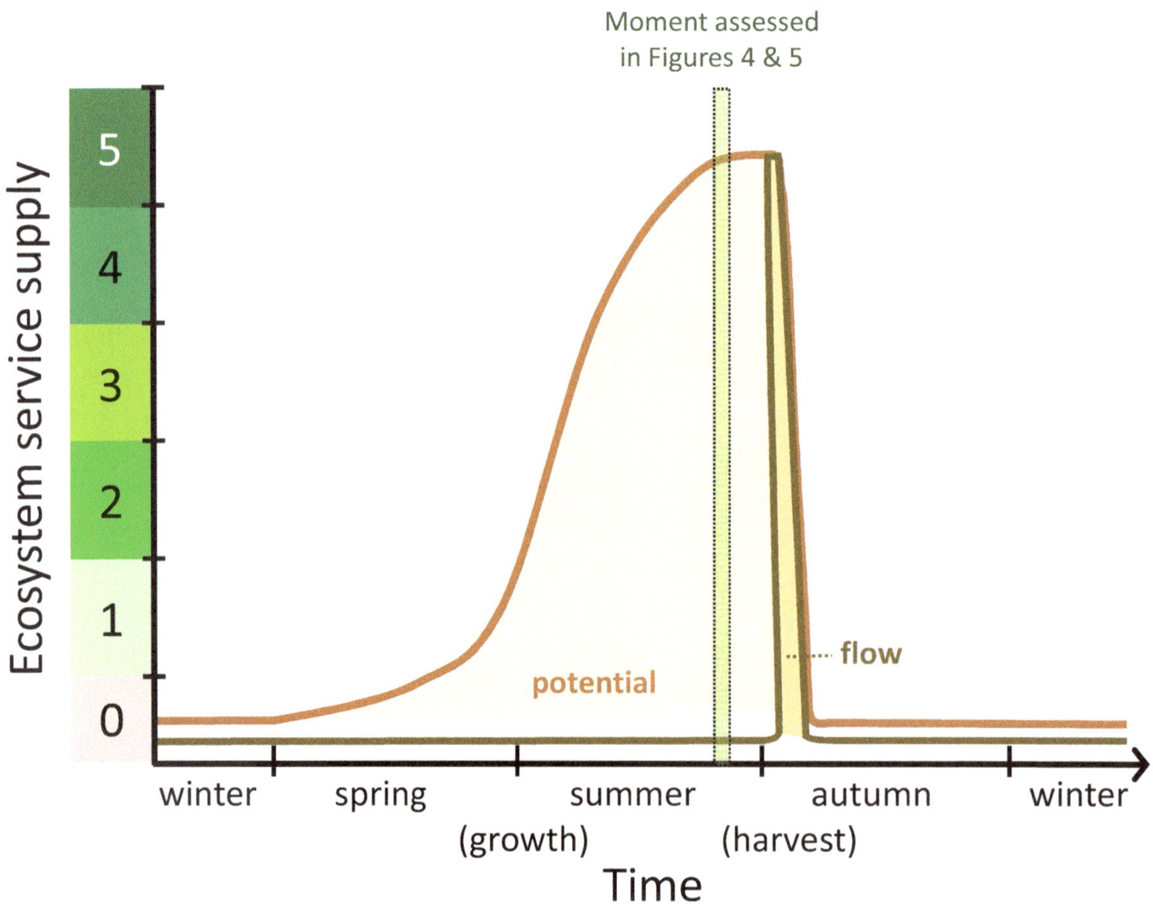

Figure 3: Ecosystem service potential and flow. Example of monoculture summer annual (e.g. wheat) crop-related food provisioning ecosystem service during the growth and harvest period (compare values with Figures 4 and 5).

3.3 *Cultural ecosystem services*

Cultural ecosystem services are difficult to assess due to their intangible nature and due to several methodological concerns. Therefore, they are rarely fully considered in ecosystem service assessments (Plieninger et al. 2013). Most of related quantifications and indicators reduce cultural ecosystem services to marketable services such as (eco)tourism and recreation (see review of Hernández-Morcillo et al. 2013). More intangible cultural ecosystem services, such as landscape aesthetics, spiritual experience or knowledge systems, have been assessed based on interviews, questionnaires and hedonic price models. More recent studies tend to integrate further assessment tools like landscape metrics to assess landscape aesthetics (Frank et al. 2013), Participatory GIS (PGIS; Palomo et al. 2013; Plieninger et al. 2013; Fagerholm et al. 2012) or Delphi surveys (Scolozzi et al. 2012).

A clear distinction between potentials and flows is not trivial in the case of cultural ecosystem services. Schröter et al. (2014) assessed recreational residential amenity service potentials by delineating suitability of SPUs for providing a location for second homes. Related ecosystem service flows were indicated based on existing cabins. Following our ecosystem services definition (Box 1), we considered existing facilities such as cabins, hotels or recreational areas as well as occurring events as ecosystem service potentials (stocks). The actual utilisation of these facilities or events is used to indicate cultural ecosystem service flows (see Table 3). Nahuelhual et al. (2013) applied a comparable method to calculate and map recreation and ecotourism potentials versus recreation and ecotourism opportunities. Potentials were calculated based on tourism attraction capacity for locations with touristic facilities and natural attractions. Opportunities were indicated by tourism carrying capacities (maximum visitor number in the study area depending on physical and biological settings and management). Visitor or consumer carrying capacities can also be used to estimate ecosystem service flows related to potentials for several other cultural ecosystem services. Here, biophysical carrying capacities need to be considered, as well as the sensibility of visitors towards a maximum number of other visitors. This fact can become relevant in crowded areas or periods (Brandt et al. 2013).

Nevertheless, appropriate SPU and SBA delineations remain delicate for cultural ecosystem services. For most of them, spatial mismatches exist between the ecosystems that supply services and people that enjoy them (Bagstad et al. 2013b). Lautenbach et al. (2011) evaluated demands for recreation services as the possibility for people to access the services. However, access is not always easy to specify for intangible services such as landscape aesthetics. Is the SPU located where the aesthetic landscape is or where the consumer (observer) is at a good location to look at it? One approach to solve this dilemma is creating viewsheds. Viewsheds can be lines of sight to connect view paths between source locations (e.g. aesthetic landscape features) and locations for human use (areas of potential enjoyment; Bagstad et al. 2013b). Calculations are mainly carried out using Digital Elevation Models (DEM) in GIS. The whole viewshed (the area that can be seen from a given location) would be the SPU for landscape aesthetic cultural ecosystem service potential. If other data are not obtainable, methods such as questionnaires, interviews, surveys, travel cost estimations, hedonic price models or willingness to pay are regularly applied for cultural ecosystem service assessments.

The demand for spatially confined ecosystem services was recommended to be mapped at the location where the final beneficiary uses the ecosystem service by Schröter et al. (2012). For the time being, we decided to denote all benefitting entities, such as touristic infrastructure, educational and spiritual facilities, their visitors, communities or households (at home location). as service benefitting areas (or sites). However, we prefer to localise demands for cultural ecosystem services at the consumer's home site, i.e. the place where she/he spends the majority of the year. This strict definition still needs to be proven further. For example, it is not clear for how long recreational or spiritual experiences are "taken home" by the consumer or whether the demand should better be located within the SPUs.

4 The Matrix

The ecosystem service matrices consist of ecosystem services (currently 11 regulating, 14 provisioning and 6 cultural services; according to Tables 1-3) on the x-axis and geobiophysical spatial units (e.g. the 44 CORINE[8] land cover types used here) on the y-axis. At the intersections (here altogether 1364), the different spatial units' ecosystem service potentials (Figure 4), flows (Figure 5) or demands (Figure 6) were assessed on a scale from 0 (no relevant supply or demand) to 5 (maximum relevant supply or demand) for a hypothetical 'normal' European landscape at one time point in summer before harvest. For further description of the method see Burkhard et al. (2009 and 2012a). The normalisation to this relative 0-5 scale aims at making different ecosystem services (measured and assessed by various indicators and units) comparable with each other. For calculations of the single values per land cover type, statistics (Kandziora et al. 2013b; Kroll et al. 2012), model results (Nedkov and Burkhard 2012), expert knowledge (Vihervaara et al. 2010 and 2012), interview results (Kaiser et al. 2013), monitoring or other data sources (Baral et al. 2013) can be used. Respective data then have to be classified to the six categories using appropriate class breaks[9]. Furthermore, it should be noted that the 0-class indicates no 'relevant' ecosystem supply or demand. Depending on the ecosystem service type, on land cover type, as well as on spatial and temporal scale, this can mean that the ecosystem service is not supplied or demanded at all or at a low amount. We know from recent case studies (e.g. Kandziora et al. 2013a and 2013b; Kaiser et al. 2013; Müller et al. subm.) that earlier published versions of the ecosystem service matrix (Burkhard et al. 2012a and 2009) did not consequently differentiate between ecosystem service potentials and flows (as defined in Chapter 2.2; Box 1; Tables 1-3). Moreover, the set of assessed ecosystem services had to be changed. Therefore, we now provide two different matrices: one for ecosystem service potentials (Figure 4) and one for ecosystem service flows (Figure 5).

It is important to note that assessed potentials and flows as well as demand values are exemplary numbers for hypothetical central European 'normal' landscapes, based on our experience from the work in different case studies. Like all expert-based assessments, the values are strongly dependent on the evaluator's experience, knowledge and objectivity (Burkhard et al. 2012a). We defined a day in summer, before the main harvest period, as appropriate time step for the evaluation. This is especially relevant for the assessment of many provisioning and several cultural ecosystem services. The given values are exemplary and illustrate the application potential of the matrix method. In other studies, the evaluations have to be modified according to the specific conditions, points in time and data obtained. Modification may also be needed for the matrix' axes to reflect peculiarities in land cover, to include other geobiophysical characteristics (y-axis) or specific relevant ecosystem services (x-axis).

4.1 *Ecosystem service potential matrix*

The matrix in Figure 4 shows a clear pattern of high potentials in the land cover types with less human impacts. Forests, wetlands and water bodies received especially high rankings. The more anthropogenically influenced land cover types (most of them in the upper part of the matrix) have considerably lower ecosystem service potentials, except for some cultural ecosystem services available in urban areas. Many agricultural land cover types show high potentials for food-related ecosystem service supply. This is typical for agricultural areas before harvest (see Chapter 3.2). As mentioned before (Chapter 3.2), the linkage of freshwater supply from groundwater sources to above-ground land cover types is not always feasible. Therefore, only above-ground water bodies were considered for freshwater supply here.

[8] http://www.eea.europa.eu/data-and-maps/figures/corine-land-cover-types-2006

[9] Equal intervals classification methods should be used to group the data into the 0-5 classes. Other methods such as natural breaks or quantiles would manipulate results and are less suitable to make the different classes and their values comparable with each other.

	Regulating services	Global climate regulation	Local climate regulation	Air quality regulation	Water flow regulation	Water purification	Nutrient regulation	Erosion regulation	Natural hazard regulation	Pollination	Pest and disease control	Regulation of waste	Provisioning services	Crops	Biomass for energy	Fodder	Livestock (domestic)	Fibre	Timber	Wood Fuel	Fish, seafood & edible algae	Aquaculture	Wild foods & resources	Biochemicals & medicine	Freshwater	Mineral resources*	Abiotic energy sources*	Cultural services	Recreation & tourism	Landscape aesthetics & inspiration	Knowledge systems	Religious & spiritual experience	Cultural heritage & cultural diversity	Natural heritage & natural diversity
Continuous urban fabric		0	0	0	0	0	0	2	0	0	1	0		0	0	0	0	0	0	0	0	0	0	0	0	0	1		3	3	2	2	1	0
Discontinuous urban fabric		0	0	0	0	0	0	1	0	1	1	0		1	0	0	0	0	0	0	0	0	0	0	0	0	1		3	2	2	2	2	0
Industrial or commercial units		0	0	0	0	0	0	2	0	0	1	0		0	0	0	0	0	0	0	0	0	0	0	0	0	1		0	0	0	0	2	0
Road and rail networks		0	0	0	0	0	0	1	0	0	0	0		0	0	0	0	0	0	0	0	0	0	0	0	0	0		0	0	0	0	1	0
Port areas		0	0	0	0	0	0	3	3	0	1	0		0	0	0	0	0	0	0	0	0	0	0	0	0	0		1	2	0	0	1	0
Airports		0	0	0	0	0	0	1	0	0	1	0		0	0	0	0	0	0	0	0	0	0	0	0	0	0		0	0	0	0	0	0
Mineral extraction sites		0	0	0	0	0	0	0	0	0	0	0		0	0	0	0	0	0	0	0	0	0	0	0	5	3		0	0	2	0	1	0
Dump sites		0	0	0	0	0	0	0	0	0	0	2		0	1	0	0	0	0	0	0	0	0	0	0	0	0		0	0	0	0	0	0
Construction sites		0	0	0	0	0	0	0	0	0	0	0		0	0	0	0	0	0	0	0	0	0	0	0	0	0		0	0	0	0	2	0
Green urban areas		2	2	2	2	2	2	2	1	2	2	2		0	0	0	0	0	0	0	0	0	0	0	0	0	0		3	3	1	0	2	1
Sport and leisure facilities		1	1	1	1	1	1	1	0	0	1	1		0	0	0	0	0	0	0	0	0	0	0	0	0	0		5	1	0	0	1	0
Non-irrigated arable land		1	2	1	2	0	1	0	1	1	2	2		5	5	5	0	5	0	0	0	0	1	3	0	0	2		1	1	2	0	3	0
Permanently irrigated land		1	3	1	1	0	1	0	1	1	2	2		5	1	2	0	4	0	0	0	0	1	3	0	0	1		1	1	2	0	3	0
Ricefields		0	2	1	1	0	1	0	0	1	1	2		5	1	2	0	0	0	0	0	0	0	0	0	0	0		1	1	2	0	3	0
Vineyards		1	1	1	1	0	1	1	0	1	1	1		4	1	0	0	0	0	1	0	0	0	0	0	0	0		3	2	3	0	5	0
Fruit trees and berries		2	2	2	2	1	2	2	2	5	3	2		4	1	0	0	0	2	2	0	0	0	2	0	0	0		3	2	2	0	4	1
Olive groves		1	1	1	1	1	1	1	0	1	2	2		4	1	0	0	0	2	2	0	0	0	2	0	0	0		2	2	2	0	4	0
Pastures		2	1	0	1	0	1	1	1	0	2	4		0	1	5	5	0	0	0	0	0	2	0	0	0	5		2	2	2	0	3	1
Annual and permanent crops		1	2	1	1	0	1	2	1	1	2	2		4	2	4	1	5	0	0	0	0	1	1	0	0	2		1	1	2	0	3	0
Complex cultivation patterns		1	2	1	1	0	1	1	1	2	3	2		4	2	2	1	4	0	1	0	0	1	2	0	0	1		2	2	2	0	3	0
Agriculture & natural vegetation		2	3	2	2	2	2	2	1	2	3	2		3	3	2	2	4	1	1	0	0	2	1	0	0	1		2	2	3	1	3	3
Agro-forestry areas		2	2	2	2	2	2	3	1	3	3	3		2	3	2	3	2	3	3	0	0	2	1	0	0	0		2	2	2	0	3	2
Broad-leaved forest		5	5	5	3	5	5	5	4	4	4	4		0	1	1	0	1	5	5	0	0	5	3	0	0	0		5	5	5	3	4	5
Coniferous forest		5	5	5	3	5	5	5	4	4	4	4		0	1	1	0	1	5	5	0	0	5	3	0	0	0		5	5	5	3	4	4
Mixed forest		5	5	5	3	5	5	5	4	4	5	5		0	1	1	0	2	5	5	0	0	5	3	0	0	0		5	5	5	3	4	5
Natural grassland		5	2	0	1	3	4	5	1	1	1	2		0	1	2	3	0	0	0	0	0	5	1	0	0	2		3	4	5	1	3	3
Moors and heathland		3	4	0	2	3	3	2	2	2	2	3		0	1	1	1	0	0	2	0	0	2	1	0	0	0		4	4	5	1	2	4
Sclerophyllous vegetation		2	2	1	1	1	2	1	1	2	2	3		0	1	1	1	1	2	2	0	0	1	3	0	0	1		2	3	4	1	2	4
Transitional woodland shrub		2	2	1	1	1	2	1	1	2	2	3		0	2	1	1	1	1	2	0	0	1	1	0	0	1		2	3	4	1	2	2
Beaches, dunes and sand plains		0	0	0	1	1	1	0	5	0	1	1		0	0	0	0	0	0	0	0	0	0	1	0	1	0		5	4	4	1	3	2
Bare rock		0	0	0	0	1	0	2	1	0	0	0		0	0	0	0	0	0	0	0	0	0	0	0	1	0		2	3	3	2	2	1
Sparsely vegetated areas		0	1	0	1	1	1	1	1	0	1	1		0	0	0	1	0	0	0	0	0	1	0	0	0	2		1	1	3	0	2	1
Burnt areas		0	1	0	0	0	0	0	0	0	0	1		0	0	0	0	0	0	0	0	0	0	0	0	0	0		0	0	2	0	0	0
Glaciers and perpetual snow		3	4	0	5	0	0	0	0	0	1	1		0	0	0	0	0	0	0	0	0	0	0	5	0	0		5	5	4	0	0	1
Inland marshes		2	2	0	3	2	4	1	4	1	2	3		0	0	4	2	0	0	0	0	0	1	0	0	0	0		1	2	3	0	2	2
Peatbogs		5	4	0	4	4	4	2	3	2	3	4		0	2	0	0	0	0	0	0	0	1	2	1	0	0		3	2	3	0	2	4
Salt marshes		1	1	0	1	1	2	1	4	1	2	2		0	0	2	2	0	0	0	0	0	1	0	0	0	0		3	2	3	0	2	2
Salines		0	3	0	0	0	0	0	0	0	1	1		0	0	0	0	0	0	0	0	0	0	2	0	2	0		2	2	3	0	4	0
Intertidal flats		1	1	0	1	1	1	1	5	0	2	3		0	1	0	0	0	0	0	0	0	1	0	0	0	0		4	2	3	0	2	2
Water courses		0	1	0	3	3	3	0	3	0	3	5		0	2	0	0	0	0	0	3	0	4	0	5	0	3		4	4	4	2	3	3
Water bodies		1	2	0	5	2	3	0	3	0	3	5		0	1	0	0	0	0	0	4	5	4	0	5	0	1		5	4	4	2	3	3
Coastal lagoons		1	1	0	4	2	3	0	4	0	3	5		0	1	0	0	0	0	0	4	5	4	1	0	0	0		3	4	4	0	2	3
Estuaries		1	0	0	3	3	3	0	3	0	3	5		0	2	0	0	0	0	0	4	5	4	1	0	0	1		3	4	4	0	2	3
Sea and ocean		3	3	0	1	2	3	0	0	0	3	5		0	4	3	0	0	0	0	5	5	4	3	0	1	3		4	5	5	2	3	3

*abiotic outputs from natural systems (after CICES)

Figure 4: Exemplary ecosystem service potential matrix. The exemplary evaluation refers to a hypothetical European „normal" landscape in summer (before the harvest period). Scale from 0/rosy = no relevant potential; 1/grey green = low relevant potential; 2/light green = relevant potential; 3/yellow green = medium relevant potential; 4/blue green = high relevant potential; and 5/dark green = very high (maximum) relevant potential (after Burkhard et al. 2009 and 2012).

4.2 *Ecosystem service flow matrix*

	Regulating services	Global climate regulation	Local climate regulation	Air quality regulation	Water flow regulation	Water purification	Nutrient regulation	Erosion regulation	Natural hazard regulation	Pollination	Pest and disease control	Regulation of waste	Provisioning services	Crops	Biomass for energy	Fodder	Livestock (domestic)	Fibre	Timber	Wood Fuel	Fish, seafood & edible algae	Aquaculture	Wild foods & resources	Biochemicals & medicine	Freshwater	Mineral resources*	Abiotic energy sources*	Cultural services	Recreation & tourism	Landscape aesthetics & inspiration	Knowledge systems	Religious & spiritual experience	Cultural heritage & cultural diversity	Natural heritage & natural diversity
Continuous urban fabric		0	0	0	0	0	0	2	0	1	1	0		0	0	0	0	0	0	0	0	0	0	0	0	0	0		3	2	2	2	1	0
Discontinuous urban fabric		0	0	0	0	0	0	1	0	2	1	0		0	0	1	0	0	0	0	0	0	0	0	0	0	1		2	1	2	2	2	0
Industrial or commercial units		0	0	0	0	0	0	2	0	1	1	0		0	0	0	0	0	0	0	0	0	0	0	0	0	1		0	0	0	0	2	0
Road and rail networks		0	0	0	0	0	0	1	0	0	0	0		0	0	0	0	0	0	0	0	0	0	0	0	0	0		0	0	0	0	1	0
Port areas		0	0	0	0	0	0	3	3	0	1	0		0	0	0	0	0	0	0	0	0	0	0	0	0	0		1	2	0	0	1	0
Airports		0	0	0	0	0	0	1	0	0	1	0		0	0	0	0	0	0	0	0	0	0	0	0	0	0		0	0	0	0	0	0
Mineral extraction sites		0	0	0	0	0	0	0	0	0	0	0		0	0	0	0	0	0	0	0	0	0	0	0	3	1		0	0	1	0	0	0
Dump sites		0	0	0	0	0	0	0	0	0	0	2		0	0	0	0	0	0	0	0	0	0	0	0	0	0		0	0	0	0	0	0
Construction sites		0	0	0	0	0	0	0	0	0	0	0		0	0	0	0	0	0	0	0	0	0	0	0	0	0		0	0	0	0	1	0
Green urban areas		2	2	2	2	2	2	2	1	2	2	2		0	0	1	0	0	0	0	0	0	0	0	0	0	0		3	3	1	0	2	1
Sport and leisure facilities		1	1	1	1	1	1	1	0	0	1	1		0	0	0	0	0	0	0	0	0	0	0	0	0	0		4	1	0	0	1	0
Non-irrigated arable land		1	2	1	2	0	3	0	1	3	2	2		0	0	0	0	0	0	0	0	0	0	0	0	0	0		1	1	1	0	1	0
Permanently irrigated land		1	3	1	3	0	3	0	1	3	2	2		0	0	0	0	0	0	0	0	0	0	0	0	0	0		1	1	1	0	1	0
Ricefields		0	2	1	4	0	3	0	0	1	1	2		0	0	0	0	0	0	0	0	0	0	0	0	0	0		1	1	1	0	2	0
Vineyards		1	1	1	1	0	1	1	0	2	1	1		0	0	0	0	0	0	0	0	0	0	0	0	0	0		2	1	2	0	4	0
Fruit trees and berries		2	2	2	2	1	2	2	2	3	3	2		0	0	0	0	0	0	0	0	0	0	0	0	0	0		2	1	1	0	3	1
Olive groves		1	1	1	1	1	1	1	0	2	2	2		0	0	0	0	0	0	0	0	0	0	0	0	0	0		2	2	2	0	3	0
Pastures		1	1	0	1	0	1	1	1	1	2	4		0	0	4	2	0	0	0	0	0	0	0	0	0	0		2	2	2	0	2	1
Annual and permanent crops		1	2	1	2	0	2	2	1	2	2	2		0	0	0	0	0	0	0	0	0	0	0	0	0	0		1	1	1	0	2	0
Complex cultivation patterns		1	2	1	2	0	2	1	1	3	3	2		0	0	1	0	0	0	0	0	0	0	1	0	0	0		2	2	1	0	2	0
Agriculture & natural vegetation		1	3	2	2	1	2	2	1	2	3	2		0	0	1	0	1	0	1	0	0	0	1	0	0	1		2	2	2	1	2	1
Agro-forestry areas		1	2	2	2	1	1	3	1	2	3	3		0	0	0	1	1	1	1	0	0	0	1	0	0	0		2	2	1	0	2	1
Broad-leaved forest		4	5	5	3	4	5	5	3	1	4	4		0	1	1	0	1	1	1	0	0	1	1	0	0	0		4	4	4	2	2	4
Coniferous forest		4	5	5	3	4	5	5	3	1	4	4		0	1	1	0	1	1	1	0	0	1	1	0	0	0		4	4	4	2	2	3
Mixed forest		4	5	5	3	4	5	5	3	1	5	5		0	1	1	0	1	1	1	0	0	1	1	0	0	0		4	4	4	2	2	4
Natural grassland		2	2	0	1	3	4	5	1	2	1	2		0	0	2	1	0	0	0	0	0	1	1	0	0	0		3	4	4	1	2	2
Moors and heathland		2	4	0	2	3	3	2	2	2	2	3		0	1	1	0	0	0	1	0	0	1	1	0	0	0		3	4	4	1	1	3
Sclerophyllous vegetation		1	2	1	1	1	2	1	1	1	2	3		0	0	1	0	0	1	1	0	0	1	0	0	0	0		2	3	4	1	1	3
Transitional woodland shrub		1	2	1	1	1	2	1	1	1	2	3		0	1	1	0	1	1	1	0	0	1	1	0	0	0		2	3	4	1	1	1
Beaches, dunes and sand plains		0	0	0	1	1	1	0	5	0	1	1		0	0	0	0	0	0	0	0	0	0	1	0	0	0		5	4	4	1	1	1
Bare rock		0	0	0	0	1	0	2	1	0	0	0		0	0	0	0	0	0	0	0	0	0	0	0	0	0		2	3	3	1	1	1
Sparsely vegetated areas		0	1	0	1	1	1	1	1	0	1	1		0	0	0	1	0	0	0	0	0	0	0	0	0	0		1	1	2	0	1	1
Burnt areas		0	1	0	0	0	0	0	0	0	0	1		0	0	0	0	0	0	0	0	0	0	0	0	0	0		0	0	2	0	0	0
Glaciers and perpetual snow		3	4	0	3	0	0	0	0	0	1	1		0	0	0	0	0	0	0	0	0	0	0	1	0	0		2	4	3	0	0	1
Inland marshes		2	2	0	2	2	4	1	4	1	2	3		0	0	2	1	0	0	0	0	0	0	0	0	0	0		1	2	3	0	1	1
Peatbogs		4	4	0	3	4	4	2	3	2	3	4		0	2	0	0	0	0	0	0	0	0	1	0	0	0		3	2	3	0	1	3
Salt marshes		0	1	0	1	1	2	1	4	1	2	2		0	0	2	1	0	0	0	0	0	0	0	0	0	0		3	2	3	0	1	1
Salines		0	2	0	0	0	0	0	0	0	1	1		0	0	0	0	0	0	0	0	0	0	2	0	2	0		2	2	3	0	3	0
Intertidal flats		1	1	0	1	1	1	1	5	0	2	3		0	0	0	0	0	0	0	0	0	0	0	0	0	0		4	2	3	0	1	1
Water courses		0	1	0	3	3	3	0	3	0	3	5		0	0	0	0	0	0	0	2	0	2	0	2	0	2		4	4	3	1	2	2
Water bodies		1	2	0	3	2	3	0	3	0	3	5		0	0	0	0	0	0	0	2	2	3	0	2	0	1		5	4	3	1	2	2
Coastal lagoons		1	1	0	2	1	3	0	4	0	3	5		0	0	0	0	0	0	0	2	2	2	0	0	0	0		3	4	3	0	1	2
Estuaries		1	0	0	2	3	3	0	3	0	3	5		0	0	0	0	0	0	0	2	2	2	0	0	0	0		3	4	3	0	1	2
Sea and ocean		2	3	0	1	1	3	0	0	0	3	5		0	1	1	0	0	0	0	3	2	3	2	0	1	1		4	5	3	1	2	2

*abiotic outputs from natural systems (after CICES)

Figure 5: Exemplary ecosystem service flow matrix. The exemplary evaluation refers to a hypothetical European „normal" landscape in summer (before the harvest period). Scale from 0/rosy = no relevant flow; 1/grey green = low relevant flow; 2/light green = relevant flow; 3/yellow green = medium relevant flow; 4/blue green = high relevant flow; and 5/dark green = very high (maximum) relevant flow (after Burkhard et al. 2009 and 2012).

The ecosystem service flow matrix (Figure 5) clearly shows that most provisioning ecosystem service potentials (as indicated in Figure 4) have not been harvested yet (evaluation time was set to be summer, before the harvest period). In this case, differences between potentials and flows become especially clear. For regulating ecosystem services, values for several service flows in agricultural systems (agro-ecosystems) and urban systems were higher compared to earlier matrix values (in Burkhard et al. 2012a and 2009). This is because, according to the ecosystem service definition in Box 1, additional inputs have been included to better reflect conditions and flows in real agro-ecosystems. This can lead to higher *de facto* used ecosystem service flows (based on additional system inputs) compared to naturally available ecosystem service potentials. However, the differentiation between purely natural and anthropogenic inputs remains complex in many cases.

The most extreme cases of human-dominated 'landscapes' are urban (eco)systems. Many studies on urban ecosystem services have been carried out recently (Larondelle & Haase 2013; Elmqvist et al. 2013; Bastian et al. 2012). The dependence of urban areas on surrounding rural regions for decoupled or directional supply of almost all regulating and provisioning ecosystem services was shown by Kroll et al. (2012) and is clearly reflected in Figure 5. Regarding cultural ecosystem services, most of the potentials were fully used during the summer period.

4.3 Ecosystem service flow/potential comparison matrix

The comparison of ecosystem service potentials versus ecosystem service flows shows clear differences that are subject to seasonal influences mainly related to time-delayed growing/harvest periods. This becomes obvious especially for agricultural provisioning ecosystem services. Other provisioning ecosystem service potentials, such as fish stocks or forest stands, will (by management intention) not be depleted to their full extent. Comparisons between potentials and flows can deliver important information for sustainable resource management (Villamagna et al. 2013) or nature protection. Data on areas where ecosystem service potentials significantly exceed actual flows should not be abused however, for further environmental exploitation or identification of areas for increasing human activities.

For most of the regulating ecosystem services, a coherent distinction between potential and flow is difficult (see Chapter 3.1), resulting in similar potential and flow scores (0-values in Figure 6). For pollination, water flow and nutrient regulation in agricultural systems, actual flows exceed potentials significantly. Here, additional service flows have been added to the system in the form of omni-directional pollination supply, added irrigation water or fertiliser application. Pollination supply flows (and demands) would be higher for several land cover types and plants during the blooming period. However, a summer period was assessed in the matrices, which means lower pollination service flows compared to earlier periods of the year. Cultural ecosystem services show a more or less balanced pattern with some exceptions where potentials have not fully been harnessed in summer. This was the case in urban areas or in cases related to cultural heritage and cultural diversity linked for example to traditional forms of land use.

	Regulating services	Global climate regulation	Local climate regulation	Air quality regulation	Water flow regulation	Water purification	Nutrient regulation	Erosion regulation	Natural hazard regulation	Pollination	Pest and disease control	Regulation of waste	Provisioning services	Crops	Biomass for energy	Fodder	Livestock (domestic)	Fibre	Timber	Wood Fuel	Fish, seafood & edible algae	Aquaculture	Wild foods & resources	Biochemicals & medicine	Freshwater	Mineral resources*	Abiotic energy sources*	Cultural services	Recreation & tourism	Landscape aesthetics & inspiration	Knowledge systems	Religious & spiritual experience	Cultural heritage & cultural diversity	Natural heritage & natural diversity
Continuous urban fabric		0	0	0	0	0	0	0	0	-1	0	0		0	0	0	0	0	0	0	0	0	0	0	0	0	1		0	1	0	0	0	0
Discontinuous urban fabric		0	0	0	0	0	0	0	0	-1	0	0		1	0	-1	0	0	0	0	0	0	0	0	0	0	0		1	1	0	0	0	0
Industrial or commercial units		0	0	0	0	0	0	0	0	-1	0	0		0	0	0	0	0	0	0	0	0	0	0	0	0	0		0	0	0	0	0	0
Road and rail networks		0	0	0	0	0	0	0	0	0	0	0		0	0	0	0	0	0	0	0	0	0	0	0	0	0		0	0	0	0	0	0
Port areas		0	0	0	0	0	0	0	0	0	0	0		0	0	0	0	0	0	0	0	0	0	0	0	0	0		0	0	0	0	0	0
Airports		0	0	0	0	0	0	0	0	0	0	0		0	0	0	0	0	0	0	0	0	0	0	0	0	0		0	0	0	0	0	0
Mineral extraction sites		0	0	0	0	0	0	0	0	0	0	0		0	0	0	0	0	0	0	0	0	0	0	0	2	2		0	0	1	0	1	0
Dump sites		0	0	0	0	0	0	0	0	0	0	0		0	1	0	0	0	0	0	0	0	0	0	0	0	0		0	0	0	0	0	0
Construction sites		0	0	0	0	0	0	0	0	0	0	0		0	0	0	0	0	0	0	0	0	0	0	0	0	0		0	0	0	0	1	0
Green urban areas		0	0	0	0	0	0	0	0	0	0	0		0	0	-1	0	0	0	0	0	0	0	0	0	0	0		0	0	0	0	0	0
Sport and leisure facilities		0	0	0	0	0	0	0	0	0	0	0		0	0	0	0	0	0	0	0	0	0	0	0	0	0		1	0	0	0	0	0
Non-irrigated arable land		0	0	0	0	0	-2	0	0	-2	0	0		5	5	5	0	5	0	0	0	0	1	3	0	0	2		0	0	1	0	2	0
Permanently irrigated land		0	0	0	-2	0	-2	0	0	-2	0	0		5	1	2	0	4	0	0	0	0	1	3	0	0	1		0	0	1	0	2	0
Ricefields		0	0	0	-3	0	-2	0	0	0	0	0		5	1	2	0	0	0	0	0	0	0	0	0	0	0		0	0	1	0	1	0
Vineyards		0	0	0	0	0	0	0	0	-1	0	0		4	1	0	0	0	0	1	0	0	0	0	0	0	0		1	1	1	0	1	0
Fruit trees and berries		0	0	0	0	0	0	0	0	2	0	0		4	1	0	0	0	2	2	0	0	0	2	0	0	0		1	1	1	0	1	0
Olive groves		0	0	0	0	0	0	0	0	-1	0	0		4	1	0	0	0	2	2	0	0	0	2	0	0	0		0	0	0	0	1	0
Pastures		1	0	0	0	0	0	0	0	-1	0	0		0	1	1	3	0	0	0	0	0	2	0	0	0	5		0	0	0	0	1	0
Annual and permanent crops		0	0	0	-1	0	-1	0	0	-1	0	0		4	2	4	1	5	0	0	0	0	1	1	0	0	2		0	0	1	0	1	0
Complex cultivation patterns		0	0	0	-1	0	-1	0	0	-1	0	0		4	2	1	1	4	0	1	0	0	1	1	0	0	1		0	0	1	0	1	0
Agriculture & natural vegetation		1	0	0	0	1	0	0	0	0	0	0		3	3	1	2	3	1	0	0	0	2	0	0	0	0		0	0	1	0	1	2
Agro-forestry areas		1	0	0	0	1	1	0	0	1	0	0		2	3	2	2	1	2	2	0	0	2	0	0	0	0		0	0	1	0	1	1
Broad-leaved forest		1	0	0	0	1	0	0	1	3	0	0		0	0	0	0	0	4	4	0	0	4	2	0	0	0		1	1	1	1	2	1
Coniferous forest		1	0	0	0	1	0	0	1	3	0	0		0	0	0	0	0	4	4	0	0	4	2	0	0	0		1	1	1	1	2	1
Mixed forest		1	0	0	0	1	0	0	1	3	0	0		0	0	0	0	1	4	4	0	0	4	2	0	0	0		1	1	1	1	2	1
Natural grassland		3	0	0	0	0	0	0	0	-1	0	0		0	1	0	2	0	0	0	0	0	4	0	0	0	2		0	0	1	0	1	1
Moors and heathland		1	0	0	0	0	0	0	0	0	0	0		0	0	0	1	0	0	1	0	0	1	0	0	0	0		1	0	1	0	1	1
Sclerophyllous vegetation		1	0	0	0	0	0	0	0	1	0	0		0	1	0	1	1	1	1	0	0	0	3	0	0	1		0	0	0	0	1	1
Transitional woodland shrub		1	0	0	0	0	0	0	0	1	0	0		0	1	0	1	0	0	1	0	0	0	0	0	0	1		0	0	0	0	1	1
Beaches, dunes and sand plains		0	0	0	0	0	0	0	0	0	0	0		0	0	0	0	0	0	0	0	0	0	0	0	1	0		0	0	0	0	2	1
Bare rock		0	0	0	0	0	0	0	0	0	0	0		0	0	0	0	0	0	0	0	0	0	0	0	1	0		0	0	0	1	1	0
Sparsely vegetated areas		0	0	0	0	0	0	0	0	0	0	0		0	0	0	0	0	0	0	0	0	1	0	0	0	2		0	0	1	0	1	0
Burnt areas		0	0	0	0	0	0	0	0	0	0	0		0	0	0	0	0	0	0	0	0	0	0	0	0	0		0	0	0	0	0	0
Glaciers and perpetual snow		0	0	0	2	0	0	0	0	0	0	0		0	0	0	0	0	0	0	0	0	0	0	4	0	0		3	1	1	0	0	0
Inland marshes		0	0	0	1	0	0	0	0	0	0	0		0	0	2	1	0	0	0	0	0	1	0	0	0	0		0	0	0	0	1	1
Peatbogs		1	0	0	1	0	0	0	0	0	0	0		0	0	0	0	0	0	0	0	0	1	1	1	0	0		0	0	0	0	1	1
Salt marshes		1	0	0	0	0	0	0	0	0	0	0		0	0	0	1	0	0	0	0	0	1	0	0	0	0		0	0	0	0	1	1
Salines		0	1	0	0	0	0	0	0	0	0	0		0	0	0	0	0	0	0	0	0	0	0	0	0	0		0	0	0	0	1	0
Intertidal flats		0	0	0	0	0	0	0	0	0	0	0		0	1	0	0	0	0	0	0	0	1	0	0	0	0		0	0	0	0	1	1
Water courses		0	0	0	0	0	0	0	0	0	0	0		0	2	0	0	0	0	0	1	0	2	0	3	0	1		0	0	1	1	1	1
Water bodies		0	0	0	2	0	0	0	0	0	0	0		0	1	0	0	0	0	0	2	3	1	0	3	0	0		0	0	1	1	1	1
Coastal lagoons		0	0	0	2	1	0	0	0	0	0	0		0	1	0	0	0	0	0	2	3	2	1	0	0	0		0	0	1	0	1	1
Estuaries		0	0	0	1	0	0	0	0	0	0	0		0	2	0	0	0	0	0	2	3	2	1	0	0	1		0	0	1	0	1	1
Sea and ocean		1	0	0	0	1	0	0	0	0	0	0		0	3	2	0	0	0	0	2	3	1	1	0	0	2		0	0	2	1	1	1

*abiotic outputs from natural systems (after CICES)

Figure 6: Exemplary ecosystem service potential - flow comparison matrix. The evaluations' (from Figures 4 and 5) comparison refers to a hypothetical European „normal" landscape in summer (before the harvest period). Scale from −5/brown red = flow (actual use) exceeds potential (potential supply) significantly; via 0/rosy = flow = potential; to 5/dark green = potential exceeds flow significantly.

4.4 Ecosystem service demand matrix

Demand calculations are mainly based on population numbers and average consumption patterns but also on land use activities and on their demands for certain services (Kroll et al. 2012; Burkhard et al. 2012a). For example, all agricultural activities show high demands for whole bundles of regulating ecosystem services (Figure 7). If they cannot be fulfilled by the ecosystem service potentials, additional inputs (as mentioned before) come into play, increasing ecosystem service flows in order to meet increased demands. This differentiation between natural functions, additional inputs, as well as between service flows and demands in the case of specific land use forms is complicated and no satisfying solution has been found yet.

The ecosystem service demand matrix (Figure 7) clearly illustrates that land cover types with the highest population numbers and high human activities (uppermost parts of the matrix) have the highest demands for multiple ecosystem services. More near-natural land cover types show much lower or no relevant demands for ecosystem services because normally less people are present there.

4.5 Ecosystem service flow and demand budget matrix

Figure 8 shows the comparison between ecosystem service flows (as evaluated in Figure 5) in relation to demands for ecosystem services (Figure 7). The compared entities need to be quantified in the same units or need to be normalised into the relative 0–5 scale (Burkhard et al. 2012a). Figure 8 shows an obvious undersupply of ecosystem services in the human-dominated land cover types (upper part of the matrix), whereas several of the more near-natural land cover types show ecosystem services oversupply. For a more detailed description of the method and exemplary applications see Kroll et al. (2012); Nedkov & Burkhard (2012); Burkhard et al. (2012a).

With regard to environmental management and related land use decisions, regional budgets for individual ecosystem service flows and demands do not necessarily need to be neutral (0-values in Figure 8). Some ecosystem services with decoupled supply patterns may be better and more sustainably provided by other regions. It is one task of future-oriented environmental management to optimise land use decisions toward sustainability based on ecosystem services. However, global supply-demand budgets have to be zero in the long-term if sustainable management is to be achieved and a depletion of natural capital is to be avoided.

A matrix comparing ecosystem service potentials with actual demands could easily be estimated using the same approach. However, exposing ecosystem service potentials directly related to human demands holds the risk of being abused for further exploitation of natural resources. Particularly spatial designation of ecosystem service potentials in maps can foster further or new land conversions towards more intensive forms of land use or even support land-grabbing activities. Therefore, information on ecosystem service supply has to be prepared and documented carefully and has to fulfil certain criteria for end-use.

	Regulating services	Global climate regulation	Local climate regulation	Air quality regulation	Water flow regulation	Water purification	Nutrient regulation	Erosion regulation	Natural hazard regulation	Pollination	Pest and disease control	Regulation of waste	Provisioning services	Crops	Biomass for energy	Fodder	Livestock (domestic)	Fibre	Timber	Wood Fuel	Fish, seafood & edible algae	Aquaculture	Wild foods & resources	Biochemicals & medicine	Freshwater	Mineral resources*	Abiotic energy sources*	Cultural services	Recreation & tourism	Landscape aesthetics & inspiration	Knowledge systems	Religious & spiritual experience	Cultural heritage & cultural diversity	Natural heritage & natural diversity
Continuous urban fabric		4	5	5	4	5	1	1	5	1	5	3		5	5	1	5	3	3	1	5	5	5	5	5	4	2		4	4	3	4	4	4
Discontinuous urban fabric		3	5	5	5	4	2	1	4	2	4	2		4	4	2	4	3	3	1	4	4	4	5	5	3	3		4	4	3	3	2	3
Industrial or commercial units		5	2	5	4	5	3	1	5	1	3	4		3	5	1	3	5	5	4	4	4	4	5	5	5	5		1	1	4	1	3	1
Road and rail networks		4	2	4	4	0	0	3	4	0	2	1		1	4	0	1	0	2	0	0	0	0	0	1	2	0		2	2	1	1	1	0
Port areas		3	2	2	5	3	0	4	5	0	5	2		1	5	0	1	2	5	2	2	2	1	1	3	3	1		2	2	2	1	2	0
Airports		5	2	4	1	2	1	1	5	0	5	1		1	5	0	1	1	1	0	1	1	1	1	3	2	1		1	1	1	1	1	0
Mineral extraction sites		1	0	0	2	4	0	4	3	0	0	3		0	3	0	0	1	2	0	0	0	0	0	2	1	0		0	0	0	0	0	0
Dump sites		2	2	2	0	3	0	0	5	0	4	5		0	1	0	0	0	0	0	0	0	0	0	2	0	0		0	0	0	0	0	0
Construction sites		1	2	1	2	2	2	2	3	0	1	2		0	4	0	0	4	4	0	0	0	0	0	2	4	1		0	0	0	0	0	0
Green urban areas		0	3	2	2	0	0	0	2	2	3	1		0	1	0	1	0	0	0	0	0	0	0	2	0	0		4	4	2	0	2	1
Sport and leisure facilities		0	3	3	1	1	0	0	3	0	3	1		0	1	1	1	1	1	0	2	2	2	3	3	1	1		3	3	1	0	2	0
Non-irrigated arable land		2	2	1	2	0	3	3	2	3	4	3		0	1	0	0	0	0	0	0	0	0	1	0	0	0		0	0	1	0	1	0
Permanently irrigated land		2	2	1	2	5	3	2	2	3	4	3		0	1	0	0	0	0	0	0	0	0	1	5	0	0		0	0	1	0	1	0
Ricefields		3	2	1	5	5	3	2	2	1	4	3		0	1	0	0	0	0	0	0	0	0	1	5	0	0		0	0	2	0	3	0
Vineyards		1	3	1	1	4	3	5	3	2	4	2		0	1	0	0	0	1	0	0	0	0	2	4	0	0		0	0	2	0	3	0
Fruit trees and berries		1	2	1	1	2	3	1	3	3	4	2		0	1	0	0	0	1	0	0	0	0	2	3	0	0		0	0	2	0	2	0
Olive groves		1	2	1	1	2	3	1	3	2	4	2		0	1	0	0	0	0	0	0	0	0	2	1	0	0		0	0	2	0	2	0
Pastures		3	1	1	1	2	3	1	2	1	4	4		0	1	5	0	0	1	0	0	0	0	1	2	0	0		0	0	1	0	1	0
Annual and permanent crops		1	1	1	1	2	5	1	3	2	3	2		0	1	0	0	0	0	0	0	0	0	1	1	0	0		0	0	1	0	1	0
Complex cultivation patterns		1	1	1	1	2	5	1	2	3	3	2		0	1	0	0	0	0	0	0	0	0	1	1	0	0		0	0	1	0	1	0
Agriculture & natural vegetation		1	1	1	1	2	3	1	1	2	3	2		0	1	0	0	0	0	0	0	0	0	1	2	0	0		0	0	1	0	1	0
Agro-forestry areas		1	1	1	1	2	3	1	2	2	3	2		0	1	0	0	0	0	0	0	0	0	1	2	0	0		0	0	1	0	1	0
Broad-leaved forest		0	0	0	0	0	0	0	0	0	0	0		0	0	0	0	0	0	0	0	0	0	0	0	0	0		0	0	0	0	0	0
Coniferous forest		0	0	0	0	0	0	0	0	0	0	0		0	0	0	0	0	0	0	0	0	0	0	0	0	0		0	0	0	0	0	0
Mixed forest		0	0	0	0	0	0	0	0	0	0	0		0	0	0	0	0	0	0	0	0	0	0	0	0	0		0	0	0	0	0	0
Natural grassland		0	0	0	0	0	0	0	0	0	0	0		0	0	0	0	0	0	0	0	0	0	0	0	0	0		0	0	0	0	0	0
Moors and heathland		0	0	0	0	0	0	0	0	0	0	0		0	0	0	0	0	0	0	0	0	0	0	0	0	0		0	0	0	0	0	0
Sclerophyllous vegetation		0	0	0	0	0	0	0	0	0	0	0		0	0	0	0	0	0	0	0	0	0	0	0	0	0		0	0	0	0	0	0
Transitional woodland shrub		0	0	0	0	0	0	0	0	0	0	0		0	0	0	0	0	0	0	0	0	0	0	0	0	0		0	0	0	0	0	0
Beaches, dunes and sand plains		0	0	0	0	0	0	0	0	0	0	0		0	0	0	0	0	0	0	0	0	0	0	0	0	0		1	0	0	0	0	0
Bare rock		0	0	0	0	0	0	0	0	0	0	0		0	0	0	0	0	0	0	0	0	0	0	0	0	0		0	0	0	0	0	0
Sparsely vegetated areas		0	0	0	0	0	0	0	0	0	0	0		0	0	0	0	0	0	0	0	0	0	0	0	0	0		0	0	0	0	0	0
Burnt areas		0	0	0	0	0	0	0	0	0	0	0		0	0	0	0	0	0	0	0	0	0	0	0	0	0		0	0	0	0	0	0
Glaciers and perpetual snow		0	0	0	0	0	0	0	0	0	0	0		0	0	0	0	0	0	0	0	0	0	0	0	0	0		0	0	0	0	0	0
Inland marshes		0	0	0	0	0	0	0	0	0	0	0		0	0	0	0	0	0	0	0	0	0	0	0	0	0		0	0	0	0	0	0
Peatbogs		0	0	0	0	0	0	0	0	0	0	0		0	0	0	0	0	0	0	0	0	0	0	0	0	0		0	0	0	0	0	0
Salt marshes		0	0	0	0	0	0	0	0	0	0	0		0	0	0	0	0	0	0	0	0	0	0	0	0	0		0	0	0	0	0	0
Salines		0	0	0	0	0	0	0	0	0	0	0		0	0	0	0	0	0	0	0	0	0	0	0	0	0		0	0	0	0	0	0
Intertidal flats		0	0	0	0	0	0	0	0	0	0	0		0	0	0	0	0	0	0	0	0	0	0	0	0	0		0	0	0	0	0	0
Water courses		0	0	0	0	0	0	0	0	0	0	0		0	0	0	0	0	0	0	0	0	0	0	0	0	0		0	0	0	0	0	0
Water bodies		0	0	0	0	0	0	0	0	0	0	0		0	0	0	0	0	0	0	0	0	0	0	0	0	0		0	0	0	0	0	0
Coastal lagoons		0	0	0	0	0	0	0	0	0	0	0		0	0	0	0	0	0	0	0	0	0	0	0	0	0		0	0	0	0	0	0
Estuaries		0	0	0	0	0	0	0	0	0	0	0		0	0	0	0	0	0	0	0	0	0	0	0	0	0		0	0	0	0	0	0
Sea and ocean		0	0	0	0	0	0	0	0	0	0	0		0	0	0	0	0	0	0	0	0	0	0	0	0	0		0	0	0	0	0	0

*abiotic outputs from natural systems (after CICES)

Figure 7: Exemplary ecosystem service matrix showing demands for ecosystem services within different land cover classes. The exemplary evaluation refers to a hypothetical European „normal" landscape in summer (before the harvest period). Scale from: 0/rosy = no relevant demand; 1/dark rosy = low relevant demand; 2/light red = relevant demand; 3/red = medium relevant demand; 4/dark red = high relevant demand; and 5/brown red = very high relevant demand (after Burkhard et al. 2012).

	Regulating services	Global climate regulation	Local climate regulation	Air quality regulation	Water flow regulation	Water purification	Nutrient regulation	Erosion regulation	Natural hazard regulation	Pollination	Pest and disease control	Regulation of waste	**Provisioning services**	Crops	Biomass for energy	Fodder	Livestock (domestic)	Fibre	Timber	Wood Fuel	Fish, seafood & edible algae	Aquaculture	Wild foods & resources	Biochemicals & medicine	Freshwater	Mineral resources*	Abiotic energy sources*	**Cultural services**	Recreation & tourism	Landscape aesthetics & inspiration	Knowledge systems	Religious & spiritual experience	Cultural heritage & cultural diversity	Natural heritage & natural diversity
Continuous urban fabric		-4	-5	-5	-4	-5	-1	1	-5	0	-4	-3		-5	-5	-1	-5	-3	-3	-1	-5	-5	-5	-5	-5	-4	-2		-1	-2	-1	-2	-3	-4
Discontinuous urban fabric		-3	-5	-5	-5	-4	-2	0	-4	0	-3	-2		-4	-4	-1	-4	-3	-3	-1	-4	-4	-4	-5	-5	-3	-2		-2	-3	-1	-1	0	-3
Industrial or commercial units		-5	-2	-5	-4	-5	-3	1	-5	0	-2	-4		-3	-5	-1	-3	-5	-5	-4	-4	-4	-4	-5	-5	-5	-4		-1	-1	-4	-1	-1	-1
Road and rail networks		-4	-2	-4	-4		0	-2	-4		-2	-1		-1	-4		-1		-2						-1	-2			-2	-2	-1	-1	0	
Port areas		-3	-2	-2	-5	-3		-1	-2		-4	-2		-1	-5		-1	-2	-5	-2	-2	-2	-1	-1	-3	-3	-1		-1	0	-2	-1	-1	0
Airports		-5	-2	-4	-1	-2	-1	0	-5		-4	-1		-1	-5		-1	-1	-1		-1	-1	-1	-1	-3	-2	-1		-1	-1	-1	-1	-1	0
Mineral extraction sites		-1			-2	-4		-4	-3			-3			-3			-1	-2						-2	2	1				1			
Dump sites		-2	-2	-2		-3			-5		-4	-3			-1										-2									
Construction sites		-1	-2	-1	-2	-2	-2	-2	-3		-1	-2			-4			-4	-4						-2	-4	-1						1	
Green urban areas		2	-1	0	0	2	2	2	-1	0	-1	1			-1	1	-1								-2				-1	-1	-1		0	0
Sport and leisure facilities		1	-2	-2	0	0	1	1	-3		-2	0			-1	-1	-1	-1	-1		-2	-2	-2	-3	-3	-1	-1		1	-2	-1		-1	
Non-irrigated arable land		-1	0	0	0		0	-3	-1	0	-2	-1			-1									-1					1	1	0		0	
Permanently irrigated land		-1	1	0	1	-5	0	-2	-1	0	-2	-1			-1									-1	-5				1	1	0		0	
Ricefields		-3	0	0	-1	-5	0	-2	-2	0	-3	-1			-1									-1	-5				1	1	-1		-1	
Vineyards		0	-2	0	0	-4	-2	-4	-3	0	-3	-1			-1				-1					-2	-4				2	1	0		1	
Fruit trees and berries		1	0	1	1	-1	-1	1	-1	0	-1	0			-1				-1					-2	-3				2	1	-1		1	1
Olive groves		0	-1	0	0	-1	-2	0	-3	0	-2	0			-1									-2	-1				2	2	0		1	
Pastures		-2		-1	0	-2	-2	0	-1	0	-2	0			-1	-1	2		-1					-1	-2				2	2	1		1	1
Annual and permanent crops		0	1	0	1	-2	-3	1	-2	0	-1	0			-1									-1	-1				1	1	0		1	
Complex cultivation patterns		0	1	0	1	-2	-3	0	-1	0	0	0			-1	1								0	-1				2	2	0		1	
Agriculture & natural vegetation		0	2	1	1	-1	-1	1	0	0	0	0			-1	1		1		1				0	-2		1		2	2	1	1	1	1
Agro-forestry areas		0	1	1	1	-1	-2	2	-1	0	0	1			-1		1	1	1	1				0	-2				2	2	0		1	1
Broad-leaved forest		4	5	5	3	4	5	5	3	1	4	4			1	1		1	1	1			1	1					4	4	4	2	2	4
Coniferous forest		4	5	5	3	4	5	5	3	1	4	4			1	1		1	1	1			1	1					4	4	4	2	2	3
Mixed forest		4	5	5	3	4	5	5	3	1	5	5			1	1		1	1	1			1	1					4	4	4	2	2	4
Natural grassland		2	2		1	3	4	5	1	2	1	2				2	1						1	1					3	4	4	1	2	2
Moors and heathland		2	4		2	4	3	2	2	2	2	3			1	1				1			1	1					3	4	4	1	1	3
Sclerophyllous vegetation		1	2	1	1	1	2	1	1	1	2	3				1			1	1			1						2	3	4	1	1	3
Transitional woodland shrub		1	2	1	1	1	2	1	1	1	2	3			1	1		1	1	1			1	1					2	3	4	1	1	1
Beaches, dunes and sand plains					1	1	1		5		1	1												1					4	4	4	1	1	1
Bare rock						1		2	1																				2	3	3	1	1	1
Sparsely vegetated areas			1		1	1	1	1	1		1	1					1												1	1	2		1	1
Burnt areas			1				1					1																			2			
Glaciers and perpetual snow		3	4		3						1	1													1				2	4	3			1
Inland marshes		2	2		2	2	4	1	4	1	2	3				2	1												1	2	3		1	1
Peatbogs		4	4		3	4	4	2	3	2	3	4			2									1					3	2	3		1	3
Salt marshes			1		1	1	2	1	4	1	2	2				2	1												3	2	3		1	1
Salines			2								1	1												2		2			2	2	3		3	
Intertidal flats		1	3		1	1	1	1	5		2	3																	4	2	3		1	1
Water courses		1	1		3	3	3		3		3	5									2		2		2		2		4	4	3	1	2	2
Water bodies		1	2		3	2	3		3		3	5									2	2	3		2		1		5	4	3	1	2	2
Coastal lagoons		1	1		2	1	3		4		3	5									2	2	2						3	4	3		1	2
Estuaries		1			2	3	3		3		3	5									2	2	2						3	4	3		1	2
Sea and ocean		2	3		1	1	3				3	5			1	1					3	2	3	2		1	1		4	5	3	1	2	2

*abiotic outputs from natural systems (after CICES)

Figure 8: Exemplary ecosystem service flow - demand budget matrix. The exemplary evaluation refers to a hypothetical European „normal“ landscape in summer (before the harvest period). Scale from –5/brown red = demand exceeds flow significantly = strong undersupply; via 0/rosy = demand = flow = neutral balance; to 5/dark green = flow exceeds the demand significantly = strong oversupply. Empty fields = neither a relevant flow of nor a relevant demand for the particular ecosystem service (after Burkhard et al. 2012).

5 Discussion

5.1 Potentials, flows, demands and their localization, indication and quantification

Ecosystem service flow evaluations (as defined here) are more closely linked to actual flows of goods and services. Therefore, the information they provide can be better connected to demands for ecosystem goods and services and related trade-off assessments. This is where added values for current environmental management and landscape planning can be found. Until now, most landscape planning approaches mainly consider landscape functions without sufficiently and spatially explicitly integrating demands for goods and services (de Groot et al. 2010). The opportunity and technology to provide spatially explicit ecosystem service evaluations and supply-demand budgets on different scales has the potential to make ecosystem services a focal tool for environmental management.

More elaborated analyses of demand localisations and demand flows as well as quantification examples are essential next steps in spatially connecting ecosystem service potentials, flows and demands. Improvement of ecosystem service demand indicators is needed to distinguish between consuming motivations, consumption rates and actual needs. A distinction between demand potentials versus demand flows is conceivable here. A better consideration of ecosystem service flows from remote regions, for example by integrating ecosystem service or ecological footprint calculations, life cycle analyses or supply chain analysis, are important further steps for the integration of the concept into decision-making and environmental management in broader contexts and on larger scales. The distinction of ecosystem service potentials, flows and demands is especially difficult for regulating ecosystem services. This dilemma became apparent when deriving the exemplary indicators (shown in Table 1) and assigning the ecosystem service matrix scores. A stronger integration of regulating ecosystem services by more detailed analyses of actual ecosystem service 'production', interrelations and societal needs (demand) is required. This includes interpretations of legal and administrative constraints. Until now, no direct demands from human societies could be indicated for most of the regulating ecosystem services.

5.2 The ecosystem service matrix concept

Applications of the approach in case studies and related ecosystem service map compilations delivered expressive results (e.g. Kandziora et al. 2013b; Nedkov and Burkhard 2012; Vihervaara et al. 2010). The matrices presented in this article show clear patterns when applying a land cover-based evaluation of ecosystem service potentials, flows and demands. The quality and depth of such analyses are of course strongly dependent on the quality and suitability of the data and information used. An accounting concept on a relative normalised scale (like the 0-5 scale applied here) can be used as representative for any appropriate quantification unit such as GJ, tons, currency units, species numbers, NPP, °C, turnover rates, availabilities or abundances.

Concerning the score estimation patterns, the ecosystem service potential and flow matrices of regulating services show considerable similarities with high scores from more nature-near land cover types and low scores from more human-influenced or inhabited land cover types. This phenomenon also applies to most cultural ecosystem services, except to those generated by urban areas. Urban areas can provide high recreation and tourism services as well as knowledge and religious experiences. Nevertheless, the two matrices in Figures 4 and 5 show notably different score patterns for provisioning ecosystem services provided by agricultural and more natural lands, as the estimation time was set to summer before the harvest period. This setting led to a high potential surplus for most agricultural products in the ecosystem service potential - flow comparison matrix (Figure 6). The differences between the potential and flow matrices reveal the importance of defining an appropriate point in time to assess ecosystem service flows. With respect to the ecosystem service demand matrix (Figure 7), a generally opposing pattern to the potential matrix can be observed. In other words, greater demands come from more human-inhabited land cover types (mainly in the upper part of the matrix) with more natural land covers showing negligible demands.

The ecosystem service flow - demand budget matrix (Figure 8) shows significant spatial mismatches between ecosystem service flows and demands, especially between urbanised areas and non-urban regions (as shown in Radford & James 2013; Kroll et al. 2012).

However, the notion of relevant supply or demand introduces uncertainty into the method because 'relevant' can be interpreted differently. Moreover, appropriate reference values to calibrate the assessed values in space and time need to be defined. What if, for example, today's assessed maximum relevant ecosystem service supply (assessed with 5) is not the maximum anymore in a future state and theoretically a higher class (6) needs to be introduced? We recommend using an adaptive approach, where the maximum values are always in class 5. This means that the classification may need to be adapted if values show an increase or decrease or if different regions are to be compared with each other. For comparison and budget estimations of ecosystem service potentials, flows and demands, all 0-5 normalisations should preferably be based on the same quantification units and dimensions to make the values better comparable.

The integration of temporal supply and demand dynamics (seasonal aspects, medium- and long-term dynamics, "hot moments") is important in the assessment of certain ecosystem services as their potentials, flows and demands can vary significantly through time, especially in regions with distinct seasonal fluctuations. For other ecosystem services, average annual values may provide sufficient information. In case of the matrix method, the third dimension Z (time) (see Figure 9) should be included and a clearly defined temporal assessment scale (cp. summer period before harvest as in Figures 3-8) is inevitable for a successful application of the approach.

Furthermore, the cross-comparisons of ecosystem service assessment values need to be improved: a) vertically within ecosystem service categories and across different spatial units (e.g. land cover types; comparisons are relatively reliable), and b) horizontally within individual spatial units across different ecosystem services (comparisons still need to be improved; see horizontal and vertical arrows in Figure 9). The values shown in the different matrices here are meant to explain the method and to demonstrate its application potential. The proper quantification and testing of the values should be done within distinct case studies. The results will feed back into the refinement of the method.

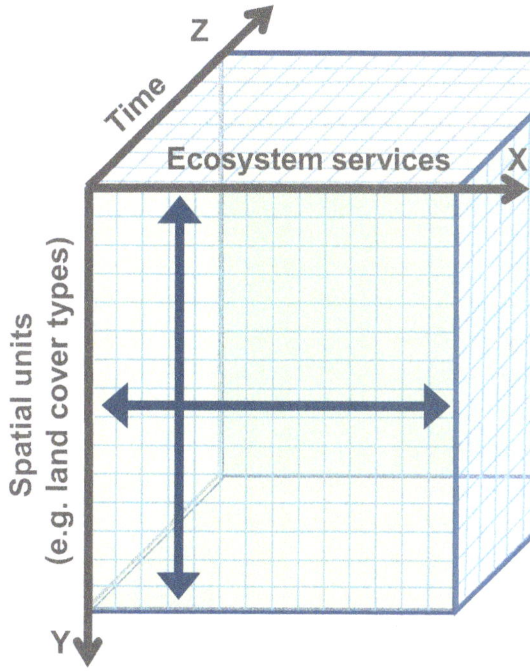

Figure 9: Integration of temporal aspects into the matrix by including a third dimension (3D-Matrix) and horizontal (within spatial units) as well as vertical (within ecosystem service types) cross-comparisons of assessment values.

Most of the evaluation values in the matrices presented here are based on expert opinions. The integration and combination of more comprehensive quantification methods (models, measurements, statistics, surveys) will improve the reliability of evaluation values for the different classes. Moreover, land cover is only one aspect of ecosystems, and therefore only a proxy indicator. Further integration of geobiophysical (biotic, abiotic), land use (intensity) and socio-economic data for quantification and localization will foster a more comprehensive assessment. Uncertainty analyses of evaluation methods and results are required as further important steps in the matrix assessment method.

6 Conclusions

Summing up the questions raised in the introduction, we can conclude that a distinction between ecosystem service potentials, flows and demands is needed and practical for the integration of this concept into research, decision making and environmental management. Appropriate indicators for service potentials, flows and demands were proposed, which need to be applied and further developed in case studies. The indicators presented here are examples and further ones should be added to the tables. Ecosystem service indicators have to reflect case study peculiarities and are depending on specific research/management questions. The demand for ecosystem services is considered to be best localized in the areas where people live and spend most of their time. However, additional development is needed in the conceptualisation of ecosystem service demand but also for supply assessments, especially with regard to regulating ecosystem services.

Information on ecosystem service potentials should be used to assess future options for long-term landscape planning and environmental management in terms of sustainable flows of ecosystem services. Ecosystem service maps are useful for sustainable decision making, for example by identifying supply-demand mismatches across landscapes and their changes over time. They should not be used to enhance human exploitation of natural resources. Therefore, appropriate institutions to sustainably manage ecosystem services on spatial and temporal scales that match the scales of the service supply and demand should be established.

The concepts, indicators and exemplary evaluated numbers provided here are open for discussion. We hope to stimulate scientific debate in these highly relevant and dynamic research topics. We are looking forward to future discussions in order to jointly improve concepts and methods for ecosystem service assessments to be used for sustainable environmental management. Better linkages between knowledge and method providers and end-users in science, decision and policy making must be established.

7 Acknowledgement

The work on this article has mainly been funded by the LEGATO project (German Ministry of Research and Education BMBF within the BMBF-Funding Measure "Sustainable Land Management" (http://nachhaltiges-landmanagement.de/en/; Funding No. 01LL0917F).

References

Alkemade, R.; Burkhard, B.; Crossman, N.; Nedkov, S. & K. Petz (Eds.) (2014): Quantifying ecosystem services and indicators for science, policy and practice. Special Issue. Ecological Indicators 37, 161-266

Bagstad, K.J.; Semmens, D.J.; Waage, S. & R. Winthrop 2013a. A comparative assessment of decision-support tools for ecosystem services quantification and valuation. Ecosystem Services 5, 27–39.

Bagstad, K.J.; Johnson, G.W.; Voigt, B. & F. Villa 2013b. Spatial dynamics of ecosystem service flows: a comprehensive approach to quantifying actual services. Ecosystem Services 4, 117–125.

Baral, H.; Keenan, R.J.; Fox, J.C.; Stork, N.E. & S. Kasel 2013. Spatial assessment of ecosystem goods and services in complex production landscapes: A case study from south-eastern Australia. Ecological Complexity 13, 35–45.

Baral, H.; Keenan, R.J.; Sharma, S.K.; Stork, N.E. & S. Kasel 2014. Spatial assessment and mapping of biodiversity and conservation priorities in a heavily modified and fragmented production landscape in north-central Victoria, Australia. Ecological Indicators 36, 552–562.

Bastian, O.; Haase, D. & K. Grunewald, 2012. Ecosystem properties, potentials and services – The EPPS conceptual framework and an urban application example. Ecological Indicators 21, 7–16.

Bastian, O.; Syrbe, R.-U.; Rosenberg, M.; Rahe, D. & K. Grunewald 2013. The five pillar EPPS framework for quantifying, mapping and managing ecosystem services. Ecosystem Services 4, 15–24.

Boyd, J. & S. Banzhaf 2007. What are ecosystem services? The need for standardized environmental accounting units. Ecological Economics 63, 616–626.

Brandt, J.; Aagaard Christensen, A.; Roar Svenningsen, S. & E. Holmes 2013. Landscape practise and key concepts for landscape sustainability. Landscape Ecology 28 (6), 1125-1137.

Burkhard, B.; Kroll, F.; Müller, F. & W. Windhorst 2009. Landscapes' capacities to provide ecosystem services – a concept for land-cover based assessments. Landscape Online 15, 1–22.

Burkhard, B.; Fath, B.D. & F. Müller 2011. Adapting the adaptive cycle: Hypotheses on the development of ecosystem properties and services. Ecological Modelling 222: 2878– 2890.

Burkhard, B.; Kroll, F.; Nedkov, S. & F. Müller 2012a. Mapping supply, demand and budgets of ecosystem services. Ecological Indicators 21, 17–29.

Burkhard, B.; de Groot, R.; Costanza, R.; Seppelt, R.; Jørgensen, S.E. & M. Potschin 2012b. Solutions for sustaining natural capital and ecosystem services. Ecological Indicators 21, 1–6.

Burkhard, B.; Crossman, N.; Nedkov, S.; Petz, K. & R. Alkemade 2013. Mapping and Modelling Ecosystem Services for Science, Policy and Practice. Ecosystem Services 4: 1-3.

Chan, K.; Satterfield, T. & J. Goldstein 2012. Rethinking ecosystem services to better address and navigate cultural values. Ecological Economics 74, 8–18.

Clerici, N.; Paracchini, M.L. & J. Maes 2014. Land-cover change dynamics and insights into ecosystem services in European stream riparian zones. Ecohydrology & Hydrobiology http://dx.doi.org/10.1016/j.ecohyd.2014.01.002.

Costanza, R. 2008a. Ecosystem services: multiple classification systems are needed. Biological Conservation 141, 350–352.

Costanza, R. 2008b. Natural capital. http://www.eoearth.org/view/article/154791 (Date: 29.10.2013).

Costanza, R. & , H.E. Daly 1992. Natural Capital and Sustainable Development. Conservation Biology 6 (1), 37-46.

Crossman, N.; Burkhard, B. & S. Nedkov 2012. Quantifying and Mapping Ecosystem Services. International Journal of Biodiversity Science, Ecosystem Services & Management 8 (1-2). 1-4.

Crossman, N.D.; Burkhard, B.; Nedkov, S.; Willemen, L.; Petz, K.; Palomo, I.; Drakou, E.G.; Martín-Lopez, B.; McPhearson, T.; Boyanova, K.; Alkemade, R.; Egoh, B.; Dunbar, M. & J. Maes 2013. A blueprint for mapping and modelling ecosystem services. Ecosystem Services 4: 4-14.

Daily, G.C.; Polasky, S.; Goldstein, J.; Kareiva, P.M.; Mooney, H.A.; Pejchar, L.; Ricketts, T.H.; Salzman, J. & R. Shallenberger 2009. Ecosystem services in decision-making: time to deliver. Frontiers in Ecology and the Environment 7 (1), 21–28.

de Groot, R.S.; Wilson, M.A. & R.M.J. Boumans 2002. A typology for the classification, description and valuation of ecosystem functions, goods and services. Ecological Economics 41 (3), 393–408.

de Groot, R.S.; Alkemade, R.; Braat, L.; Hein, L. & L. Willemen 2010. Challenges in integrating the concept of ecosystem services and values in landscape planning, management and decision making. Ecological Complexity 7, 260–272.

Dick, J.; Maes, J.; Smith, R.I.; Paracchini, M.L. & G. Zulian 2014. Cross-scale analysis of ecosystem services identified and assessed at local and European level. Ecological Indicators 38, 20–30.

Egoh, B.; Drakou, E.G.; Dunbar, M.B.; Maes, J. & L. Willemen 2012. Indicators for mapping ecosystem services: a review. Report EUR25456EN. Publications Office of the European Union, Luxembourg.

Egoh, B.; Reyers, B.; Rouget, M.; Richardson, D.M.; Le-Maitre, D.C. & A.S. van Jaarsveld 2008. Mapping ecosystem services for planning and management. Agriculture, Ecosystems and the Environment 127, 135–140.

Elmqvist, T.; Fragkias, M.; Goodness, J.; Güneralp, B.; Marcotullio, P.J.; McDonald, R.I.; Parnell, S.; Schewenius, M.; Sendstad, M.; Seto, K.C. & C. Wilkinson (Eds.) 2013. Urbanization, Biodiversity and Ecosystem Services: Challenges and Opportunities - A Global Assessment. Springer.

Fagerholm, N.; Käyhkö, N.; Ndumbaro, F. & M. Khamis 2012. Community stakeholders' knowledge in landscape assessments e mapping indicators for landscape services. Ecological Indicators, 18, 421-433.

Fisher, B.; Turner, R.K. & P. Morling 2009. Defining and classifying ecosystem services for decision making. Ecological Economics 68 (3), 643–653.

Frank, S.; Fürst, C.; Koschke, L.; Witt, A. & F. Makeschin 2013. Assessment of landscape aesthetics - Validation of a landscape metrics-based assessment by visual estimation of the scenic beauty. Ecological Indicators 32, 222-231.

García-Nieto, A.P.; García-Llorente, M.; Iniesta-Arandia, I. & B. Martín-López 2013. Mapping forest ecosystem services: From providing units to beneficiaries. Ecosystem Services 4, 126–138.

Gimona, A.; van der Horst, D. 2007. Mapping hotspots of multiple landscape functions: a case study on farmland afforestation in Scotland. Landscape Ecology 22, 1255-1264.

Haberl, H.; Steinberger, J.K.; Plutzar, C.; Erb, K.-H.; Gaube, V.; Gingrich, S. & F. Krausmann 2012. Natural and socioeconomic determinants of the embodied human appropriation of net primary production and its relation to other resource use indicators Ecological Indicators 23, 222-231.

Haines-Young, R. & M. Potschin 2010. The links between biodiversity ecosystem services and human well-being. In: Raffaelli, D. & C. Frid (Eds.). Ecosystem Ecology: A New Synthesis. Cambridge University Press, Cambridge, 110–139.

Haines-Young, R.; Potschin, M. & F. Kienast 2012. Indicators of ecosystem service potential at European scales: mapping marginal changes and trade-offs. Ecological Indicartos 21, 39–53.

Hernández-Morcillo, M.; Plieninger, T. & C. Bieling 2013. An empirical review of cultural ecosystem service indicators. Ecological Indicators 29, 434–444.

Honey-Rosés, J. & L.H. Pendleton 2013. A demand driven research agenda for ecosystem services. Ecosystem Services 5, e160–e162.

Hou, Y.; Burkhard, B. & F. Müller 2013. Uncertainties in landscape analysis and ecosystem service assessment. Journal of Environmental Management 127, S117–S131.

Kaiser, G.; Burkhard, B.; Römer, H.; Sangkaew, S.; Graterol, R.; Haitook, T.; Sterr, H. & D. Sakuna-Schwartz 2013. Mapping tsunami impacts on land cover and related ecosystem service supply in Phang Nga, Thailand. Natural Hazards and Earth System Sciences 13, 3095-3111.

Kandziora, M.; Burkhard, B. & F. Müller 2013a. Interactions of ecosystem properties, ecosystem integrity and ecosystem service indicators e a theoretical matrix exercise. Ecological Indicators 28, 54–78.

Kandziora, M.; Burkhard, B. & F. Müller 2013b. Mapping provisioning ecosystem services at the local scale using data of varying spatial and temporal resolution. Ecosystem Services 4, 47-59.

Kandziora, M.; Dörnhöfer, K.; Oppelt, N. & F. Müller (in review). Detecting land use and land cover changes in Northern German agricultural landscapes to assess ecosystem service dynamics. Landscape Online.

Kemkes, R.J.; Farley, J. & C.J. Koliba 2010. Determining when payments are an effective policy approach to ecosystem service provision. Ecological Economics 69, 2069-2074.

Kienast, F.; Bolliger, J.; Potschin, M.; de Groot, R.S.; Verburg, P.H.; Heller, I.; Wascher, D. & R. Haines-Young 2009. Assessing landscape functions with broad-scale environmental data: insights gained from a prototype development for Europe. Environmental Management 44, 1099–1120.

Kroll, F.; Müller, F.; Haase, D. & N. Fohrer 2012. Rural–urban gradient analysis of ecosystem services supply and demand dynamics. Land Use Policy 29, 521– 535.

Kumar, P. & M.D. Wood 2010. Valuation of Regulating Services of Ecosystems: Methodology and Applications. Taylor & Francis,

Larondelle, N. & D. Haase 2013. Urban ecosystem services assessment along a rural–urban gradient: A cross-analysis of European cities. Ecological Indicators 29, 179–190.

Lautenbach, S.; Kugel, C.; Lausch, A. & R. Seppelt 2011. Analysis of historic changes in regional ecosystem service provisioning using land use data. Ecological Indicators 11 (2), 676–687.

MA (Millennium Ecosystem Assessment), 2005. Ecosystems and Human Well-being: Synthesis. Island Press/World Resources Institute, Washington, DC.

Maes, J., Paracchini, M.L., Zulian, G., 2011. A European Assessment of the Provision of Ecosystem Services: Towards an Atlas of Ecosystem Services. Publications Office of the European Union, Luxembourg, doi:10.2788/63557, p. 81.

Martínez-Harms, M.J. & P. Balvanera 2012. Methods for mapping ecosystem service supply: a review. International Journal of Biodiversity Science, Ecosystem Services & Management 8, 17–25.

Maslow, A.H. 1943. A Theory of Human Motivation. Psychological Review 50, 370-396.

Mubareka, S.B.; Maes, J.; Lavalle, C. & A. De Roo 2013. Estimation of water requirements by Livestock in Europe. Ecosystem Services 4, 139–145.

Müller, F. 2005. Indicating ecosystem and landscape organisation. Ecological Indicators 5 (4), 280–294.

Müller, F. & B. Burkhard 2012. The indicator side of ecosystem services. Ecosystem Services 1, 26-30.

Müller, A.; Burkhard, B.; Müller, F.; Grescho, V. & J. Settele (subm.): Land cover-based ecosystem service assessment of irrigated rice cropping systems with different production intensities in Southeast Asia.

Nahuelhual, L.; Carmona, A.; Lozada, P.; Jaramillo, A. & M. Aguayo 2013. Mapping recreation and ecotourism as a cultural ecosystem service: An application at the local level in Southern Chile. Applied Geography 40, 71-82.

Nedkov, S. & B. Burkhard 2012. Flood regulating ecosystem services - Mapping supply and demand, in the Etropole municipality, Bulgaria. Ecological Indicators 21: 67-79.

Palomo, I.; Martín-López, B.; Potschin, M.; Roy Haines-Young, R. & C. Montes 2013. National Parks, buffer zones and surrounding lands: mapping ecosystem service flows. Ecosystem Services 4, 104–116.

Papendiek, F.; Ende, H.-P.; Steinhardt, U. & H. Wiggering 2012. Biorefineries: relocating biomass refineries to the rural area. Landscape Online 27, 1–9.

Petz, K. & A.P.E. van Oudenhoven (2012): Modelling land management effect on ecosystem functions and services: a study in the Netherlands. International Journal of Biodiversity Science, Ecosystem Services & Management 8 (1-2), 135-155.

Plieninger, T.; Dijks, S.; Oteros-Rozas, E. & C. Bieling 2013. Assessing, mapping, and quantifying cultural ecosystem services at community level. Land Use Policy 33, 118– 129.

Radford. K.G. & P. James 2013. Changes in the value of ecosystem services along a rural-urban gradient: A case study of Greater Manchester, UK. Landscape and Urban Planning 109, 117-127.

Raudsepp-Hearne, C.; Peterson, G.D.; Bennett, E.M. 2010. Ecosystem service bundles for analyzing tradeoffs in diverse landscapes. Proceedings of the National Academy of Sciences. doi/10.1073/pnas.0907284107.

Reid, W.V.; Berkes, F.; Wilbanks, T. & D. Capistrano (Eds.) 2006. Bridging scales and knowledge systems: concepts and applications in ecosystem assessment / Millennium Ecosystem Assessment. Island Press/World Resources Institute, Washington, DC.

Ruppert, K. & F. Schaffer 1969. Zur Konzeption der Sozialgeographie. Geographische Rundschau 21, 52-55.

Schneiders, A.; Van Daele, T.; Van Reeth, W. & W. Van Landuyt 2012. Biodiversity and ecosystem services: complementary visions on world's natural capital? Ecological Indicators 21, 123–133.

Schröter, M.; Barton, D.N.; Remme, R.P. & L. Hein 2014. Accounting for capacity and flow of ecosystem services: A conceptual model and a case study for Telemark, Norway. Ecological Indicators 36, 539–551.

Schröter, M.; Remme, R.P. & L. Hein 2012. How and where to map supply and demand of ecosystem services for policy-relevant outcomes? Ecological Indicators 23, 220-221.

Schulp, C.J.E.; Lautenbach, S. & P.H. Verburg 2014. Quantifying and mapping ecosystem services: Demand and supply of pollination in the European Union. Ecological Indicators 36, 131–141.

Scolozzi, R.; Morri, E. & R. Santolini 2012. Delphi-based change assessment in ecosystem service values to support strategic spatial planning in Italian landscapes. Ecological Indicators 21, 134–144.

Seppelt, R.; Fath, B.; Burkhard, B.; Fisher, J.L.; Grêt-Regamey, A.; Lautenbach, S.; Pert, P.; Hotes, S.; Spangenberg, J.; Verburg, P.H. & A.P.E. van Oudenhoven 2012. Form follows function? Proposing a blueprint for ecosystem service assessments based on reviews and case studies. Ecological Indicators 21, 145–154.

Swetnam, R.D.; Fisher, B.; Mbilinyi, B.P.; Munishi, P.K.T.; Willcock, S.; Ricketts, T.; Mwakalila, S.; Balmford, A.; Burgess, N.D.; Marshall, A.R. & S.L. Lewis 2010. Mapping socio-economic scenarios of land cover change: a GIS method to enable ecosystem service modelling. Journal of Environmental Management 92(3), 563-74.

Syrbe, R.-U. & U. Walz 2012. Spatial indicators for the assessment of ecosystem services: providing, benefiting and connecting areas and landscape metrics. Ecological Indicators 21, 80–88.

TEEB 2010. The economics of ecosystems and biodiversity: mainstreaming the economics of nature: a synthesis of the approach, conclusions and recommendations of TEEB.

Vandewalle, M.; Sykes, M.T.; Harrison, P.A.; Luck, G.W.; Berry, P.; Bugter, R.: Dawson, T.P.: Feld, C.K.; Harrington, R.; Haslett, J.R.; Hering, D.; Jones, K.B.; Jongamn, R.; & S. Lavorel 2009. Review paper on concepts of dynamic ecosystems and their services. The Rubicode Project Rationalising Biodiversity Conservation in Dynamic Ecosystems. http://www.rubicode.net/rubicode/RUBICODE_Review_on_Ecosystem_Services.pdf (Date: 17.10.2013).

van Oudenhoven, A.P.E.; Petz, K.; Alkemade, R.; de Groot, R.S. & L. Hein 2012. Indicators for assessing effects of management on ecosystem services. Ecological Indicators 21, 110–122.

Vigerstol, K.L. & J.E. Aukema 2011. A comparison of tools for modeling freshwater ecosystem services. Journal of Environmental Management 92, 2403-2409.

Vihervaara, P.; Kumpula, T.; Tanskanen, A. & B. Burkhard 2010. Ecosystem services – A tool for sustainable management of human–environment systems. Case study Finnish Forest Lapland. Ecological Complexity 7/3: 410-420.

Vihervaara, P.; Kumpula, T.; Ruokolainen, A.; Tanskanen, A. & B. Burkhard 2012. Using detailed biotope data for Ecosystem service assessment in Natural protection areas. International Journal of Biodiversity Science, Ecosystem Services & Management 8 (1-2). 169-185.

Villamagna, A.M.; Angermeier, P.L. & E.M. Bennet 2013. Capacity, pressure, demand, and flow: A conceptual framework for analyzing ecosystem service provision and delivery. Ecological Complexity 15, 114–121.

Willemen, L.; Drakou, E.G.; Dunbar, M.B.; Mayaux, P. & B.N. Egoh 2013. Safeguarding ecosystem services and livelihoods: understanding the impact of conservation strategies on benefit flows to society. Ecosystem Services 4, 95–103.

Wood, D.; Fels, J. & John Krygier 2010. Rethinking the Power of Maps. Guilford Pubn.

Table 1: Regulating ecosystem services: definitions, service potential (stock) indicators, service flow indicators, demand indicators, exemplary service providing units (hotspots) and service benefitting areas, suggested spatial and temporal assessment scales (based on Kandziora et al. 2013a; Burkhard et al. 2009 and 2012; Schröter et al. 2014; Syrbe and Walz 2012; de Groot et al. 2010).

Regulating ecosystem service	**Definition**	Exemplary **Service potential indicators**	Exemplary **Service flow indicators**	Exemplary **Demand indicators***	Exemplary **Service providing units (SPU)** (hotspots)	Exemplary **Service benefitting areas (SBA)**	**SPU - SBA spatial relations**	**Rival**	**Spatial assessment scale**	**Temporal scale** (hot moments)
Global climate regulation	Long-term storage of potential greenhouse gases in ecosystems.	Amount of methane, carbon dioxide and water vapour stored in vegetation, soils and marine systems (t C/ha)	Amount of methane, carbon dioxide and water vapour taken up by vegetation, soils and marine systems (t CO_2/ha per year)	Greenhouse gas emissions by industry, traffic, households (t CO_2/ha per year)	Soils, forests (standing biomass), peatlands, oceans	The world, climate change-affected regions, agriculture	In situ, omni-directional	Yes	Global, continental	Long-term
Local climate regulation	Changes in local climate components like wind, precipitation, temperature, radiation due to ecosystem properties.	Temperature (°C); albedo (%); precipitation (mm); wind (Bft); evapotranspiration (mm); shaded areas (ha; %)	Temperature amplitudes (K); precipitation, wind or evapotranspiration deviation from surrounding areas (%)	Excess heat, rain or storm performance (°C, mm. Bft) or periods (d/a); Air conditioning use (kWh/a)	Forests, wetlands, lakes, oceans, (urban) green areas, air-circulation corridors	Residential and recreation areas	In situ, omni-directional, directional	No	Local, regional	Medium-term, annual, seasonal
Air quality regulation	Capturing/filtering of dust, chemicals and gases from air.	Leaf area index, difference between open land and throughfall deposition (kg/ha); immission concentrations (ppm)	Aerosols or pollutants removed (kg/ha per year); air quality standards amplitudes (ppb)	Level of pollutants in the air (ppb); air quality standards deviation (ppb); critical loads exceedance (kg/ha per year)	Woods, hedges, green areas	Residential and recreation areas	In situ, omni-directional, directional	No	Local, regional	Medium-term, annual, seasonal
Water flow regulation	Water cycle feature maintenance (e.g. water storage and buffer, natural drainage, irrigation and drought prevention).	Water storage capacity (m^3/ha); groundwater recharge rate (mm/ha per year)	Water released for hydrological process use, e.g. plant or animal uptake, soil processes (m^3/ha per year); available water content (v%);amount of excess water (m^3/ha per year)	Periods at permanent wilting point (d/a); soil field capacity (v%); periods of excess water or floods (d/a)	Water bodies, wetlands, forests, glaciers	Agricultural areas, residential areas, industrial areas	In situ, omni-directional, directional	Yes (agri-culture) No (floods)	Regional	Medium-term, annual, seasonal
Water purification	Ecosystem ability to purify water, e.g. from sediments, pollutants, nutrients, pesticides, disease-causing microbes and pathogens.	Water quality indicators: sediment load (g/l); total dissolved solids (mg/l)	Elements removed from water (kg/m^3 per year); water quality standards amplitudes (ppb; mg/l)	Level of pollutants in the water (ppb); water quality standard deviation (ppb; mg/l)	Water bodies, aquatic flora, riparian strips, filtrating soils, forests, wetlands, grasslands	Residential or recreation areas, agriculture, industry	In situ, omni-directional, directional	Yes	Local, regional, catchment	Medium-term, annual, seasonal
Nutrient regulation	Ecosystem ability to recycle nutrients, e.g. N, P.	Nutrient turnover rates of, e.g. N, P (y^{-1}); water quality indicators, e.g. N (mg/l), P (mg/l); electrical conductivity (µS/cm); total dissolved solids (mg/l); soil potentials (CEC; SOC; texture)	Nutrients available for plant uptake (kg/ha per year); amount of excess nutrients (kg/ha per year); nutrients filtered or adsorbed (kg/ha per year)	Periods of nutrient deficit or excess (d/a); fertilizer needs (kg/ha per year); periods of eutrophication (d/a)	Forests, grasslands, wetlands, marshes, water bodies, oceans	Agricultural areas, communities	In situ, omni-directional, directional	Yes	Local, regional, catchment	Medium-term, annual, seasonal
Erosion regulation	Soil retention and the ability to prevent and mitigate soil erosion and landslides.	Vegetation cover (%); loss of soil particles by water and wind (kg/ha per year); USLE factors for assessment of potential soil loss and landslide frequency (n/ha per year)	Amount of soil retained or sediment captured (kg/ha per year); amount of prevented erosion events (n/a)	Number of erosion events (n/ha per year); soil loss by erosion (kg/ha per year)	Forests, hedges, groves around and between acre fields, pastures, grasslands	Agricultural fields, infra-structure, residential areas	In situ, omni-directional, directional	No	Local, regional	Short-term (events), long-term (regulation)
Natural hazard protection	Protection and mitigation of floods, storms, fires and avalanches.	Water-storage potential (m^3/ha); natural barriers (dunes, mangroves, wetlands, coral reefs, forests) (%; m/ha; ha)	Number of prevented hazards (n/a); Prevented fatalities, damage to property or infrastructure (n/a; €/a)	Number of hazards and fatalities (n/a); damage costs (€/a)	Forests, mangroves, beaches, coral reefs, wetlands, water bodies	Built areas, land uses, infra-srtucture and industry within hazard-prone zones	In situ, omni-directional, directional	No	Local, regional	Short-term (events), long-term (regulation)
Pollination **	Bees, birds, bats, moths, flies, wind, non-flying animals contributing to pollen transfer and reproduction of plants.	Species numbers and amount of pollinators (n/ha); potential habitats for pollinators (ha/ha; %; n/ha)	Amount of pollinated plants (n/ha per year; %/a; kg/ha per year)	Amount of agricultural, garden or wild plants demanding pollination (n/ha per year; %/a; kg/ha per year)	Gardens, fruit and berry plantations, forests, wetlands, agricultural areas	Agricultural, garden and wild plant areas, fruit tree plantations, farmers	Omni-directional	No	Regional	Annual
Pest and disease control**	Ecosystem ability to control pests and diseases due to genetic variations of plants and animals making them less prone to diseases and actions of predators and parasites.	Populations of biological disease and pest control agents (n/ha); Potential habitats for control agents (ha/ha; %; n/ha)	Number of prevented pest and disease outbreaks or predator and parasite actions (n/ha per year; %/a)	Number of pest and disease outbreaks (n/ha per year); Plants and animals damaged (%/a; n/a); Yield losses (%/a; €/a)	Forests, wetlands, water bodies, gardens, agricultural areas	Com-munities, transport facilities, agricultural fields, farms, stables, crops, animals, farmers	In situ, omni-directional, directional	No	Local, regional	Annual, long-term
Regulation of waste**	Ecosystem ability to filter and decompose organic material in water and soils.	Amount and number of decomposers (n/ha); immobilization potential in plants and soils	Decomposition rate (kg/ha per year); Pollutants recycled or Immobilized (kg/ha per year)	Level of organic material in water and soils (ppb); environmental standards deviation(ppb)	Soils, forests, pastures, wetlands, water bodies, oceans	Com-munities, industry, dump sites, agriculture	In situ, omni-directional, directional	Yes	Local, regional	Annual, long-term

* Demand for regulating ecosystem services is problematic to indicate because direct ecosystem service-human benefit relations are often very complex. Indicators suggested here relate to relevant ecosystem states or regulating processes instead. The demand is mostly oriented toward a reduction of the indicator values or the indicating concentrations.

** For these ecosystem services a strong overlap with ecosystem functions and a high potential of double-counting must be noted.

Table 2: Provisioning ecosystem services: definitions, servicepotential (stock) indicators, service flow indicators, demand indicators, exemplary service providing units (hotspots) and service benefitting areas, suggested spatial and temporal assessment scales (based on Kandziora et al. 2013a; Burkhard et al. 2009 and 2012; Schröter et al. 2014; Syrbe and Walz 2012; de Groot et al. 2010).

Provisioning ecosystem service	**Definition**	Exemplary **Service potential indicators**	Exemplary **Service flow indicators**	Exemplary **Demand indicators**	Exemplary **Service providing units (SPU)** (hotspots)	Exemplary **Service benefitting areas (SBA)**	**SPU - SBA spatial relations**	**Rival**	**Spatial assessment scale**	**Temporal scale** (hot moments)
Crops	Plants usable for human nutrition.	Standing stock +/or net primary production (t C/ha + t C/ha per year; kJ/ha + kJ/ha per year)	Harvested crops (t/ha per year, kJ/ha per year); Yield (€/ha per year)	Crop product consumption (kg/person per year; kJ/person per year)	Agricultural fields, gardens, fruit and berry plantations	Farms, food industry, communities, households	Decoupled	Yes	Regional, local	Annual, seasonal, short-term (harvest rhythm)
Biomass for energy	Plants usable for energy conversion (e.g. sugar cane, maize).	Standing stock +/or net primary production (t C/ha + t C/ha per year; kJ/ha + kJ/ha per year)	Harvested plants (t/ha per year, kJ/ha per year); Yield (€/ha per year)	Energy use based on biomass (kWh/person per year)	Agricultural fields, short rotation coppice, oceans	Farms, industry, communities, households	Decoupled	Yes	Regional, local	Annual
Fodder	Nutritional substances for domestic animals.	Standing stock +/or net primary production (t C/ha + t C/ha per year; kJ/ha + kJ/ha per year)	Fodder plant harvest (t/ha, kJ/ha per year); Yield (€/ha per year); Area used for harvesting fodder (ha)	Fodder use for domestic animals (kg/livestock per year)	Graslands, pastures, agroforestry, marshlands	Farms, stables, pastures	Decoupled	Yes	Regional, local	Annual
Livestock (domestic)	Domestic animals useable for nutrition and related products (dairy, wool).	Number of animals (n/ha; kJ/ha); Animal production (t C/ha per year; kJ/ha per year)	Respective animal products (t/ha per year); Yield (€/ha per year)	Meat consumption (kg/person per year); Related products consum-ption (kg/person per year)	Pastures, farms, stables, grassland, agroforestry	Farms, communities, households	Decoupled	Yes	Regional	Annual
Fibre	Natural fibre (e.g. cotton, jute sisal, silk, cellulose) usable for e.g. cloths, fabric, paper.	Biomass +/or growth of fibre (t/ha + t/ha per year)	Harvested fibre (t/ha per year; kJ/ha per year); Yield (€/ha per year)	Fibre use (t/region per year)	Agricultural fields, farms, natural vegetation	Farms, industry, construction, communities, households	Decoupled	Yes	Regional	Annual
Timber	Wood useable for human purposes (e.g. construction).	Standing stock +/or net primary production (t C/ha + t C/ha per year; kJ/ha + kJ/ha per year)	Harvested wood (solid m^3/a; volume/a); Yield (€/ha per year)	Timber use (t/region per year)	Forests, silvicultural areas, fruit and berry plantations, agroforestry	Forester, sawmills, wood industry, construction, communities, households	Decoupled	Yes	Regional, local	Long-term
Wood fuel	Wood suitable for energy conversion and/or heat production.	Standing stock +/or net primary production (t C/ha + t C/ha per year; kJ/ha + kJ/ha per year)	Harvested wood fuel (m^3/ha per year); Yield (€/ha per year)	Wood used as fuel (m^3/person per year)	Forests, short rotation coppice, hedgerows, agroforestry	Forester, industry, communities, households	Decoupled	Yes	Regional, local	Medium-term
Fish, seafood and edible algae	Seafood, algae useable for food, fish meal and fish oil.	Fish stock +/or growth (t C/ha + t C/ha per year; kJ/ha + kJ/ha per year)	Caught fish/seafood/algae (t/ha per year, kJ/ha per year); Yield (€/ha per year)	Seafood/algae consumption (kg/person per year)	Water bodies and courses, coastal lagoons, oceans	Fishermen, food industry, communities, households	Decoupled	Yes	Regional, local	Medium-term
Aquaculture	Seafood/algae in marine and terrestrial aquaculture farms.	Animal stock +/or growth (t C/ha + t C/ha per year; kJ/ha + kJ/ha per year)	Harvest of seafood/algae (t/ha per year, kJ/ha per year); Yield (€/ha per year)	Aquaculture product consumption (kg/person per year)	Aquaculture farms, fish ponds, water bodies, rice fields, coastal lagoons, estuaries, oceans	Fish farmer, food industry, communities, households	Decoupled	Yes	Local	Medium-term, annual
Wild food, semi-domestic livestock and ornamental resources	Berries, mushrooms, (edible) plants, wild animals, fish and natural ornaments available for recreational fishing, hunting or collection; semi-domestic animal husbandry.	Amount of respective items available; stock +/or growth of respective wild species (n/ha; kg/ha; kg/ha + kg/ha per year; kJ/ha + kJ/ha per year)	Catch of fish; game taken (kg/ha per year); Harvested plant biomass (t C/ha per year); Yield (€/ha per year)	Wild food consumption (kg/person per year); Ornamental item sale (n/region per year); Business volumes (€/a)	Forests, grasslands, agricultural fields, water bodies and courses, mountains	Forester, hunter-gatherers, angler, herder, industry, communities	Decoupled	Yes	Local, regional	Annual
Biochemicals and medicine	Natural products useable as biochemicals, medicine and/or cosmetics.	Amount or number of substances useable for medicine, biochemical, cosmetics (kg/ha; n/ha); Stock +/or net primary production (t C/ha + t C/ha per year; kJ/ha + kJ/ha per year)	Yield of respective products (€/ha per year)	Substances used (kg/ha per year); Products sale (€/region per year)	Forests, gardens	Gatherer, gardener, pharmacy, beauty industry, consumer	Decoupled	Yes	Local, regional	Annual
Freshwater	Fresh and process water available for e.g. drinking, domestic use, industrial use, irrigation.	Fresh- and/or process water availability (l/ha per year; m^3/ha per year); Total amount of water (m^3/ha); Groundwater recharge rate (m^3/ha)	Water withdrawal (l/region per year; m^3/region per year)	Water use (l or m^3 /person per year; l or m^3/industrial sector per year)	Water reservoirs, water bodies and courses, glaciers, groundwater	Water supply companies, agriculture, industry, communities, households	In situ; directional; (decoupled)	Yes	Local, regional	Annual, medium-term
Mineral resources* **	Minerals extractable close from surface or above surface (e.g. sand for construction, lignite, gold, salts).	Minerals available for extraction (t/ha)	Excavated minerals (t/ha per year); Earnings (€/a)	Minerals used (t/person per year; t/industrial sector per year)	Coal beds, ore veins, moraines, eskers, sea bed, salines	Mining companies, industry, construction, communities, households	Decoupled	Yes	Local	Annual, long-term
Abiotic energy sources* **	Abiotic energy sources useable for conversion (e.g. solar, wind, water and geothermic power).	Areas and natural settings potentially suitable for energy conversion (ha/ha; n/ha; GW/ha)	Converted energy (kWh/ha per year); Produced electricity (kWh/ha per year); Yields (€/ha per year)	Energy use (kWh/person per year; kWh/industrial sector per year)	Open spaces on- and offshore, water bodies, geothermal fields	Wind or solar farmer, energy companies, communities, households	Decoupled (electricity); in-situ (water mill; geothermic)	No	Local, regional	Annual

*** Abiotic outputs from natural systems (after CICES); often not acknowledged as ecosystem services, but of high relevance for policy decisions and land use/resource management.

Table 3: Cultural ecosystem services: definitions, service potential (stock) indicators, service flow indicators, demand indicators, exemplary service providing units (hotspots) and service benefitting areas, suggested spatial and temporal assessment scales (based on Kandziora et al. 2013a; Burkhard et al. 2009 and 2012; Schröter et al. 2014; Syrbe and Walz 2012; de Groot et al. 2010).

Cultural ecosystem service	Definition	Exemplary Service potential indicators	Exemplary Service flow indicators	Exemplary Demand indicators	Exemplary Service providing units (SPU) (hotspots)	Exemplary Service benefitting areas (SBA)	SPU - SBA spatial relations	Rival	Spatial assessment scale	Temporal scale (hot moments)
Recreation and tourism	Outdoor activities and tourism relating to the local environment or landscape, including forms of sports, leisure and outdoor pursuit.	Number of facilities (e.g. hotels, restaurants, hiking paths, parking lots; n/ha); Results from questionnaires on nature and leisure preferences (wildlife-viewing, hiking, fishing, sports)	Number of facility visitors (n/facility per year); Turnover from tourism (€/ha per year)	Results from question-naires on holiday plans and expectations	Forests, water bodies, beaches, mountains, urban green, gardens, leisure facilities	Touristic infrastructure, visitors, communities, households (at home location)	In situ, omni-directional, directional, decoupled	De-pen-ding on visitor car-rying capa-city	Local, regional	Seasonal, annual
Landscape aesthetic, amenity and inspiration	Visual quality of the landscape/ecosystems or parts of them influencing human well-being and the need to create something as well as the sense of beauty people obtain from looking at landscapes/ecosystems.	Evaluations from questionnaires; Scenic beauty estimation via landscape metrics	Number of paintings/illustrations, songs, products portraying the resp. landscape/ecosystem (n/landscape type); results of travel cost or willingness to pay estimations	Results from question-naires on landscape preferences and expectations	Viewsheds, seascapes, water bodies and courses, forests	Touristic infrastructure, trader, industry, visitors, communities, households (at home location)	In situ, omni-directional, directional, decoupled	No	Regional	Seasonal, annual
Knowledge systems	Environmental education based on ecosystems/landscapes and knowledge in terms of traditional knowledge and specialist expertise arising from living in this particular environment.	Number of environmental educational-related facilities (n/ha)	Number of environmental educational-related events and number of their users (n/a)	Requests for environment al education (n requests/a)	Geotopes, traditional land use systems, forests	Education facilities, research, industry, visitors, households (at home location)	In situ, decoupled	No	Regional - global	Medium-term, long-term
Religious and spiritual experience	Spiritual or emotional values that people or religions attach to local ecosystems or landscapes due to religious and/or spiritual experience.	Number of spiritual facilities or items (n/ha)	Number of visitors of spiritual facilities or items for performance of rituals and maintain the relationship with ancestors (n/facility per year)	Requests for religious and spiritual experience (n requests per year)	Forests, trees, water bodies, rocks, graveyards	Spiritual facilities, visitors, households (at home location)	In situ, decoupled	No	Regional	Seasonal, annual
Cultural heritage and cultural diversity	Values that humans place on the maintenance of historically important (cultural) landscapes and forms of land use (cultural heritage).	Areas and natural settings potentially suitable for traditional land use (ha/ha; n/ha); Results from question-naires on local people's personal preferences	Number of traditional land use forms (n/ha); Number of employees in traditional land use forms (n/ha)	Number of job appli-cations and trainees in traditional land use forms (n/a)	Agricultural fields, gardens, vineyards, terraced fields, hedgerows, silviculture, villages	Traditional land use regions, visitors, households (at home location)	In situ, decoupled	No	Local - Global	Long-term
Natural heritage and natural diversity	The existence value of nature and species themselves, beyond economic or direct human benefits.	Potential habitats for endangered, protected and/or rare species (n/ha)	Abundance of endangered, protected and/or rare species (n/ha)	Relevant guidelines for nature protection (n/ha)	Natural forests, peatlands, water bodies and courses, mountains	Nature itself, households (at home location)	In situ, decoupled	No	Regional - global	Long-term

5

Sustainable landscape development and value rigidity: the Pirsig's monkey trap

Zurlini Giovanni[1]*, Petrosillo Irene[1], Bozsik András[2], Cloud Jon[3], Aretano Roberta[1], Lincoln Noa Kekuewa[4]

1 Landscape Ecology Laboratory, Dept. of Biological and Environmental Sciences and Technology, University of Salento, Lecce, Italy

2 University of Debrecen, Department of Plant Protection, Debrecen, Hungary

3 Cloud Holding Inc., 342 Indian Road Cres Toronto, Canada

4 Emmett Interdisciplinary Program in the Environment and Resources, Stanford University, Palo Alto, CA, USA

Abstract

New broader, adaptable and accommodating sets of themes have been proposed to help to identify, understand and solve sustainability problems. However, how this knowledge will foster decisions that lead to more desirable outcomes and analyses necessary to transition to sustainability remains a critical theoretical and empirical question for basic and applied research. We argue that we are still underestimating the tendency to lock into certain patterns that come at the cost of the ability to adjust to new situations. This rigidity limits the ability of persons, groups, and companies to respond to new problems, and can make it hard to learn new facts because we pre-select facts as important, or not, in line with our established values. Changing circumstances demand to reappraise values like in the case of Pirsig's monkey and its rice. There is an urgent need to go beyond such local, static and short-term conceptions, where landscape sustainability has been incorrectly envisioned as a durable, stable condition that, once achieved, could persist for generations. We argue that to manage a global transition toward more environmentally efficient and, therefore, more sustainable land-use we have to reappraise societal values at the root of overregulation and rigidity.

Keywords:
landscape sustainability, value rigidity, rigidity traps, societal values

*Corresponding author. Email: giovanni.zurlini@unisalento.it

1 Introduction

New broader, adaptable and accommodating sets of themes have been proposed to help to identify, understand and solve sustainability problems (Leemans and Solecki 2013). They span from environmental change issues (e.g. past, present and future trends), environmental change assessments (e.g. defining, quantifying or attributing change, vulnerabilities and risks), system dynamics and sustainability (e.g. transitions, thresholds, feedbacks and tipping points), sustainability governance and transformation (e.g. societal change over time, public–private partnerships, green economy), to sustainability challenges (e.g. planetary boundaries, solutions, vulnerabilities, adaptation versus mitigation and environmental security), and sustainability science (e.g. analytical frameworks to assess sustainability, interdisciplinarity and transdisciplinarity, capacity building, and science communication and outreach).

However, how this knowledge will foster decisions that lead to more desirable outcomes and analyses of the processes necessary for the transition to sustainability remains a critical theoretical and empirical question for basic and applied research (Miller et al. 2014). We are still underestimating the tendency to lock into certain patterns that come at the cost of the ability to adjust to new situations and occur on levels varying from the cell and the mind to societies (Scheffer and Westley 2007). This resulting rigidity limits the ability of persons, groups, and companies to respond to new problems, and some of them may have contributed to the collapse of ancient societies. This is well surmised by the famous Albert Einstein quote "we cannot solve our problems with the same thinking we used when we created them."

While sustainability scientists acknowledge the importance of the societal values as a subject of inquiry of sustainability research (Clark 2007), those values are, however, widely neglected (Miller et al. 2014).

The need to consider changing societal values at the root of rigidity is well illustrated in Pirsig's (1974) book "Zen and the Art of Motorcycle Maintenance: An Inquiry into Values". There, the South Indian monkey trapper drills a hole in a coconut, puts a ball of rice inside and chains the coconut to a stake. The monkey smells the rice, inserts its hand to grasp the rice, and becomes trapped since its fist with the ball of rice is now too big to pass through the hole and it will not let go of the rice. Pirsig calls this trap "value rigidity". The usually high value the monkey places on rice needs re-evaluation in this life-threatening situation. If the monkey gave up a bit of rice it would save its life, but because of its consolidate value rigidity the monkey does not and results captured. In this metaphor, value rigidity skews the value we attach to facts and, because of value rigidity we might get stuck in a "rigidity trap" (Carpenter and Brock 2008). Such locks usually have an obvious evolutionary advantage for the monkey, even though locked attitudes and modes of behavior in individuals and groups can easily lead to undesirable lock-in situations (Scheffer and Westley 2007) like for the Pirsig's monkey.

The problem we presently face is to assess the implications from local to global scale of a "static" and "ordered" landscape condition in social-ecological landscapes (SELs), provided by the cross-scale intersections of land-uses, plans and norms (order), and how this can be made sustainable in face of predictable as well as unpredictable change and disturbance (disorder) (Zurlini et al. 2013). This is widely related to different lock-in situations. For instance, the current land-use planning system is insufficiently equipped to stimulate sustainable development (Diamond 1995), and the emergence of land overregulation and rigidity in some parts of the world can result in drastic changes in others, what may not be evident through traditional local and single scale-based approaches.

As land-use transformation is becoming a main global driver given the worldwide changes to forests, farmlands, waterways, and air (MEA 2005; Turner II et al. 2007), global analysis is required to determine the net effect of local land use decisions and assess implications for greenhouse gas concentrations and climate change. This is because regulations to protect natural and human-managed ecosystems and world trade policies may merely shift land-uses from one country to another, by increasing imports, and mitigate climate change by the use of biofuels in one place increasing global greenhouse gas emissions due to the parallel response of land-use changes in remote locations (Lambin and Meyfroidt 2011).

This highlights the need to go beyond such local, static and short-term conceptions of landscape planning and management, where sustainability has been incorrectly envisioned as a durable, stable condition that, once achieved, could persist for generations (Ahern 1999). This can be seen in the shifting academic consideration of "sustainable" solutions to those of adaptation and resilience (e.g., Berkes et al. 2003; Nelson et al. 2007).

In this perspective, the purpose of this paper is to present the most important facets of values rigidity in order to achieve sustainable social-ecological landscapes (SELs). In particular, we argue that in order to manage a transition toward more environmentally efficient and, therefore, more sustainable land-use we have to consider changing societal values at the root of over-regulation and rigidity.

The paper is organized as follows: the next section provides an overview of several important components of values, their use in sustainability research and the role of values in shaping the SELs. Then, the concept of rigidity trap is introduced and, in the fourth section, few examples of socio-ecological concepts, which need re-evaluation to escape rigidity traps, are proposed. Finally, a general discussion on the significant changes in human values necessary to escape rigidity traps in the light of sustainable development is presented. The paper ends proposing a possible solution to achieve sustainable landscape planning and management.

2 Defining values

Many authors have defined values as beliefs, either individual or social, about what is valuable and important in life (e.g. RCEP 1998; Slootweg at al. 2001; Stolp et al. 2002; Adger at al. 2009; O'Brien and Wolf 2010) and that function as important guiding principles and influence behavior and perceptions across a variety of situations (Feather 1996; Rokeach 1973; Swartz, 1994). One mode of defining sustainability is through the values that represent or support it (Kates et al. 2005). Various statements supporting sustainability are all expressions of values, such as the Millennium Declaration that is an explicit statement founded on a core set of important values that are essential to international relations: freedom, equality, solidarity, tolerance, respect for nature, and shared responsibility. Abstract values, such as peace, freedom, equity, justice, are of great importance to humanity, because they suggest essential principles for societies that guide our current actions and shape our vision for the future (Wu 2013) and explain the motivational bases of attitudes and behavior (Rokeach 1979; Karp 1996; Rohan 2000; Bardi and Schwartz 2003; Leiserowitz et al. 2006).

The Schwartz's value theory (Schwartz 1992) provides one of the most widely cited value frameworks identifying ten universal values (power, achievement, hedonism, stimulation, self-direction, universalism, benevolence, conformity, tradition, security) according to the motivation that underlies each of them. Although the nature of values and their structure may be universal, their hierarchy and the type of goal or motivation that they express could change among levels of organization and over time. We can say that human values are scale dependent and may be viewed in a nested context: individuals focus on local, short-term interests (e.g. power, achievement), communities focus on longer-term intergenerational interests (e.g. tradition, security) and nations focus on global issues (universalism).

Considering the temporal scale, the importance attributed to values may change when new motivations emerge. For example, at the individual level, as people grow older, they tend to become more embedded in social networks, start a family and attain stable positions in the occupational world. This implies a change in their value's priority, becoming less preoccupied with their self-enhancement values (power, achievement) and more concerned with the welfare of others (benevolence, universalism values). At the societal level, as proved by Inglehart and Welzel (2005), socioeconomic development tends to produce generation value differences and a shift from survival values toward self-expression values.

The construction of a system of values, analyzing motivations, can be also traced to Maslow (1943), who developed a holistic-dynamic theory of motivation, well-known as 'Maslow's hierarchy of needs' (Figure 1). It represents a pyramid of basic human needs in five levels (physiological needs, safety needs, love and belongingness needs, self-esteem, and self-actualization) that motivate and drive behavior and are sequenced in order of priority from lowest to highest.

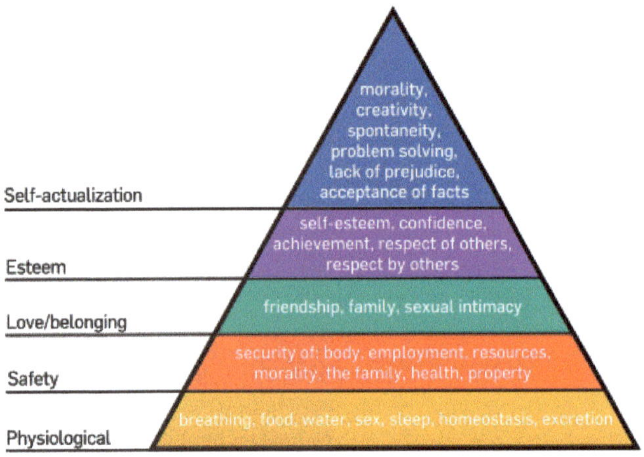

Figure 1: An interpretation of Maslow's (1943) hierarchy of needs, represented as a pyramid with the more basic needs at the bottom (from Maslow's 1943).

The Maslow's theory suggests that first we need to satisfy basic physiological needs, such as food, air, water, then our needs for safety (e.g. security of environment, employment, health, law), belongingness (e.g. love, friendship, family), esteem (e.g. achievement, self-esteem, respect,) and lastly self-actualization (e.g. personal growth, self-fulfillment, morality, creativity). In this perspective, if basic (physiological and safety) needs are not met, it is unlikely that other higher values at the top of the pyramid (democracy, social justice, equity) will be prioritized (Hagerty 1999; Tischler 1999; Yawson et al. 2009; O'Brien and Wolf 2010, Tay and Diener 2011). The ability of humans to satisfy these needs arises from the opportunities available and constructed from social, built, human and natural capital (and time) (Costanza et al., 2008, Petrosillo et al. 2013). In this viewpoint, in the sequence of Maslow's pyramid the human motivations connect the human being primarily to the natural capital and then to the built, social and human capital evoking the framework suggested by Meadows (1998) which relates natural wealth to ultimate human purpose through technology, economy, politics and ethics.

Several researchers pointed out that the opportunities for sustainability increase as more of Maslow's human needs are met (Kofinas and Chapin 2009; Sidiropoulos 2013). In the perspective of environmental sustainability, Udo and Jansson (2009) have demonstrated that poorer nations that are struggling to survive are less concerned with environmental sustainability than advanced and stable nations at higher levels in the hierarchy of needs. This confirms the hypothesis of the environmental Kuznets' curve that provides an inverted U-shape relationship between economic growth (income per capita) and the environment. In particular, at low incomes a trade-off between economic growth and environmental quality exists, while at higher incomes economic growth is complementary to improved environmental quality. This could be explained in terms of lax environmental regulations and low ability to pay for conservation during initial phases of economic development, followed by greater public concern for the environment leading to more stringent regulatory standards and greater ability to pay for environmental issues as incomes rise (Dasgupta et al. 2002). However, as mentioned by Etzioni (1998) and Stobbelaar and van Mansvelt (1999), the continuous consumption by societies long after their basic needs are met is affecting the ability to be environmentally sustainable.

The review of the sustainability literature by the U.S. National Resource Council (1999) identified that the most cited value to be sustained was "life support systems", highlighting that the environment is a source of goods and services essential for the life support of humankind. Thus, ecosystem services are inherently related to human needs and values (Wallace 2007; Dominati et al. 2010) because ecosystems provide goods and services like food, fiber for clothing, sources of energy, support for infrastructures, clean air, clean water, flood mitigation, recycling of wastes, which are useful for the fulfillment of physiological and safety needs. Moreover, aesthetics, spiritual and cultural human experiences are important components of cultural ecosystem services that are crucial for meeting human needs in the upper part of Maslow's hierarchy (Wu 2013). In addition, Termorshuizen and Opdam (2009) have re-conceptualized the idea of ecosystem services as landscape services in order to better recognize the link between values and local landscape patterns and processes. Thus, landscape becomes conceptualized as a value delivery system, through a value chain of structure–function-value.

However, societal values contribute to landscape condition by encouraging or limiting human activities (Rapport et al. 1998). For example, in communities where forests are valued more as timber production than as habitats for wildlife, one might expect an increase in deforestation. Furthermore, societal values can play a key role in judging the acceptability of landscape conditions, as in the case of permissive policies that, to achieve short-term economic benefits, have resulted in environmental consequences affecting the long-term societal goals of landscape sustainability (Rapport et al. 1998).

Finally, the same socio-ecological landscape could be valued differently by different stakeholders because of different needs and perceptions (Aretano et al. 2013). For example, the same agrarian landscape may offer utilitarian and symbolic values for a group of farmers and aesthetic values for a group of non-farmers. Also the scale can affect the values perception of stakeholders involved (MEA, 2005). Consider, for instance, the values of wetland services in the Netherlands for stakeholders at different jurisdictional levels: recreation, reed cutting and fisheries are seen to be of value at the municipal scale, whereas at the provincial scale, main concerns are recreation, but also nature conservation. At the national level, nature conservation is by far the most important service (Hein et al. 2006). However, a clear understanding of scale can lead to development of more effective landscape policies and management at all scales (Zurlini et al. 2010).

3 Values and Rigidity traps

Rigidity is a well-known psychological notion. Psychologists speak about a social trap (Cross and Guyer 1980) when individuals operating for short-term positive gain ("reinforcement") had a tendency to over-exploit a resource, which led to a long-term overall loss to society. Cross and Gruyer (1980) and Costanza (1987) recommended transforming social traps into trade-offs by charging a tariff or other financial contribution on the participants liable for initiating and causing long-term harmful and hazardous situations. Funds gathered this way can be employed to study and ameliorate environmental effects. Odum (1997) has used rigidity in an ecological sense.

Maslow's hierarchy of needs (Figure 1) is a concept to be considered here as the basis of rigidity in values. The essential needs near the bottom of the pyramid produce values that have to be correspondingly rigid. We cannot live without food, for example, and our social systems cannot exist without security. We cannot reevaluate those needs and related values because they are essential to us. However, we are also living at the top of the pyramid, in a realm where needs and related values could be much less rigid and be bent to serve our changing needs. The social puzzle considered daily by law, psychology, education and marketing experts, rest with the motivational factors and forces capable of inspiring self-realization.

In socio-ecological systems, after a phase of growth, followed by a phase in which that growth was conserved, there seemed to be the need for release. Failure to release the creativity for the next phase created a rigidity to the systems, which Holling (2001) described as the "rigidity trap" where institutions become highly connected, self-reinforcing, and inflexible and cannot change or adapt to new conditions, nor escape from a trajectory toward an undesired regime (Gunderson and Holling 2002). These systems are characterized by high capital (accumulated wealth and abundant social and natural capital), high connectivity (efficient methods of social control in which any novelty is ejected) and, in contrast to the phase where those conditions exist in an adaptive cycle, high resilience (the system has a great ability to resist external disturbances and persist, even beyond the point where it is adaptive and creative (Holling 2001). Although some level of "rigidity" is useful, and obviously essential for providing structure to society (civil rights, norms, land-use planning, maintaining values, and fostering resilience to short-term changes), it can exist at the expense of creativity and adaptability and rigidity may also result in overexploitation, or force trades-offs that may become maladaptive over time (Gunderson and Holling 2002).

Escape from this trap is always difficult, because it often means stopping doing something that we have done for years as leaving a job, ending a program, abandoning an approach. But the adaptive cycle tells us that unless we release the resources of time, energy, money and skill locked up in our routines and our institutions on a regular basis, it is hard to create anything new or to look at things from a different perspective. Without the continuous infusion of novelty and innovation in our lives, our organizations and our systems, there is an increase in rigidity. For example, Nielsen and Reenberg (2010) argued that the Fulbe ethnic group in rural Burkina Faso has been less successful in adapting to extreme climate events because they prioritize values such as personal integrity, worthiness, and individual freedom. Prioritizing such values lock them into such pattern that makes them more reliant on climate sensitive activities and gives them less access to embracing alternative successful livelihood strategies.

There are cases where fostering specified resilience of particular parts of a SEL to specific disturbances may cause the system to lose resilience in other ways (Cifdaloz et al. 2010). And this can also occur at global scale, where fostering resilience of particular countries may cause the global interconnected system to lose resilience in another. The ability of our institutions to move at multiple scales along more sustainable trajectories will depend on their aptitude to learn from experience and inform and adapt future sustainability visions, values and transition strategies (Norton 2005). Understanding why rigidity can sometimes make sense may help in finding ways to avoid traps in situations where flexible response and innovation are needed.

Moving this metaphor to the landscape perspective, a "rigidity trap" is formally characterized by low heterogeneity and higher aggregation and connectivity of entities (e.g., land-uses, land-covers), a great capacity to focus on a singular approach, and low capacity to explore alternatives, with consequent little capacity to dissipate stress, which may accumulate to high levels through the panarchy (Holling 2001; Carpenter and Brock 2008). Value rigidity of Pirsig's monkey is at the root of the rigidity trap. Value rigidity can make it hard to learn new facts and to recognize important facts because we pre-select facts as important, or not, in line with our established values. Changing circumstances demand to reappraise values like in the case of Pirsig's monkey and its rice. We have still to change beliefs, analyses or hunches that can immobilize us far more effectively than preparedness to live with uncertainty and surprises.

One interesting example of rigidity trap in SELs is highlighted through the cultural strategies for resource management within different Oceanic islands, where the small islands inevitably reached resource limitations. Initial settlement and growth occurred with little or no resource constraints. Over time growing population lead to the approach of each island's carrying capacity. Kirch (1984) laid out six theoretical models for the long term population growth of these islands, the cultural equivalent of the r/K selection model of island biogeography (MacArthur and Wilson 1967) to include (1) extinction, (2) exponential, (3) logistic, (4) overshoot, (5) oscillating, and (6) step models. While examples exist showing the range of development pathways (see Kirch 1984; Kirch and Rallu 2007), the islands that avoided extinction or overshoot developed stringent cultural standards in order to live within their resource limitations. Some directly curtailed population – such as infanticide, human sacrifice, or more frequent wars – while others indirectly controlled growth rates through strict ownership and enforcement of natural resources. In all cases the values of the populations would be considered overbearing from a modern perspective because they inherently must limit unrestricted freedom to pursue higher 'needs' in order to preserve the essential needs of people the society. While many paths were taken once the resource scarcity was acknowledged, and all of them limiting in their own way, it is very clear that those that did not react came out the worst. This is the case of Easter Island (Rapa Nui) that has become widely known as a case of "ecocide", where the ancient Polynesians recklessly destroyed their environment and, as a consequence, suffered collapse (Hunt and Lipo 2009).

Although in Collapse, Jared Diamond (2005) attributes the demise of island groups primarily to their size and resource diversity, the importance of cultural response was given little attention. The 'Law of Evolutionary Potential' – 'the more specialized and adapted a form in a given evolutionary stage, the smaller its potential for passing to the next stage (Service 1975)' – speaks to the value of rigidity in solving short term or localized problems and to the disadvantage rigid polities incur when faced with new challenges or external or competing pressure. The successful society in this model becomes locked into their adaptations under the assumption that success at adaptation breed conservatism. In Oceania the rigidity of the successful polities, particularly on the smaller islands where competition for power was limited, likely played a role in the success or failure of island populations; rigidity and conservatism resulted from the investment in controlling the current polity system. Therefore, the more complex and interconnected a society and its polities become, the more investment is needed to maintain it and the more rigid and conservative, as well as vulnerable to change, it will become (Gunderson and Holling 2002). It is to be noted that environmental limitations relate to the potential of political complexity, with greater resources allowing for increased competition and corresponding complexity. In Oceania the ability to adapt to resource limitations was an essential element to long-term stability, but what is less clear is how Oceanic islands would fare in the face of continuous change.

Success breeds rigidity and, therefore, consequently lowers potential success under different conditions and constraints. This theory is often presented as an explanation of why new regimes can often out-compete established ones. Although not presented under the same 'law,' this concept is applied to the success and failures of businesses over time, with new, innovative businesses out-competing established, but rigid, cooperation. As highlighted in the oceanic example, the moral of the paradox is that there are hard choices to make, but that those choices must be made. Restrictions of needs higher on Maslow's pyramid (Figure 1) were restricted or controlled in order to ensure the ability to provide the more essential needs at the bottom. It may be argued that the global community is approaching resource limitations for continued growth and that, at least at the decision making level, is in denial about the need to make difficult, restrictive decisions. Instead we often take decisions that reinforce the rigidity of the current system (Holling and Meffe 1996) that places equivalent value on disparate 'needs'.

4 Examples of socio-ecological concepts that need re-evaluation to escape from rigidity traps

The usually high value the monkey places on rice needs re-evaluation in this life-threatening situation. Changing circumstances demand to reappraise a number of leading concepts like biodiversity, ecological corridors, society's techno-social systems, and the privatization of value, as an example.

Humans trying to understand the current state or predict the future condition of complex systems regularly resort to simple, easily interpreted surrogates like biodiversity as parts of the whole complexity, which can be understood and even used by non-scientists to make planning and management decisions. Yet, the overall information we can gain from a set of indicators either structural and/or functional (e.g. biodiversity) will never match that of the whole system, since each individual indicator carries only partial information.

Fortunately, the complexity of living systems emerges not from a random association of a large number of interacting factors, but rather from a smaller number of key-controlling processes (Holling 2001; Gunderson and Holling 2002). Thus, despite the infinite complexity, much of the fundamental nature of systems can often be captured and described by single key-state variables, as many features of the system's state tend to shift in concert with them. We can argue whether and how biodiversity, as it is currently conceived and pursued, is still one of them. However, the problem is that each complex system is rather unique with nonlinear behavior and sudden shifts and surprises, and not always and anywhere the same key-state-variables are in operation and in the same way. Large studies of land-use and land-cover change, for example, have not found evidence for any single, ever-present driver of change (Ostrom et al. 2007). Yet the challenge is still to identify such single key-state variables, based on the best knowledge we have, as representative of the many features of the system's state that tends to shift in concert with them. In more extensive terms, the same applies to sustainability science of complex adaptive systems.

Furthermore, there is a need to recognize the evolving values in light of our changing situations; that is to not hold too tightly to our current notion of a "biodiverse" system and recognize that in our situation we may have to make hard decisions in cutting certain systems away in order to maintain anything at all. We have lost many species, and will continue to lose many more. As the systems we attempt to protect lost critical species for maintenance, such as the large mega-fauna, our notion of what that system is must also change. The notion of biodiversity is already changing from what it was just a couple decades ago. Species richness, the number of different species in an ecosystem, was for a long time the predominant measure of global biodiversity; this is now changing rapidly as researchers are realizing that rather than considering the sheer number of species, it is which specific species that are present that is crucial for ecological functions (Stuart-Smith et al. 2013). Furthermore, the realization that saving a species does not really work without the conservation of a critical mass of habitat has begun to alter the way we think about species conservation (Noss and Cooperrider 1994). We think this is an important change that needs to continue happening. While biodiversity is important it is not the right level to focus our management on. Biodiversity, at least how we think of it, is a positive benefit of healthy, intact ecosystems but is simply too complex to be managed directly.

In some cases, the rigidity of protecting individual species has resulted in the degradation of habitats that may, ultimately, further result in losses of biodiversity rather than in an increase. For example, the protection of green sea turtles in Hawaii through the endangered species act has been wildly successful, with huge increases in numbers and nesting (Balazs and Chaloupka 2004; Chaloupka et al. 2008). While the recovery of the green sea turtle has been wildly successful given the population increase, the green sea turtle is not considered "recovered" because their populations are still below historical levels (Kittenger et al. 2013). The recovery to historical levels, however, is not a realistic goal due to the concept of the "shifting baseline syndrome" (Pinnegar and Engelhard 2008; Papworth et al. 2009) given that the ecosystem has changed drastically through introduced and invasive species, degradation through human impacts and land runoff, and over-exploitation and harvesting. Thus, considering the historical population of sea turtles as a baseline is no longer relevant under the new paradigm. This is an example of how our attachment to a value (number of turtles) results in mismanagement; although the turtles are not up to historical numbers, they appear to be overly abundant in the modern ecosystem. The conservation of this single species has resulted in unexpected environmental and social consequences; sea turtles may now be over grazing native seaweeds and competing against other seaweed eating species, and an increase of shark attacks in the past few years may be another unintended side effect. Other well-known examples of mismanagement and consequences caused by shifting baselines exist such as the Kaibab deer in Canada and the US (Binkley et al. 2006).

The complete knowledge of biodiversity will remain an unknown for humanity as many other things and, however, even if we know all the elements (species), along with the interactions between taxa and with other communities, and the external forcing from environmental factors we could not yet be capable to reasonably predict what we are really interested in, that is the whole functioning, and this because of the complexity and intrinsic uncertainty of social-ecological systems. Furthermore, the values we place on biodiversity are entirely based upon the complexities of humans' cultural norms, societal structures, and the interaction with the environment. There are many layers of value consisting of both tangible and intangible benefits, and the changing levels of biodiversity or changing relationship of a society to that biodiversity can change a biota from a nuisance to a valued rarity very quickly.

As to corridors, an accepted goal of conservation is to build a conservation network that is resilient to fragmentation and environmental change (Zaccarelli et al. 2008). However, fragmentation is a relative concept as well as connectivity. Effective corridors should provide suitable and reliable connectivity among habitats across spatiotemporal scales for species mobile or less mobile for gene exchange under uncertainty and change. However, often a "static" vision of landscapes is adopted (i.e. the cartography of land-uses/covers) whereas landscapes (habitats included) are dynamic. Indeed, they do change either under different seasonal conditions, or under multiple driving forces like, for instance, climate change. As a result, what we are looking for, i.e. fragmentation or effective corridors, can systematically change on the map, and what is fragmented or suitable as corridor under certain conditions could not be suitable or fragmented when season, conditions or the set of focal species are changed. This is an example of how our attachment to valued spatially explicit mapping of land uses/covers can result in mismanagement of conservation networks.

Just because we are not so good in predicting the future and what could be a suitable network sustaining functional diversity and gene exchange, we should rely on past time series (at a suitable scale) to define the trajectory of every landscape piece to see whether it is predictable or not, that is, if it is persistent or not. Once you get a "predictability" map (e.g., Zurlini et al. 2013) then you can think of applying different modeling tools to derive, under uncertainty, what possibly could be an effective corridor network and a suitable fragmentation for the future (Zurlini et al. 2014a). So one could discover that along with "classical" green and blue ways other elements in the landscape could be crucial for the network based on their predictability. One could also discover which unpredictable landscape pieces are crucial for the maintenance of the overall connectivity in the face of climate change and search to transform them in "persistent" through planning and management efforts.

The same principle should be applied to fragmentation/connectivity for marine systems (Treml et al. 2008). Indeed, in the case of many marine species and population, connectivity is determined largely by ocean currents transporting larvae and juveniles among distant patches of suitable habitat. So, connectivity relies on the persistence of ocean currents suggesting areas that might be prioritized for marine conservation efforts and that are working like "stepping stones" in the maintenance of the overall network. In this case, you might identify "new" candidate stepping stone areas in case of predicted changes in the oceanic current pattern following the ongoing climate change. Unfortunately, most of marine biologists and ecologists involved in marine conservation do not consider the importance of ocean currents.

Another example refers to society's techno-social systems that are becoming ever faster and more computer-orientated generating faster versions of the existing human behavior, as humans lose the ability to intervene in real time in the global financial market (Johnson et al. 2013). This is a kind of global techno-social trap where a behavioral response is constantly dictated by value rigidity of pre-established competitive necessities of profit and is characterized by large numbers of sub-second extreme events. The proliferation of these sub-second events shows a significant correlation with the onset of the system-wide financial collapse in 2008 (Johnson et al. 2013).

The housing bubble in the US is one example of how the valuation that we rely on became so skewed from reality that it caused a global "catastrophe" (e.g., Baker 2008). We attempt to counterbalance private valuation with laws and statutes that protect public values. The Environmental Protection Act essentially says that we (as a society) value certain aspects of the environment as a public good over the private value that individuals might be able to extract from that good (for example Sax and DiMento 1974). Often, these are viewed in opposition (protection of environment vs. private value) rather than complimentary (balancing of two different values to maximize total value). Ostrom (2009) discusses how value derived from shared spaces is not the simple response dictated by the tragedy of the commons. The tragedy of the commons is fundamentally flawed if there is communication and cooperation; the fair and just solution to the tragedy (as it is proposed with the shared pasture) should be reached through shared values that are communicated within a group. Yet, rather than allow communities or other social blocks to find a fair and just solution, our current norm is to sell the common pasture to a single entity, who (more often than not) will manage it for individual private benefit. This illustrates how an initially beneficial concept, the privatization of value, may have become a burden and potential detriment to our society. While it may not be necessary to abandon concepts as they evolve into traps, they will need to adapt or be regulated in order to avoid the development of rigidity traps.

5 Discussion

The long-term goal for scholars of sustainability science is to recognize which combination of state variables in complex adaptive systems tends to lead to relatively sustainable and productive use of particular resource systems operating at specific spatial and temporal scales and which combination, instead, might get us stuck in the Pirsig's monkey trap leading to resource collapses and high costs for humanity.

We all just seem to be concerned with trying to grab the biggest handful of rice we can rather than thinking about how we – as individuals, communities, and a global society – are going to get our hand out of the hole. In order to escape the value rigidity trap, the very first value that would need to change is to acknowledge that we are in a trap. We perceive that this will be the hardest value to change, and that the rest would come relatively simply after that. We would argue that many people (monkeys), particularly the one's with power in the decision making process, tend to think that there is no trap. People feel this way for different reasons – be it a belief that technology will solve any confronted problem or be it an unawareness of our dependence on natural systems – which complicates the messages needed to change values. Indeed, many people feel that the trap is actually a blessing because after all they do have a handful of rice (if they can just pull their hand out).

The classic work of Donella Meadows (1941 – 2001), provides insight into how do we change the structure of systems to produce more of what we want and less of that which is undesirable in the light of sustainable development. She proposed a list of places (leverage points) to intervene in complex systems in increasing order of effectiveness (Meadows 2009) as reported in Table 1.

Table 1: A list of leverage points to intervene in complex systems in increasing order of effectiveness (modified from Meadows 2009).

Order	Leverage points to intervene in complex systems
12	**Numbers**: Constants and parameters such as subsidies, taxes, and standards.
11	**Buffers**: The sizes of stabilizing stocks relative to their flows.
10	**Stock-and-Flow Structures**: Physical systems and their nodes of intersection.
9	**Delays**: The lengths of time relative to the rates of system changes.
8	**Balancing Feedback Loops**: The strength of the feedbacks relative to the impacts they are trying to correct.
7	**Reinforcing Feedback Loops**: The strength of the gain of driving loops.
6	**Information Flows**: The structure of who does and does not have access to information.
5	**Rules**: Incentives, punishments, constraints.
4	**Self-Organization**: The power to add, change, or evolve system structure.
3	**Goals**: The purpose or function of the system.
2	**Paradigms**: The mindset out of which the system — its goals, structure, rules, delays, parameters – arises.
1	**Transcending Paradigms**: no paradigm is right.

As one can see the most effective leverage points are paradigms and transcending paradigms, apparently very difficult to change but the most effective for a real change. Mostly, the numbers are not worth the sweat put into them. Our discussion on value rigidity falls largely within those highest two leverage points. In the words of Donella "...the shared ideas in the minds of society, the great big unstated assumptions, constitute that society's paradigm, or deepest set of beliefs about how the world works. These beliefs are unstated because it is unnecessary to state them—everyone already knows them. Money measures something real and has real meaning; therefore, people who are paid less are literally worth less. Growth is good. One can "own" land. Those are just a few of the paradigmatic assumptions of our current culture, all of which have utterly dumbfounded other cultures, who thought them not the least bit obvious" (Meadows 2009). Notice that most of the current sustainability research, even the most advanced on complex systems, instead, is focused on least effective leverage points.

Societal value shifts occur in front of us and the response of public attitude to an increase in perceived problem size is expected to be abruptly discontinuous (Scheffer and Westley 2007). In societies with little difference among individuals, if the problem is perceived to be small (and the perceived pay-off of taking action is low), the attitude of most individuals is passive, but if the perceived severity of the problem has grown sufficiently to a critical point then society abruptly shifts to a predominantly active attitude (e.g. creating political pressure to solve the problem) (Scheffer and Westley 2007). What awareness or pain (quantitative and qualitative) is sufficient to provide the realization of letting go of the rice, extracting the hand from the coconut and rethinking the problem of getting to the rice without sacrifice of life? For each individual the awareness or pain (discomfort) is different. It requires a major portion of the population to realize the need for change in order to precipitate a paradigm shift. How many monkeys will need to be captured before the remaining monkeys realize the coconut with the rice in it is a trap (paradigm shift)?

The values expressed by China in 1995, for instance, were very different than those being expressed now. The melting of Mao Zedong's value rigidity gave way to a massive paradigmatic change in only one generation. What are the discomforting values being left behind, and what level of awareness or discomfort was required to birth the societal value change for a society or a troop of monkeys? The monkey population with the value flexibility should increase as the value rigid monkeys are decreased by the consequence of their rigidity and the trap; eventually, the societal value will change (behavior imitation, learning and sharing).

Self-sealing beliefs can be self-correcting when extreme events such as wildfires or hurricanes (e.g. Katrina) foster change in long-established rules and practices like planning and management (Scheffer and Westley 2007). This can give us hope on our adaptive capacity, but we must not rely on extreme events like that for a change. On the contrary, we have to foster our ability to schedule and plan in a collaborative way. However, extreme events may also provoke other feedback processes working to maintain the status quo, such as the financial and/or political support that accompanies continued crisis management. Thus, things might remain the same or even worse through the pathological cycle of resource degradation even when resource managers recognize that things would improve if they approached their work differently (Repetto and Allen, 2006). It would be really hard to escape such lock-in situation like in the Oceanic island case.

However, there are some barriers that enhance such trapped situations and impede the transition of values into concrete actions for the achievement of sustainability (Blake 1999; Kollmuss and Agyeman 2002; Leiserowitz et al. 2006), and the strength of a particular value could also affect the real attainment of a sustainability goal. For instance, environmental protection, that is widely recognized by many people as a key value, it is considered to be a low priority if compared to other values (e.g., economic growth), as evidenced by the current relationship between society and nature that is undoubtedly unsustainable. Furthermore, changing values would also require changes in lifestyles and in economic and social institutions. People often, lack the time, money, or literacy to translate their values into action or to overcome bad lifestyles and the social, economic, and political context (social norms, laws, infrastructure, available technologies) may affect how individuals order their values, giving precedence to some values over others, and obstructing sustainability across multiple contexts.

Thus, it is clear that significant changes in human values and priorities are required to achieve sustainability. Societal value shifts occur in front of us, and thus we can only hope that the degree of awareness will increase to make paradigms shifts with the fewest possible pains of any kind.

6 Conclusions

Human values and motivations that are experienced in landscapes determine the value system that guides land use. The reflection on the way society and we ourselves handle our motivations is particularly relevant when the shift to a sustainable development of the landscape is at stake, a shift that requires turning from maximum tolerable consumption levels to minimum required consumption levels of all limited resources (VanMansvelt e Van der Lubbe, 1999). Especially in the rich countries, the perspectives for a change to achieve sustainable landscapes as the basis for human livelihood should be taken seriously.

In this perspective, adopting management practices and interventions over the short term only to meet current societal needs may be short-sighted. For example, if for a given region a current primary societal value is agricultural production, maximizing production, although can be perceived to be beneficial, may compromise future ecosystem services, such as a diversity of gene pools and water quality, indicating a clearly unsustainable pathway of development.

As highlighted by Zurlini et al. (2014b) an overregulated planning and management of land use along with intensive agricultural systems can lead in the Mediterranean region to a rigidity trap. This can occur through a pathological cycle of resource degradation and stress (soil and water), followed by social-economical response aimed at reestablishing or maintaining productivity of the resource-degrading activity, with consequent further degradation and erosion of system adaptive capacity to cope with shocks and surprises. This can dramatically enhance desertification processes and be detrimental to the necessary adaptability of landscape elements and biodiversity and can alter the vegetation and the patterns of regional climate variables like temperatures and precipitation (Pielke, 2005; Makarieva et al., 2014) with adverse effects on the ability of ecosystems to support the water cycle on land. These are examples in which the adoption of strategies for adaptability that are socially desirable may lead to vulnerable social–ecological systems and persistent undesirable and unsustainable states. In a rigidity trap there is a tendency to lock management and governance into their existing attitudes or worldviews, making it difficult to respond to changing conditions. Moving out of this trap requires not only a shift in the social and economic dimension, but also active stewardship of ecosystem processes (Folke et al. 2009). In this context, the well-informed landscape managers and decision makers should focus particular attention on ways to sustain the all forms of capital, without which actual and future generations cannot meet their needs (Folke et al. 2009) and must take into account that they inevitably change over time and that people differ through time and across space in the value they plays on different forms of capital. For this reason, an effective landscape management is a critical issue that requires both resilience and adaptability, because the losses of many forms of human, social and natural capital are especially problematic because of the impossibility or extremely high cost of providing appropriate substitutes.

Furthermore, landscape managers must consider that societal values fluctuate over time in response to population patterns, economic opportunities, ethics and environmental conditions and are scale dependent. For this reason, over the long term, successful sustainable landscape management should include planning and design considering local and broader-scale perspectives (Jones et al. 2013) that must be flexible to respond to changing ecosystem processes and societal needs and values.

Moreover, stakeholders have different, and often conflicting, values and worldviews that they bring to decision-making processes. In this respect, we need improved social networking and we do believe that a potential solution could be the development of cooperation through hierarchies of stakeholders and decision makers (monkeys). Cooperation will be necessary for multiple stakeholders in the panarchy of SELs in social networks within and between organizational levels for managing SEL resilience under uncertainty and change (Walker et al. 2002; Olsson et al. 2004) to make adaptive plans and management of landscape patterns (Jones et al. 2013). The ultimate goal is to determine if there are a set of landscape design options that will help sustain multiple ecosystem services, but especially in areas under rapid land-use conversion or that are being affected by broad-scale drivers such as climate change. Can a landscape with relatively large amounts of developed/converted land be still designed to function more like a landscape dominated by natural land cover? What landscape features and elements provide the greatest benefit to ecosystem services? How do those opportunities change with biophysical setting and scale?

However, cooperation to this end is not easy, the emergence of cooperation in nature and human societies depends crucially on how the benefits from cooperation are divided and whether individuals have complete information about their payoffs (Fischer et al. 2013). A "social trap" can emerge where individuals, groups or organizations are unable to cooperate owing to mutual distrust and lack of social capital, even where cooperation would benefit all. Examples include civil strife, pervasive corruption, ethnic discrimination, depletion of natural resources, and misuse of social insurance systems (Rothstein 2005).

Cross-scale collaborative planning networks such as the U.S. Fire Learning Network (Butler and Goldstein 2010) can facilitate overcoming the rigidity traps that prevent resource management agencies from responding to complex cross-scalar problems in SELs. Yet, changing circumstances demand to reappraise values like in the case of Pirsig's monkey and its rice. The intentional induction of cooperation could be promoted across the panarchy of SELs through the establishment of social initiatives that increase the perception of similarity within and among stakeholders to reach a minimal level that makes cooperation advisable (Fischer et al. 2013). In other words, we must be fully aware that we might get stuck in a rigidity trap to be aware of the similarity of our common condition, to overcome mutual distrust to change some paradigmatic assumptions of our current culture, and to start real cooperation regardless of which kind of monkey we are.

A great help for fostering this awareness is coming from Internet and Social Media exchanging information and new ideas that can possibly change our way of decision-making, for example, Avaaz (http://www.avaaz.org) with over 27 million members worldwide. We have become networked minds, social decision-makers, more than ever before, and this has several fundamental implications. Our social values, and economic theories and practices must change accordingly in our SELs, and new institutional requirements must be developed for global-scale network interactions and highly interdependent decisions to support the social decision-maker, the "homo socialis", rather be tailored to the perfect egoist called "homo economicus" (Helbing 2013).

Acknowledgements

Many of these ideas have emerged during a discussion that took place among the authors, but that is still open, in ResearchGate (www.researchgate.net/post/Sustainable_development_and_the_Pirsigs_monkey_trap), on some aspects of sustainability related to value rigidity.

We sincerely thank the anonymous reviewers for their careful attention to the manuscript. Their comments and suggestions helped us to greatly improve the previous version.

References

Adger WN, Dessai S, Goulden M, Hulme M, Lorenzoni I, Nelson DR, Naess LO, Wolf J, Wreford A (2009). Are there social limits to adaptation to climate change? Climatic Change 93(3-4):335–354. DOI 10.1007/s10584-008-9520-z

Ahern J (1999) Spatial concepts, planning strategies, and future scenarios: a framework method for integrating landscape ecology and landscape planning. In: Klopatek JM, Gardner RH (eds) Landscape ecological analysis: issues and applications, Springer-Verlag, New York, pp 175–201

Aretano R, Petrosillo I, Zaccarelli N, Semeraro T, Zurlini G, 2013. People perception of landscape change effects on ecosystem services in small Mediterranean islands: A combination of subjective and objective assessments. Landscape Urban Plan. 112:63-73, DOI 10.1016/j.landurbplan.2012.12.010

Baker D (2008) The housing bubble and the financial crisis. Real-World Econ Review 46:73–81

Balazs GH, Chaloupka M (2004) Thirty-year recovery trend in the once depleted Hawaiian green sea turtle stock. Biol Conserv 117(5):491–498. DOI 10.1016/j.biocon.2003.08.008

Bardi A, Schwartz SH, (2003) Values and behavior: Strength and structure of relations. Pers Soc Psychol B 29(10): 1207-1220. DOI 10.1177/0146167203254602

Berkes F, Colding J, Folke C (2003) Navigating Social-Ecological Systems: Building resilience for complexity and change. Cambridge University Press, Cambridge

Binkley D, Moore MM, Romme WH, Brown PM (2006) Was Aldo Leopold right about the Kaibab deer herd? Ecosystems 9(2):227–241. DOI 10.1007/s10021-005-0100-z

Blake J (1999) Overcoming the 'valueaction gap' in environmental policy: tensions between national policy and local experience. Local Env 4(3):257–278. DOI 10.1080/13549839908725599

Butler WH, Goldstein BE (2010) The US Fire Learning Network: springing a rigidity trap through multiscalar collaborative networks. Ecol Soc 15(3), 21. [online] URL: http://www.ecologyandsociety.org/vol15/iss3/art21/

Carpenter SR, Brock WA (2008) Adaptive capacity and traps. Ecol Soc 13(2), 40. [online] URL: http://www.ecologyandsociety.org/vol13/iss2/art40/

Chaloupka M, Bjorndal K, Balazs GH, Bolten AB, Ehrhart LM, Limpus CJ, Suganuma H, Troëng S, Yamaguchi M (2008) Encouraging outlook for recovery of a once severely exploited marine megaherbivore. Global Ecol Biogeogr 17(2):297–304. DOI 10.1111/j.1466-8238.2007.00367.x

Cifdaloz O, Regmi A, Anderies JM, Rodriguez AA (2010) Robustness, vulnerability, and adaptive capacity in small-scale social-ecological systems: The Pumpa Irrigation System in Nepal. Ecol Soc 15(3), 39. [online] URL: http://www.ecologyandsociety.org/vol15/iss3/art39/

Clark WC (2007) Sustainability science: a room of its own. Proc Natl Acad Sci USA 104(6):1737–1738. DOI 10.1073/pnas.0611291104

Costanza R (1987) Social traps and environmental policy. Bioscience 37(6):407–412. DOI 10.2307/1310564

Costanza R, Fisher B, Ali S, Beer C, Bond L, Boumans R, Danigelis NL, Dickinson J, Elliott C, Farley J, Elliott Gayer D, MacDonald Glenn L, Hudspeth TR, Mahoney DF, McCahill L, McIntosh B, Reed B, Turab Rizvi A, Rizzo DM, Simpatico T, Snapp R, 2008. An Integrative Approach to Quality of Life Measurement, Research, and Policy. S.A.P.I.EN.S., 1(1). Online since 19.12.08. http://sapiens.revues.org/169

Cross JG, Guyer MJ (1980) Social traps. University of Michigan Press, Ann Arbor, MI

Dasgupta S, Laplante B, Wang H, Wheeler D (2002) Confronting the environmental Kuznets curve. J Econ Perspect 16(1):147–168. DOI 10.1016/j.apgeog.2010.07.004

Diamond JM (2005) Collapse: How Societies Choose to Fail or Succeed. Viking Penguin, Penguin Group (USA) Inc., New York

Dominati E, Patterson M, Mackay A (2010) A framework for classifying and quantifying the natural capital and ecosystem services of soils. Ecol Econ 69(9):1858–1868. DOI 10.1016/j.ecolecon.2010.05.002

Etzioni A (1998) Voluntary simplicity: Characterization, select psychological implications, and societal consequences. J Econ Psychol 19(5): 619–643. DOI 10.1016/S0167-4870(98)00021-X

Feather NT (1996) Values, deservingness, and attitudes toward high achievers: Research on tall poppies. In: Seligman C, Olson JM, Zanna MP (Eds.) The psychology of values: The Ontario Symposium Volume 8, Lawrence Erlbaum Associates, Mahwah NJ, pp 215–251.

Fischer I, Frid A, Goerg SJ, Levin SA, Rubenstein DI, Selten R (2013) Fusing enacted and expected mimicry generates a winning strategy that promotes the evolution of cooperation. Proc Natl Acad Sci USA 110 (25):10229-10233. DOI 10.1073/pnas.1308221110

Folke C, Chapin FS III, Olsson P (2009). Transformations in Ecosystem Stewardship In: Chapin FS III, Kofinas GP, Folke C (Eds) Principles of Ecosystem Stewardship: Resilience-Based Natural Resource Management in a Changing World. Springer, New York, pp: 103-128

Gunderson LH, Holling CS (2002) Panarchy: Understanding Transformations in Human and Natural Systems. Island Press, Washington, DC

Hagerty MR (1999) Testing Maslow's Hierarchy of Needs: National Quality-of-Life Across Time. Soc Ind Res 46(3): 249-271. DOI 10.1023/A:1006921107298

Hein L, van Koppen K, de Groot RS, van Ierland EC (2006). Spatial scales, stakeholders and the valuation of ecosystem services. Ecol Econ 57(2):209-228. DOI 10.1016/j.ecolecon.2005.04.005

Helbing D (2013) Economics 2.0: The Natural Step towards a Self-Regulating, Participatory Market Society. Evol Inst Econ Rev 10(1):3–41. DOI 10.14441/eier.D2013002

Holling CS (2001) Understanding the Complexity of Economic, Ecological, and Social Systems. Ecosystems 4(5):390–405. DOI 10.1007/s10021-001-0101-5

Holling CS, Meffe GK (1996) Command and control, and the pathology of natural-resource management. Conserv Biol 10(2):328–337. DOI 10.1046/j.1523-1739.1996.10020328.x

Hunt TL, Lipo CP (2009) Revisiting Rapa Nui (Easter Island) ''Ecocide''. Pac Sci 63(4):601–616. DOI 10.2984/049.063.0407

Inglehart R, Welzel C (2005) Modernization, Cultural Change, and Democracy: The Human Development Sequence. Cambridge University Press, Cambridge, MA

Johnson N, Zhao G, Hunsader E, Qi H, Johnson N, Meng J, Tivnan B (2013) Abrupt rise of new machine ecology beyond human response time. Nature Publishing Scientific Reports 3, Article number: 2627. DOI 10.1038/srep02627

Jones KB, Zurlini G, Kienast F, Petrosillo I, Edwards T, Wade TG, Li B-L, Zaccarelli N (2013) Informing landscape planning and design for sustaining ecosystem services from existing spatial patterns and knowledge. Landsc Ecol 28 (6):1175–1192. DOI 10.1007/s10980-012-9794-4

Karp DG (1996) Values and their effect on pro-environmental behavior. Environ Behav 28(1):111–133. DOI 10.1177/0013916596281006

Kates RW, Parris TM, Leiserowitz A (2005) What is Sustainable Development? Goals, Indicators, Values, and Practice. Environment 47(3): 8–21

Kirch PV (1984) The Evolution of the Polynesian Cheifdoms. Cambridge University Press, Cambridge

Kirch PV, Rallu J (2007) The Growth and Collapse of Pacific Island Societies. University of Hawaii Press, Honolulu

Kittinger JN, Van Houtan KS, McClenachan LE, Lawrence AL (2013) Using historical data to assess the biogeography of population recovery. Ecography 36(8):868–872. DOI 10.1111/j.1600-0587.2013.00245.x

Kofinas GP, Chapin FS (2009) Sustaining livelihoods and human well-being during social–ecological change. In: Chapin FS, Kofinas GP, Folke C (eds) Principles of ecosystem stewardship: resilience-based natural resource management in a changing world, Springer, New York, pp 55–75.

Kollmuss A, Agyeman J (2002) Mind the gap: Why do people act environmentally and what are the barriers to proenvironmental behavior? Environ Edu Res 8(3):239–60. DOI 10.1080/13504620220145401

Lambin EF, Meyfroidt P (2011) Global land use change, economic globalization, and the looming land scarcity. Proc Natl Acad Sci USA 108(9):3465–3472. DOI 10.1073/pnas.1100480108

Leemans R, Solecki W (2013) Redefining environmental sustainability. Curr Opin Environ Sustainability 5(3–4): 272–277. DOI 10.1016/j.cosust.2013.07.006

Leiserowitz AA, Kates RW, Parris TM (2006) Sustainability values, attitudes and behaviors: A review of multinational and global trends. Annu Rev Env Resour 31:413–44. DOI 10.1146/annurev.energy.31.102505.133552

MacArthur RH, Wilson EO (1967) The theory of island biography. Princeton University Press, Princeton, NJ

Makarieva AM, Gorshkov VG, Sheil D, Nobre AD, Bunyard P, Li B-L, 2014. Why does air passage over forest yield more rain? Examining the coupling between rainfall, pressure, and atmospheric moisture content. J Hydrometeorol 15(1), 411–426. DOI 10.1175/JHM-D-12-0190.1

Maslow AH (1943) A Theory of Human Motivation. Psychol Rev 50(4):370–396. DOI 10.1037/h0054346

Meadows D (1998) Indicators and Information Systems for Sustainable Development. A report to the Balaton Group, The Sustainability Institute, Hartland Four Corners VT

Meadows D (2009) Leverage Points: Places to Intervene in a System. Solutions 1(1):41–49. [online] URL: http://www.thesolutionsjournal.com/node/419

Millennium Ecosystem Assessment (MEA) (2005) Ecosystems and Human Well-Being. Island Press, Washington, DC

Miller TR, Wiek A, Sarewitz D, Robinson J, Olsson L, Kriebel D, Loorbach D (2014) The future of sustainability science: a solutions-oriented research agenda. Sustain Sci. 9(2):239–246 8(3). DOI 10.1007/s11625-013-0224-6

National Research Council (NRC) (1999) Our Common Journey: A Transition Toward Sustainability. The National Academies Press, Washington, DC

Nelson DR, Adger WN, Brown K (2007) Adaptation to environmental change: contributions of a resilience framework. Annu Rev Environ Res 32:395–419. DOI 10.1146/annurev.energy.32.051807.090348

Nielsen JØ, Reenberg A (2010) Cultural barriers to climate change adaptation: A case study from Northern Burkina Faso. Global Environmental Change 20(1):142–152. DOI 10.1016/j.gloenvcha.2009.10.002

Norton BG (2005) Sustainability: a philosophy of adaptive ecosystem management. University of Chicago Press, Chicago

Noss RF, Cooperrider AY (1994) Saving Natures Legacy. Island Press, Washington, DC

O'Brien KL, Wolf J (2010). A values-based approach to vulnerability and adaptation to climate change. WIREs Clim Change 1(2):232–242. DOI 10.1002/wcc.30

Odum EP (1997) A bridge between science and society. Sinauer Associates, Inc., Sunderland

Olsson P, Folke C, Berkes F (2004) Adaptive comanagement for building resilience in social–ecological systems. Environ Manage 34(1):75–90. DOI 10.1007/s00267-003-0101-7

Ostrom E (2009) A General Framework for Analyzing Sustainability of Social-Ecological Systems. Science 325(5939):419–422. DOI 10.1126/science.1172133

Ostrom E, Janssen MA, Anderies JM (2007) Going beyond panaceas. Proc Natl Acad Sci USA 104(39):15176–15178. DOI 10.1073/pnas.0701886104

Papworth SK, Rist J, Coad L, Milner-Gulland EJ (2009) Evidence for shifting baseline syndrome in conservation. Conserv Lett 2(2):93–100. DOI 10.1111/j.1755-263X.2009.00049.x

Petrosillo I, Costanza R, Aretano R, Zaccarelli N, Zurlini G, 2013. The use of subjective indicators to assess how natural and social capital support residents' quality of life in a small volcanic island. Ecol Indic 24:609–620. DOI 10.1016/j.ecolind.2012.08.021

Pielke RA, (2005). Land use and climate change. Science 310(5754):1625–1626. DOI 10.1126/science.1120529

Pinnegar JK, Engelhard GH (2008) The "shifting baseline" phenomenon: a global perspective. Rev Fish Biol Fisher 18(1):1–16. DOI 10.1007/s11160-007-9058-6

Pirsig RM (1974) Zen and the Art of Motorcycle Maintenance. An Inquiry into Values. William Morrow and Co., Inc., New York, NY

Rapport DJ, Gaudet C, Karr JR, Baron JS, Bohlen C, Jackson W, Jones B, Naiman RJ, Norton B, Pollock MM, (1998) Evaluating landscape health: integrating societal goals and biophysical process. J Environ Manage 53(1):1–15. DOI 10.1006/jema.1998.0187

Repetto R, Allen RB (2006) On social traps and lobster traps: choppy waters on the voyage toward fisheries' harvesting rights. In: Repetto R (ed), Punctuated equilibrium and the dynamics of US environmental policy. Yale University Press, New Haven, pp 110–136

Rohan MJ (2000) A rose by any name? The values construct. Pers Soc Psychol Rev 4(3):255–277. DOI 10.1207/S15327957PSPR0403_4

Rokeach M (1973) The nature of human values. The Free Press, New York.

Rokeach M (1979) From individual to institutional values: With special reference to the values of science. In: Rokeach M (Ed.) Understanding human values: Individual and societal, Free Press, New York, pp. 47–70

Rothstein B (2005) Social Traps and the Problem of Trust. Cambridge University Press, Cambridge

Royal Commission on Environmental Protection (RCEP) (1998). Setting Environmental standards. Twenty-first Report of the Royal Commission on Environmental Protection. The Stationery Office, London.

Sax JL, DiMento JF (1974) Environmental citizen suits: Three years' experience under the Michigan Environmental Protection Act. Ecol Law Quarterly 4(1):1–18.

Scheffer M, Westley FR (2007) The Evolutionary Basis of Rigidity: Locks in Cells, Minds, and Society. Ecol Soc 12(2): 36. [online] URL: http://www.ecologyandsociety.org/vol12/iss2/art36/

Schwartz SH (1992).Universals in the content and structure of values: Theoretical advances and empirical tests in 20 countries. Adv Exp Soc Psychol 25:1–65. DOI 10.1016/S0065-2601(08)60281-6

Schwartz SH (1994). Are there universal aspects in the structure and contents of human values? J Soc Issues 50(4): 19–45. DOI 10.1111/j.1540-4560.1994.tb01196.x

Schwartz SH, Bilsky W (1987) Toward a universal psychological structure of human values. J Pers Soc Psychol, 53(3), 550–562. DOI 10.1037/0022-3514.53.3.550

Service ER (1975) Origins of the State and Civilization. Norton, New York.

Sidiropoulos E (2013) Education for Sustainability in Business Education: A Question of Value. J Clean Prod in press. DOI 10.1016/j.jclepro.2013.10.040.

Slootweg R, Vanclay F, vanSchooten M (2001) Function evaluation as a framework for the integration of social and environmental impact assessment. Impact Assess Proj Apprais 19(1):19–28. DOI 10.3152/147154601781767186

Stobbelaar DJ, van Mansvelt JD (1999) The landscape and nature production capacity of organic/sustainable types of agriculture. Agr Ecosyst Environ 77(1/2): 1–15. DOI 10.1016/S0167-8809(99)00088-2

Stolp A, Groen W, vanVliet J, Vanclay F (2002) Citizen values assessment: Incorporating citizens value judgements in environmental impact assessment. Impact Assess Proj Apprais 20(1):11–23. DOI 10.3152/147154602781766852

Stuart-Smith RD, Bates AE, Lefcheck JS, J. Duffy E, Baker SC, Thomson RJ, Stuart-Smith JF, Hill NA, Kininmonth SJ, Airoldi L, Becerro MA, Campbell SJ, Dawson TP, Navarrete SA, Soler GA,. Strain EMA, Willis TJ, Edgar GJ (2013) Integrating abundance and functional traits reveals new global hotspots of fish diversity. Nature 501(7468): 539–542. DOI 10.1038/nature12529

Tay L, Diener E (2011) Needs and Subjective Well-Being Around the World. J Pers Soc Psychol 101(2): 354–365. DOI 10.1037/a0023779

Termorshuizen J, Opdam P (2009) Landscape services as a bridge between landscape ecology and sustainable development. Landscape Ecol 24(8):1037–1052. DOI 10.1007/s10980-008-9314-8

Tischler L (1999) The growing interest in spirituality in business: A long-term socio-economic explanation. J Organ Change Manage 12(4): 273–279. DOI 10.1108/09534819910282117

Treml EA, Halpin PN, Urban DL, Pratson LF (2008) Modeling population connectivity by ocean currents, a graph-theoretic approach for marine conservation. Landscape Ecol 23(1):19–36. DOI 10.1007/s10980-007-9138-y

Turner II BL, Lambin EF, Reenberg A (2007) The emergence of land change science for global environmental change and sustainability. Proc Natl Acad Sci USA 104(52):20666–20671. DOI 10.1073/pnas.0704119104

Udo VE, Jansson PM (2009) Bridging the gaps for global sustainable development: A quantitative analysis. J Environ Manage 90(12): 3700–3707. DOI 10.1016/j.jenvman.2008.12.020

Van Mansvelt JD, Van der Lubbe MJ (1999) Checklist for sustainable landscape management. Elsevier, 181 pp.

Walker B, Carpenter S, Anderies J, Abel N, Cumming G, Janssen M, Lebel L, Norberg J, Peterson GD, Pritchard R (2002) Resilience management in social-ecological systems: a working hypothesis for a participatory approach. Conserv Ecol 6(1):14. [online] URL: http://www.consecol.org/vol6/iss1/art14

Wallace KJ (2007) Classification of ecosystem services: problems and solutions. Biol Conserv 139(3–4):235–246. DOI 10.1016/j.biocon.2007.07.015

Wu J (2013) Landscape sustainability science: ecosystem services and human well-being in changing landscapes. Landscape Ecol 28(6):999–1023. DOI 10.1007/s10980-013-9894-9

Yawson DO, Armah FA, Pappoe NM (2009) Enabling Sustainability: Hierarchical Need-Based Framework for Promoting Sustainable Data Infrastructure in Developing Countries. Sustainability 1: 946–959. DOI 10.3390/su1040946

Zaccarelli N, Riitters KH, Petrosillo I, Zurlini G (2008) Indicating disturbance content and context for preserved areas. Ecol Indic 8(6):841–853. DOI 10.1016/j.ecolind.2007.01.010

Zurlini G, Jones K, Li B-L, Petrosillo I (2010). Potentials of ecosystem service accounting at multiple scales. http://www.eoearth.org/view/article/155326

Zurlini G, Petrosillo I, Aretano R, Castorini I, D'Arpa S, De Marco A, Pasimeni MR, Semeraro T, Zaccarelli N, (2014a). Key fundamental aspects for mapping and assessing ecosystem services: predictability of ecosystem service providers at scales from local to global. Ann Bot 4:53–63.

Zurlini G, Jones K, Riitters KH, Li B-L, Petrosillo I, (2014b) Early warning signals of regime shifts from cross-scale connectivity of land-cover patterns. Ecol Indic 45:549–560. DOI 10.1016/j.ecolind.2014.05.018

Zurlini G, Petrosillo I, Jones KB, Zaccarelli N (2013) Highlighting order and disorder in social–ecological landscapes to foster adaptive capacity and sustainability. Landscape Ecol 28(6):1161–1173. DOI 10.1007/s10980-012-9763-y

6

Higher-than-present Medieval pine (Pinus sylvestris) treeline along the Swedish Scandes

Leif Kullman

Department of Ecology and Environmental Science, Umeå University, SE-901 87 Umeå, Sweden.

Abstract

The upper treeline of Scots pine (*Pinus sylvestris L.*) is renowned as a sensitive indicator of climate change and variability. By use of megafossil tree remains, preserved exposed on the ground surface, treeline shift over the past millennium was investigated at multiple sites along the Scandes in northern Sweden. Difference in thermal level between the present and the Medieval period, about AD 1000-1200, is a central, although controversial, aspect concerning the detection and attribution of anthropogenic climate warming. Radiocarbon-dated megafossil pines revealed that the treeline was consistently positioned as much as 115 m higher during the Medieval period than today (AD 2000-2010), after a century of warming and substantial treeline upshift. Drawing on the last-mentioned figure, and a lapse rate of 0.6 °C/100 m, it may be inferred that Medieval summer temperatures were about 0.7 °C warmer than much of the past 100 years. Extensive pine mortality and treeline descent after the Medieval warming peak reflect substantially depressed temperatures during the Little Ice Age. Warmer-than-present conditions during the Medieval period concur with temperature reconstructions from different parts of northern Fennoscandia, northwestern Russia and Greenland. Modern warming has not been sufficient to restore Medieval treelines. Against this background, there is little reason to view further modest warming as unnatural.

Keywords:
treeline, Pinus sylvestris, climate change, megafossils, Little Ice Age, Medieval Warm period, Swedish Scandes

*Corresponding author. Email: leif.kullman@emg.umu.se

1 Introduction

The current scientific and public concern about future climate change and associated geoecological transformations draws on unvalidated General Circulation Models (IPCC 2007). Their failure to convincingly account for stalling global temperature evolution since the late 1990s (Akasofu 2013; Tollefson 2014) indicates incomplete systemic understanding, which argues for more extensive inclusion of robust historical data as a prerequisite for realistic modelling of climate evolution and associated biological consequences (Humlum et al. 2011). The importance of putting present-day change into a wider temporal context of previous climate and ecosystem variation cannot be overstated (cf. Zwiers & Hegerl 2008). Ideally, such an effort can be accomplished by studying the history of the upper treeline, whose natural position and dynamism in principle relates to prevailing temperatures (Tranquillini 1979; Holtmeier 2003; Körner & Paulsen 2004).

The sensitivity of the treeline to climate change is empirically demonstrated by broad-scale upshift in close association with predominant climate warming over the past 100 years in widely different parts of the world (e.g. Moiseev & Shiyatov 2003; Lloyd 2005; Kullman & Öberg 2009; Harsch et al. 2009; Mazepa et al. 2011; Kirdyanov et al. 2012). Treeline rise by about 200 m (1915-2007) in the Swedish Scandes, as estimated by Kullman & Öberg (2009), is virtually what should be expected from recorded summer temperature rise by 1.4 °C and assuming a lapse rate of 0.6 °C/100 m. Thus, there is compelling evidence that the position of the upper treeline performs in a sub-centennial-scale dynamic equilibrium with climate evolution, indicative also of profound effects on high-mountain biota (Kullman 2010a, 2012).

By focusing on elevational treeline change over the past 1000 years, an assumedly wide spectrum of natural climate variability may be assessed. This time span goes from the Medieval Warm Period around AD 1000 (Lamb 1982; Ljungqvist et al. 2012) to the tail of the Little Ice Age about 100 years ago (Lamb 1982; Grove 1988; Matthews & Briffa, 2005). Climate change and variability over this entire interval, in particular its possible global extent, homogeneity and magnitude, are central and somewhat controversial aspects with bearing on the current discourse concerning the range and impacts of human-induced future climate change (Crowley & Lowery 2000; Broecker 2001; Bradley et al. 2003; Esper & Frank 2009; Melvin et al. 2012). Aside of a political/ideological side (cf. Bradley et al. 2003; Esper & Frank 2009), this debate suffers from lack of robust paleoclimate methods and global coherence of available data. In particular, extensively used paleoclimate reconstructions from northern Fennoscandia are based on tree-ring data, analyzed by different and complex methods, without consideration for sample altitude, which tend to underestimate natural climate variability (Lindholm et al. 2010; Esper et al. 2012). These efforts have yielded conflicting inferences concerning the thermal level of the Medieval Warm Period and the Little Ice Age in this region as well as globally, relative to the past century (Briffa et al. 1990, 1992; Crowley 2000; Grudd 2008; Gunnarson et al. 2011; Melvin et al. 2012; Kaufmann et al. 2009; Kobashi et al. 2011; Ljungqvist et al. 2012). If science cannot agree on regional and global climate change over the past 1000 years and the underlying mechanisms, then it would be premature to launch models prescribing the course of climate throughout the 21st century.

Against this background, the present paper is a regional contribution to the building of a global-scale understanding of climate and ecosystem evolution over the past millennium. In addition, a paleotreeline reconstruction for this particular time slice may serve as an analogue for the dynamics and structure of high-mountain ecosystem in a potentially warmer future.

Drawing on different lines of proxy data (Hiller et al. 2001; Grudd 2008; Kullman 2013; Rinne et al. 2014), the main hypothesis to be specifically tested is that the pine (*Pinus sylvestris L.*) treeline (defined below) in northern Fennoscandia and adjacent regions was consistently higher than present (AD 2000-2010) about AD 1000-1200 (Medieval Warm Period). Thereafter it is assumed to have declined substantially in concert with Little Ice Age climate cooling.

2 Study area and modern treeline zonation patterns

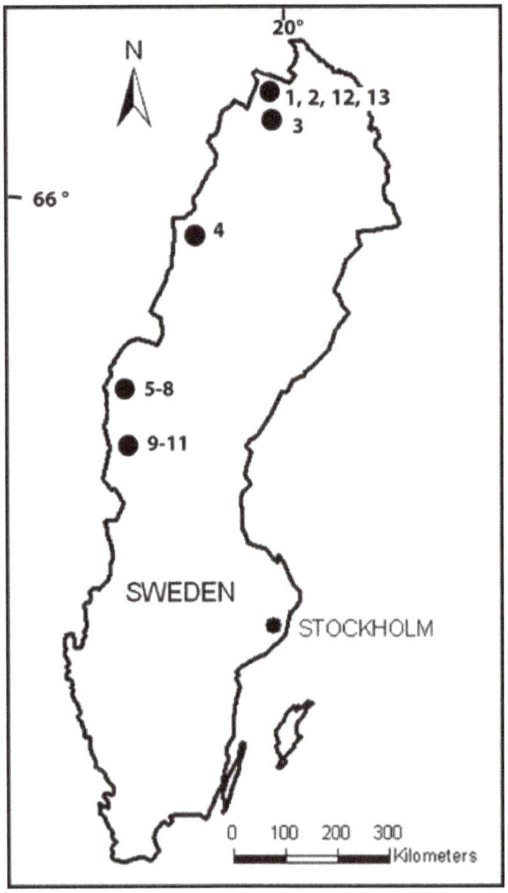

Figure 1: Location map showing the positions of the sampling sites along the Scandes in the northern half of Sweden. Numbers refer to Table 1.

Sampling spans a latitudinal range (68°24´N - 62°03´E) covering virtually the entire north-south extension of the Scandes in Sweden, between the provinces of Lapland, in the north, and Dalarna, in the south (Fig. 1). The valleys and mountain sides are covered with an upper subalpine forest belt and a treeline of mountain birch (*Betula pubescens* ssp. *czerepanovii*) adjacent to the alpine tundra. With great local variations, scattered specimens of Norway spruce (*Picea abies*) and Scots pine (*Pinus sylvestris*) form their respective treelines, around 100 and 50 m below that of the mountain birch and usually as outliers within the birch belt (Kjällgren & Kullman 2002). The present-day (2010-2013) local treelines of pine, set by the uppermost trees > 2 m tall (Fig 2), grade between c. 500 m a.s.l. in the north and 1000 m a.s.l. in the south. As a rule, the ground cover around the pine treeline is dwarf-shrub heath, with *Empetrum hermaphroditum*, *Betula nana* and *Calluna vulgaris* as dominating species, sometimes intermixed with reindeer lichens (*Cladina* spp.). In contrast to the upper limit of closed pine stands, the treeline is insignificantly altered by past and present land use or fire (Kullman 1981; Kjällgren & Kullman 2002). Since the first decade of the 20th century, the pine treeline has shifted upslope by a maximum of about 200 m (Fig. 2), in concert with a summer warming trend of 1.4 °C (Kullman & Öberg 2009).

Figure 2: Present and past pine treeline of Mt. Djupgravstöten (province of Dalarna). A. Pine trees established in the late 1920s and representing treeline advance by 120 m in elevation. B. The early-20th century treeline, reconstructed at the present-day, in the form of and old-growth stout pine. Photo: Leif Kullman 27 June, 2013.

3 Methods

It is widely recognized that treeline positions primarily relate to summer temperature (e.g. Körner and Paulsen, 2004). Thus, shifts in treeline elevations are used in this study as proxy indicators of low-frequency paleoclimate history (cf. Karlén & Kuylenstierna 1996; Tinner & Kaltenrieder 2005; Kirdyanov et al. 2012; Kullman 2013), drawing on a conventional temperature lapse rate of 0.6 ºC per 100 m in altitude (Laaksonen 1976). However, the discontinuous nature of megafossil records precludes high-resolution time series. A particular caveat associated with historical treeline reconstructions based on megafossils is that any record represents a minimum elevation constraint for the contemporary treeline position. Another complication, according to conservative generalizations based on ring counts, is that large treeline pines may attain individual ages of 700-800 years, but usually somewhat less (Bartholin & Karlén 1983; Engelmark & Hofgaard 1985; Kullman 1994). This implies that radiocarbon-dating may yield widely different ages, depending on which part of a trunk that is dated. In most of the present cases, only the outhermost and youngest parts were available for dating. Thus, the obtained ages are likely to post-date the establishment phase by some hundred years.

Primarily, this study is based on analyses of radiocarbon-dated megafossil remnants of Scots pine (*Pinus sylvestris L.*), identified within a zone 100-115 m above the modern (AD 2013) treeline along the entire Swedish Scandes. Megafossils refer to remains of trunks of formerly tree-sized specimens, which are found very close to their growing sites. Specimens lying more or less exposed on dry ground were specifically sampled, since this category of remnants is experienced to yield dates within the time interval concerned in this study (Bartholin and Karlén 1983; Kullman 2013).

The upper treeline (both past and present) is a central aspect of this study and is defined as the extreme elevation of erect pine trees, more than 2 m in height. For further elaboration of this concept and its ecological correlates, see Kullman (2010b).

As a complement to the study of upper pine treelines, "inverted" treelines (cf. Holtmeier 2003) against treeless subalpine heaths in valley bottoms, were scrutinized for the presence of megafossils. These conspicuous elements of the mountain landscape are characterized by a dense pattern of cryogenic non-sorted circles and underlying permafrost. The nature and genesis of these treelines and their relations and responses to climate and climate change are debated (Josefsson 1988; Becher et al. 2013). It is hypothesized here that their late Holocene performance mirrors that of the upper treeline as climate-driven components of the cold-marginal landscape.

Eye-witness accounts of stand-level demise near and in a zone below the treeline, during the final phase of the Little Ice Age, i.e. 19th and early 20th century, was obtained from systematic survey of contemporary geobotanical literature and itineraries.

Radiocarbon dating was performed by Beta Analytic Inc., Miami (USA). Radiocarbon ages are calibrated to calendar years before present (cal yr BP), with "present" = 1950 AD. Calibration was conducted by use of the INTCAL09 database (Reimer et al. 2009). In the text, the calibrated ages are quoted as the intercept values of radiocarbon ages with the calibration curve. Dating was carried out after pretreatment with standard laboratory procedures, including acid/alkali/acid wash.

Species identification of the recovered megafossils was unambiguous, due to remaining bark fragments. Altitudes (m a.s.l.) and geographical coordinates for the sites of all retrieved tree remains were obtained by a GPS navigator (Garmin 60CS), repeatedly calibrated against distinct points on the topographical map. Reported altitudes are rounded off to the nearest 5 m.

4 Results

A total of 37 pine megafossils, meeting the prescribed criteria, were recovered within the focused elevation band. Of these, a subsample of 13 specimens was dated, embracing the entire extent of the Swedish Scandes from north to south (Fig. 1.). It is virtually unknown whether these tree remains represent widely scattered trees or more dense stands. In some districts, however, wood of this kind has been systematically gathered by local residents for fire wood (e.g. Bartholin and Karlén, 1983). This circumstance could argue for stands denser than indicated by the current sparse spacing of megafossils exposed on the ground surface.

Both Medieval and present-day treeline positions display a strong rising tendency, north to south, in the study region, as expected for a treeline, constrained by macroclimatic heat deficiency (Cogbill & White 1991; Kjällgren & Kullman 2002; Holtmeier 2003).

Figure 3: Examples of dated pine megafossils recovered above the current treeline. Numbers refer to Table1. A. Njulla (1). B. Dalfàlcorru (3), C. Njerubäuive (5). D. Tjallingklumpen (8), F. Täljstensvalen (6).

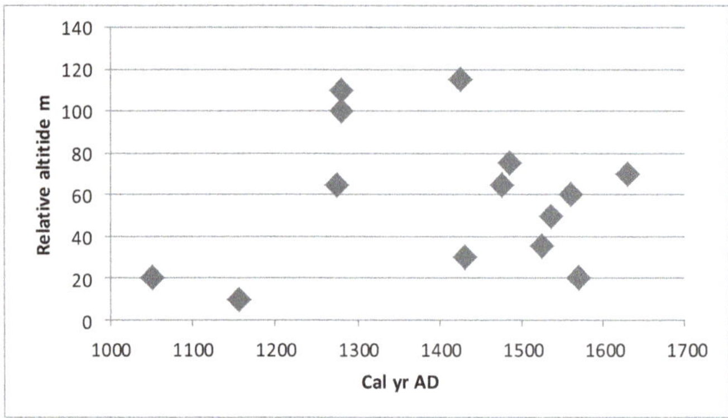

Figure 4: Radiocarbon-dates (intercept values) of all recovered pine megafossils and their relations to present-day altitude (m a.s.l.) of the pine treeline.

The obtained dates illustrate that higher-than-present treelines prevailed between AD 1050 and 1650, with an overweight in the record at the mid of this interval (Table 1, Figs. 3, 4). By the early 14th and early 15th century, the treeline peaked 100-120 m above the present position. Given a modal pine life-expectancy of 400-500 years (see above), it is reasonable to assume that the dated pines became established during the Medieval period, AD 1000-1300. This inference is supported by the heavily decayed nature of most specimens (Fig. 3).

Table 1: Radiocarbon dates of megafossil pine tree remains and their altitudes relative to the present-day (2013) treeline at the respective locality.

Age (^{14}C yr BP)	Cal yr AD 2δ	Cal yr AD Mid-point 2δ	Sample altitude (m a.s.l.)	Treeline 2013 (m a.s.l.)	Relative altitude (m)	Locality (no.)	Coordinates N lat.; E long.
270±30	1510-1750	1630	580	510	70	Njulla (1)	68 °21.829′; 18° 44.780′
710±30	1270-1290	1280	610	510	100	Slåttatjåkka (2)	68 °20.435′; 18° 42.971′
370±30	1450-1520	1485	585	510	75	Slåttatjåkka (2)	68° 20.209′; 18° 42.649′
480±30	1410-1440	1425	685	570	115	Darfálčorru (3)	67° 52.072′; 18° 38.147′
720±30	1260-1295	1280	680	570	110	Kebnetjåkka (4)	67° 52.041′; 18° 40.147′
320±30	1470-1650	1560	630	570	60	Kebnetjåkka (4)	67° 51.945′; 18° 37.057′
490±30	1410-1445	1430	720	690	30	Njereubäive (5)	65° 45.972′; 15° 20.790′
850±60	1030-1280	1155	770	760	10	Täljstensvalen (6)	63° 14.568′; 12° 27.564′
380±30	1440-1610	1525	870	835	35	Laptentjakke (7)	63° 08.351′; 12° 25.002′
410±30	1440-1510	1475	870	805	65	Tjallingklumpen (8)	63° 05.089′; 12° 28.158′
1000±70	900-1200	1050	780	760	20	N. Tväråklumpen (9)	63° 12.272′; 12° 22.151′
720±30	1260-1290	1275	1015	950	65	Ö. Barfredhågna (10)	62° 03.709′; 12° 24.447′
390±30	1440-1630	1535	1000	950	50	Ö. Barfredhågna (11)	62° 03.654′; 12° 24.277′
300±30	1490-1650	1570	1010	990	20	Djupgravstöten (12)	62° 03.187′; 12° 28.021′
830±30	1160-1260	1210	360	510	-150	Abisko-heath (13)	68° 21.527′; 18° 48.706′

Following this period, trees died consistently during subsequent centuries, particularly during the 14th century, and virtually without producing offspring, which could attain tree-size and preservation until the present. The overall outcome of this course was a general treeline lowering to a position, 2013 on average 60 m (range 10 - 115 m) lower than during the Medieval Period. Analogously, this study revealed a pine megafossil emerging by frost heave in the center of a non-sorted polygon, in a treeless subalpine heath (inverted treeline) well below the upper upper pine treeline (Fig. 5, Table 1). Radiocarbon-dating yielded an age of AD 1220 and a presumed establishment during the12th century or even earlier.

Figure 5: Subalpine heath with non-sorted circles close to Lake Torneträsk by Abisko in northern Lapland. Near the centre of the circle, s small megafossil pine remnant is exposed by frost heave. Radiocarbon-dating yielded AD 1220: Photo Leif Kullman, 14 September 2013.

5 Literature survey

Appendix I is a list of 33 published papers from northern Sweden and adjacent regions of Finland and Norway, which quite unanimously account for treeline and upper forest demise and elevational retraction during the final phase of the Little Ice Age and somewhat earlier. A preponderance of dead and dying trees, in combination with virtual no regeneration, is stressed in most of the reports. In many cases, a colder climate was explicitly assumed to be the primary reason.

Swedish forest authorities were well aware of the situation and feared an imminent and widespread alpine tundra expansion over high-lying forest regions, which prompted implementation of restrictive legislation concerning logging in marginal forests (Andersson 1906). An expression of the dire situation was that one of the first national parks in the Swedish Mountains, Sonfjället, was established (1909) with the outspoken purpose to create an arena for scientific study of ongoing treeline descent (Öberg 2009). Still today, large pine snags, reasonably representing Little Ice Age mortality and slow demise, following a more favourable climatic situation than present, are met with in the alpine tundra or upper sparse mountain birch forest of this national park and other districts (Fig. 6).

Fig. 6. Pine snag with more than 500 tree-rings and which obviously died during the Little Ice Age. Today, living pines of this dimension do not exist, neither at this elevation nor at the current treeline. Mt. Sonfjället, 875 m a.s.l. Photo: Leif Kullman, 10 June 2008.

6 Discussion

The relative small number of dated megafossils precludes any detailed discussion and regional differentiation of the obtained results.

Consistently for the entire range of the Swedish Scandes, the pine treeline, represented by the dated megafossils, was significantly higher than present during the Medieval period, although declining in altitude throughout the subsequent Little Ice Age. Over the latter epoch, the uppermost pine trees died commonly and the final of this protracted process could be witnessed about 100 years ago, as explicitly evident from the literature survey. Assuming that most of the dated pines had become established 200-300 years prior to the assigned death year (Table 1) (cf. Karlén 1996) makes it reasonable to assume that the majority of the demised trees had originated from the Medieval period around AD 1000-1300. Thus, the altitude (m a.s.l.) of the uppermost radiocarbon-dated pines is tentatively taken as a minimum elevation for the Medieval treeline. Presumably, pine trees may have prevailed at even higher elevations without leaving visible traces. The climatic favourability of this period is further indicated by concurrent pine growth in treeless environments (inverted treeline), 150 m below the upper pine treeline, currently characterized by non-sorted polygons and permafrost and totally devoid of pines and tree birches.

Obviously the Medieval epoch offered the most conducive climate for high-elevation pine tree growth and regeneration during the past 1000 years or so. Summer climate warming by c. 1.4 °C and common regional treeline rise since the early 20th century (Kullman & Öberg 2009) has not been enough to push the treeline the whole way everywhere back to its Medieval position (Kullman 2013, 2014).

Since the present-day treeline has been shifted upslope in near-equilibrium with 20th century climate warming along the Swedish Scandes (Kullman & Öberg 2009; Öberg & Kullman 2011), it is reasonable to assume that modern treelines are in a dynamic near-equilibrium with the prevailing climate. Thus, there is no substantial lag in the sense that the present-day treeline should be located far below its potential position. This responsiveness is a prerequisite for accurate summer temperature reconstructions concerning the difference between Medieval and modern summer temperatures.

As indicated above, the field-sampling procedure tends to yield minimum values of past treeline elevations. Moreover, the Medieval treeline position may be underestimated since relatively small and young trees representing short-term warming spikes have possibly not been preserved until the present. Therefore, and for the purpose of climate reconstruction, the largest recorded difference between Medieval and present treeline is used, i.e. 115 m. By drawing on a lapse rate of 0.6 °C per 100, a Medieval period, at least 0.7°C warmer than the past few decades may be computed. A similar magnitude of change is inferred from treeline and dendroecological studies at a site in the central Scandes (Linderholm et al. 2014).

Extensive tree mortality and poor regeneration in high mountain forests and treeline ecotones during the Little Ice Age, contrasting with a period of higher-than-present treeline and summer temperature around 1000 years ago, is inferred from megafossil case studies in different parts of northern Fennoscandia and northwestern Russia (Kullman 1987, 2005; Selsing 1991; Shiyatov 1993, 2003; Karlén & Kuylenstierna, 1996; Kremenetski et al. 2004; Hiller et al. 2001; Mazepa 2005; 2011; Kulti et al. 2006; Kharuk et al. 2013). Further support for the regional coherence of the view outlined above comes from dendroecological and multi-proxy studies carried out in the same region, including Greenland (Maasch et al. 2005; MacDonald et al. 2008; Shemesh et al. 2001; Grudd 2008; Lindholm et al. 2010; Vinter et al. 2010; Hanhijarvi et al. 2013; Rinne et al. 2014; Linderholm et al. 2014). Certain multi-proxy studies do recognize the Medieval Warm Period, but do not find it distinctly warmer than the past few decades (e.g. Moberg et al. 2005; Loehle 2007; Melvin et al. 2012; McCarrol et al. 2013; Briffa et al. 2013).

It may appear from Table 1 that treeline recession throughout the Little Ice Age was substantially larger in the northernmost part of the study area than in more southerly regions. Aside of regionally somewhat different climate histories, this discrepancy may relate to a more snow rich climate in the north, which has specifically counteracted pine treeline expansion in this region (Kullman & Öberg 2009; Kullman 2010).

Positional treeline change is an indicator of a fundamental reorganization of the entire ecosystem (Fagre et al. 2003; Kullman 2007; Behringer 2010). Thus, higher-than-present treelines and temperatures for some centuries around AD 1000 could imply a generally biologically richer and more productive subalpine /alpine landscape at that time, which was gradually pauperized and structurally eroded during the Little Ice Age. The modest warming of the past century may have initiated a reversal of this process (Kullman 2010a, 2012), but the structure and composition of the Medieval biological landscape has not been fully re-established, as evidenced in widely different parts of the northern hemisphere (Payette 2007; MacDonald 2010; Kullman 2014).

In perspective of the late Holocene, there is nothing particularly unnatural, unprecedented or alarming about the modern warming and associated ecological change in high mountain Fennoscandia and adjacent northern regions (cf. Kobashi et al. 2011; Kullman 2013).

7 Conclusions

Radiocarbon-dated megafossil remains of Scots pine (*Pinus sylvestris L.*) were systematically searched for and recovered on the ground surface well above the present-day treeline along the entire Swedish Scandes. These findings reveal that the treeline was positioned as much as 115 m higher than today during the apogee of the Medieval period, about AD 1000-1200.

Since treeline positions and their dynamics are mainly constrained by temperature deficiency, the obtained elevational difference of 115 m implies that summer temperatures, for purely natural reasons, were about 0.7 °C higher than present during the Medieval period. This inference is consistent with data from various parts of the northern hemisphere.

Climate warming over the past century has not yet been sufficient to restore the pine treeline to its Medieval level.

Given the obtained range of treeline change and inferred associated climate change over the past 1000 years, there is little to view the rate of extent of global warming as unnatural.

References

Akasofu, S.-I. 2013: On the present halting of global warming. *Climate 1*, 4-11.

Andersson, G. 1906: Förslaget till skyddsskogar inom Jämtlands län. *Skogsvårdsföreningens Tidskrift 3*, 57-84.

Bartholin, T. & Karlén, W. 1983: Dendrokronologii Lapland. *Dendrokronologiska Sällskapet Meddelanden 5*, 1-16.

Becher, M., Olid, C. & Klaminder, J. 2013: Buried soil organic inclusions in non-sorted circles fields in northern Sweden: Age and paleoclimatic context. *Journal of Geophysical Research: Biogeosciences 118° (11)*, 104-111.

Behringer, W. 2010: *A cultural history of climate*. 280 pp. Polity Press, London.

Bradley, R.S., Hughes, M.K. & Diaz, H.F. 2003: Climate in Medieval Time. *Science 302*, 404-405.

Briffa, K.R., Bartholin, T.S., Eckstein, D., Jones, P.D., Karlén, W., Schweingruber, F.H. & Zetterberg, P. 1990: A 1,400 year tree-ring record of summer temperatures in Fennoscandia. *Nature 346*, 434-439.

Briffa, K., Jones, P.D., Bartholin, T.S., Eckstein, D., Schweingruber, F.H., Karlén, W., Zetterberg, P. & Eronen, M. 1992: Fennoscandian summers from AD 500: temperature changes on short and long timescales. *Climate Dynamics 7*: 111-119.

Briffa, K.R., Melvin, T.M, Osborn, T.J., Hantemirov, R.M., Kirdyanov, A., Mazepa, V., Shiyatov, S.G. & Esper, J. 2013: Reassessing the evidence for tree-growth and inferred temperature change during the Common Era in Yamalia, northwest Siberia. *Quaternary Science Reviews 72*, 83-1007.

Broecker,W.S. 2001: Was the Medieval Warm Period global? *Science 291*, 1497-1499.

Crowley, T.J. 2000: Causes of climate change over the past 1000 years. *Science 289*, 270-277.

Crowley, T.J. & Lowery, T.S. 2000: How warm was the Medieval Warm Period? *Ambio 29*, 51-54.

Engelmark, O. & Hofgaard A. 1985: Sveriges äldsta tall. *Svensk Botanisk Tidskrift 79*, 415-416.

Esper, J. & Frank, D.C. 2009: IPCC on a heterogeneous Medieval Warm Period. *Climatic Change 94*, 267-273.

Esper, J., Frank, D.C., Tomonen, M., Zorita, E., Wilson, R.J.S., Luterbacher, J., Holzkämper, S., Fischer, N., Wagner, S., Nievergelt, D., Verstege, A. & Büntgen, U. 2012: Orbital forcing of tree-ring data. *Nature Climate Change 2*, 862-866.

Fagre, D.B., Peterson, D.L. & Hessl, A.E. 2003: Taking the pulse of mountains: ecosystem responses to climatic variability. *Climatic Change 59*, 263-282.

Grove, J.M., 1988: *The Little Ice Age*. 498 pp. Methuen, New York.

Grudd, H. 2008: Torneträsk tree-ring width and density AD 500-2004: A test of climatic sensitivity and a new 1500-year reconstruction of north Fennoscandian summers. *Climate Dynamics 31*, 843-857.

Gunnarson, B.E., Linderholm, H.W. & Moberg, A. 2011: Improving a tree-ring reconstruction from west-central Scandinavia: 900 years of warm-season temperatures. *Climate Dynamics 36*, 97-108.

Hanhijarvi, S., Tingley, M.P. & Korhola, A. 2013: Pairwise comparisons to reconstruct mean temperature in the Arctic Atlantic Region over the last 2,000 years. *Climate Dynamics 41*, 2039-2060.

Harsch, M., Hulme, P., McGlove, M. & Duncan, R. 2009: Are tree lines advancing? A global meta-analysis of tree line response to climate warming. *Ecology Letters 12*, 10140-1049

Hiller, A., Boettger, T. & Kremenetski, K. 2001: Mediaeval climate warming recorded by radiocarbon dated alpine tree-line shift on the Kola Peninsula, Russia. *The Holocene 11*: 491-497.

Holtmeier, F.-K. 2003: *Mountain timberlines: ecology, patchiness, and dynamics*. 437 pp. Kluwer Academic Publishers, Dordrecht.

Hughes, M.K. & Diaz, H.F. 1994: Was there a "Medieval Warm Period", and if so, where and when? *Climatic Change 26*, 109-142.

Humlum, O., Solheim, J.-E. & Stordahl, K. 2011: Identifying natural contributions to late Holocene climate change. *Global and Planetary Change 79*, 145-156.

IPCC 2007: *Climate Change 2007. The Physical Science Basis*. 996 pp. Cambridge University Press, Cambridge.

Josefsson, M. 1990: Ground temperature variations in a subarctic mountain valley, Abisko, northern Sweden. *Geografiska Annaler 72A*, 172-190.

Karlén, W. & Kuylenstierna, J. 1996: On solar forcing of Holocene climate: evidence from Scandinavia. *The Holocene 6*, 359-365.

Kaufmann, D.S., Schneider, D.P., McKay, N.P. & 8 others 2009: Recent warming reverses long-term Arctic cooling. *Science 325*, 1236-1239.

Kharuk, V.I., Ranson, K.J., Im, S.T., Oskorbin, P.A., Dvinskaya, M.L. & Ovchinnikov, D.V. 2013: Tree-line structure and dynamics at the northern limit of larch forest: Anabar Plateau, Siberia, Russia. *Arctic, Antarctic, and Alpine Research 45*, 526-537.

Kirdyanov, A.V., Hagedorn, F., Knorre, A., Fedotova, E.V., Vaganov, E.A., Naurzbaev, M.M., Moiseev, P.A. & Rigling, A. 2012: 20th century tree-line advance and vegetation changes along an altitudinal transect in the Putorana Mountains, northern Siberia. *Boreas 41*, 56-67.

Kjällgren, L. & Kullman, L. 2002: Geographical patterns of tree-limits of Norway spruce and Scots pine in the southern Swedish Scandes. *Norwegian Journal of Geography 56*, 237-245.

Kobashi, T., Kawamura, K., Severighaus, J.P., Barnola, J.-M., Nakaegawa, T., Vinther, B.M., Johnsen, S.J. & Box, J.E., 2011: High variability of Greenland surface temperatures over the past 4000 year estimated from trapped air in an ice core. *Geophysical Research Letters 38*, doi 10.1029/2011GL049444.

Körner, C. & Paulsen, J. 2004: A world-wide study of high altitude treeline temperatures. *Journal of Biogeography 31*, 713-732.

Kremenetski, K.V., Boettger, T., MacDonald, G.M., Vaschalova, T., Sulerzhitsky, L. & Hiller, A. 2004: Medieval climate warming and aridity as indicated by multiproxy evidence from the Kola Peninsula, Russia. *Palaeogeography, Palaeoclimatology, Palaeoecology 209*, 113-125.

Kullman, L. 1981. Recent tree-limit dynamics of Scots pine (Pinus sylvestris L.) in the southern Swedish Scandes. *Wahlenbergia 8*, 1-67.

Kullman, L. 1987: Little Ice Age decline of a cold marginal Pinus sylvestris forest in the Swedish Scandes. *New Phytologist 106*, 567-584.

Kullman, L. 2005: Pine (*Pinus sylvestris*) treeline dynamics during the past millennium – a population study in west-central Sweden. *Annales Botanici Fennici 42*, 95-106.

Kullman, L. 2007: Long-term geobotanical observations of climate change impacts in the Scandes of west-central Sweden. *Nordic Journal of Botany 24*, 445-467.

Kullman, L. 2010a: A richer, greener and smaller alpine world – review and projection of warming-induced plant cover change in the Swedish Scandes. *Ambio 39*, 159-169.

Kullman, L. 2010b: One century of treeline change and stability - experiences from the Swedish Scandes. *Landscape Online 17*, 1-31.

Kullman, L. 2012: The alpine treeline ecotone in the southernmost Swedish Scandes. *In* Myster, R.W. (ed.), *Ecotones between forest and grassland. New York*: Springer, 271-298.

Kullman, L. 2013: Ecological tree line history and palaeoclimate – review of megafossil evidence from the Swedish Scandes. *Boreas 42*, 555-567.

Kullman, L. 2014. Treeline (*Pinus sylvestris*) landscape evolution in the Swedish Scandes – A forty-year demographic effort viewed in a broader temporal context. *Norwegian Journal of Geography 68*, 155-167.

Kullman, L. & Öberg, L. 2009: Post-Little Ice Age tree line rise and climate warming in the Swedish Scandes – a landscape-scale perspective. *Journal of Ecology 97*, 415-429.

Kulti, S., Mikkola, K., Virtanen, T., Timonen, M. & Eronen, M. 2006: Past changes in the Scots pine forest line and climate in Finnish Lapland: a study based on megafossils, lake sediments, and GIS-based vegetation and climate data. *The Holocene 16*, 381-391.

Laaksonen, K. 1976: The dependence of mean air temperatures upon latitude and altitude in Fennoscandia. *Annales Academiae Scientiarum Fennicae A3 199*, 1-19.

Lamb, H.H., 1982: *Climate history and the modern world.* , London: Methuen, 387 pp.

Linderholm, H.W., Zhang, P., Gunnarson, B.E., Björklund, J., Farahat, E. , Fuentes, M., Rocha, E., Salo, R., Settigen, K., Stridbeck, P. & Liu, Y. 2014: Growth dynamics of treeline and lake-shore Scots pine (*Pinus sylvestris L.*) in the central Scandinavian Mountains during the Medieval climatic Anomaly and the early Little Ice Age. *Frontiers in Ecology and Evolution 22*, 1-11.

Lindholm, M., Jalkanen, R., Salminen, H., Aalto, T. & Oguitsov, M. 2010: The height-increment record of summer temperature extended over the last millennium in Fennoscandia. *The Holocene 21*, 319-326.

Ljungqvist, F.C., Krusic, P.J., Brattström, G. & Sundqvist, H.S. 2012: Northern Hemisphere temperature patterns in the last 12 centuries. *Climate of the Past 8*, 227-249.

Loehle, C. 2007: A 2000-year global temperature reconstruction based on non-treering proxies. *Energy & Environment 18*, 1049-1058.

Lloyd, A.H. 2005: Ecological histories from Alaskan tree lines provide insight into future change. *Ecology 86*, 1687-1695.

Luetscher, M., Borreguero, M., Moseley, G.E., Spötl, C. & Edwards, R.L., 2013: Alpine permafrost thawing during the Medieval Warm period identified from cryogenic cave carbonates. *Cryosphere 7*, 1073-1081.

Maasch, K.A., Mayewski, P.A., Rohling, E.J., Stager, J.C., Karlén, W., Meeker, L.D. & Meyerson. B.A. 2005: A 2000-year context for modern climate change. *Geografiska Annaler 87A*, 7-15.

MacDonald, G., Kremenetski, K.V. & Beilman, D.W. 2008: Climate change and the northern Russian treeline zone. *Philosophical Transactions of the Royal Society B 363*, 2285-2299.

MacDonald, G.M. 2010: Some Holocene palaeoclimatic and palaeoenvironmental perspectives on Arctic and Subarctic climate warming and the IPCC 4th assessment report. *Journal of Quaternary Science 25*, 39-47.

Mazepa, V.S. 2005: Stand density in the last millennium at the upper tree-line ecotone in the Polar Ural Mountains. *Canadian Journal of Forest Research 35*, 2082-2091.

Mazepa, V., Shiyatov, S., & Devi, N., 2011: Climate-driven change of stand age structure in the Polar Ural Mountains. In Bianco, J. (ed.): *Geophysical foundations and ecological effects*, 377-402. InTech, Rijeke.

McCarrol, D., Loader, N.J., Jalkanen, R. & 14 others 2013: A 1200-year multiproxy record of tree growth and summer temperature at the northern pine forest limit of Europe. *The Holocen 23*, 471-482.

Matthews, J.A. & Briffa, K.R. 2005: The Little Ice Age: Re-evaluation of an evolving concept. *Geografiska Annaler 87A*, 17-36.

Melvin, T.M., Grudd, H. & Briffa, K.R. 2012: Potential bias in "updating" tree ring chronologies using regional curve standardisation: Re-processing 1500 years of Torneträsk density and ring-width data. *The Holocene 23*, 364-373.

Moberg, A., Sonechkin, D.M., Holmgren, K., Datsenko, N.M. & Karlén, W. 2005: Highly variable Northern hemisphere temperatures reconstructed from low- and high-resolution proxy data. *Nature 433*, 613-617.

Moiseev, P. A. & Shiyatov, S. G., 2003: Vegetation dynamics at the treeline ecotone in the Ural highlands, Russia. In Nagy, L., Grabherr, G., Körner, C. & Thompson, D.B.A. (eds.), *Alpine biodiversity in Europe - A Europe-wide assessment of biological richness and change*, 423-435. Springer, Berlin.

Payette, S. 2007: Contrasted dynamics of northern Labrador tree lines caused by climate change and migrational lag. *Ecology 88*, 770-780.

Öberg, L., 2009: The heart of Härjedalen. Sonfjället. National park since 1909. *Jamtli Förlag*, Östersund, 168 pp.

Öberg, L. & Kullman, L. 2011: Recent glacier recession – a new source of postglacial treeline and climate history in the Swedish Scandes. *Landscape Online 26*, 1-38.

Reimer, P.J. & 27 others, 2009: INTCAL09 and MARINE09 radiocarbon age calibration curves, 0-50,000 years cal BP. *Radiocarbon 51*, 1111-1150.

Rinne, J., Alestalo, M. & Miettinen, A. 2014: A 70-80 year periodicity identified from tree ring temperatures AD 550-1980 in Northern Scandinavia. *Global and Planetary Change 116*, 149-155.

Selsing, L., Foldøy, O., Løken, T., Pedersen, E. & Wishman, E. 1991: A preliminary history of the Little Ice Age in a mountain area in SW Norway. *Norsk Geologisk Tidsskrift 71*, 223-228.

Shemesh, A., Rosqvist, G., Rietti-Shati, M., Rubensdotter, L., Bigler, C., Yam, R. & Karlén, W. 2001: Holocene climatic change in Swedish Lapland inferred from an oxygen-isotope record of lacustrine biogenic silica. *The Holocene 11*, 447-454.

Shiyatov, S.G. 1993: The upper timberline dynamics during the last 1100 years in the Polar Ural Mountains. *Paläoklimaforschung 9*, 195-203.

Shiyatov, S.G. 2003: Rates of change in the upper treeline ecotone in the Polar Ural Mountains. *Pages News 11*, 6-8.

Tinner, W. & Kaltenrieder, P. 2005: Rapid responses of high-mountain vegetation to early Holocene environmental changes in the Swiss Alps. *Journal of Ecology 93*, 936-947.

Tollefsson, J. 2014: The case of the missing heat. *Nature 505*, 276-278.

Tranquillini, W. 1979: *Physiological ecology of the alpine timberline*. 137 pp. Springer-Verlag, Berlin.

Vinther, B.M., Jones, P.D., Briffa, K.R., Clausen, H.B., Andersen, K.K., Dahl-Jensen, D. & Johnsen, S.J. 2010: Climatic signals in multiple highly resolved stable isotope records from Greenland. *Quaternary Science Reviews 29*, 522-538.

Zwiers, F. & Hegerl. G. 2008: Attributing cause and effect. *Nature 453*, 296-297.

Appendix I

Papers accounting for eye-witness reports of treeline recession and mountain forest decline during the final phase Little Ice Age in Fennoscandian mountain regions.

Aario, L. 1941: Polardämon und Klimaverbesserung. *Terra 53(4)*, 57-69.

Andersson, G. 1906: Förslaget till skyddsskogar inom Jämtlands län. *Skogsvårdsföreningens Tidskrift 3*, 57-84.

Ehrenheim, F. 1824: Om climaternes rörlighet. Tal hållet vid Praesidii nedläggande uti Kongl. Vetenskaps Academien. Stockholm.

Engström, J. 1835: Resa genom södra Lappland, Jemtland och Dalarne. Stockholm.

Forsslund, K.-E. 1919: *Med Dalälven från källorna till havet. Bok I. Storån*. Åhlen & Åkerlunds Förlag, Stockholm.

Forsslund, K.-E. 1921: Omkring Storvaln. Sveriges sydligaste fjäll. *Sveriges Natu*r 1921, 33-42.

Gavelin, A. 1909: Om trädgränsernas nedgång i de svenska fjälltrakterna. *Skogsvårdsföreningens Tidskrift 6*, 133-151

Gavelin, A. 1910: Trädgränsförskjutningarna inom Kamajokks vattenområde. *Sveriges Geologiska Undersökning Serie C*, 227: 1-32.

Hagem, O. 1917: Furuens og granens frøsaetning. *Meddelelse fra Vestlandets Forstlige Forsøksstation 2*, 1-188.

Hagemann, A. 1890: Altenskoven. Den Norske Forstforenings Aarbog for 1890: 1-74.

Hägerström, K.P., 1882: Bidrag till Torne lappmarks och Ofotens flora. *Botaniska Notiser 1882*, 65-96.

Holmerz, C.G. & Örtenblad, T. 1886: Om Norrbottens skogar. *Tidskrift för Skogshushållning 14*, 193-212.

Holmerz, C.G. & Örtenblad, T. 1887: Om Jemtlands lappskattefjäll. *Tidskrift för Skogshushållning 15*, 115-180.

Holmgren, A. 1904: Skogsbiologiska studier inom Arjepluogs och Jokkmokks lappmarker. *Sveriges Skogsvårdsförbunds Tidskrift 1904*, 1-22.

Kellgren, A.G. 1891: Om de skogsbildande trädens utbredning i Dalarnes fjälltrakter. *Botaniska Notiser 1891*, 182-186.

Kellgren, A.G. 1893: Några observationer öfver trädgränserna i våra sydliga fjälltrakter. *Öfversigt af Kongl. Vetenskaps-Akademiens Förhandlingar 1893*, 249-262.

Kellgren, A.G. 1895: Flyttar barrskogsgränsen nedåt? *Landtmannen 6*, 315-316.

Kranck, H. 1909: Tallformationerna i Finland. *Meddelanden från Geografiska Föreningen i Finland 8*, 1-59.

Marelius, N. 1771: Fortsättning om land- och Fjäll-ryggarne samt Gränsen mellan Sverige och Norrige. *Kongliga Vetenskapskapsacademiens Handlingar för år 1771 31*, 11-52.

Näsström, G. 1925: Glimtar från 1740-talets Jämtland och Härjedalen. *Fornvårdaren 1*, 88-112.

Nilsson, A. 1897: Om Norrbottens växtlighet med särskild hänsyn till dess skogar. *Tidskrift för Skogshushållning 25*, 139-153.

Nilsson, A. & Norling, K.G. 1895: Skogsundersökningar i Norrland och Dalarne sommaren 1894. Bihang till Kungliga Domänstyrelsens Underdåninga Berättelse. Stockholm.

Nissen, K. 1945: Et brev från Lars Levi Laestadius om vegetasjonens tilbakegang på grunn av klimaforverring. *Blyttia 3*, 38-42.

Nordfors, G.A. 1928: Fjällskogars och exponserade skogars föryngringsmöjligheter med särskild hänsyn till det producerade fröets grobarhet under extrema klimatförhållanden. *Norrlands Skogsvårdsförbunds Tidskrift 1928*, 105-165, 222-286.

Nordhagen, R. 1923. Litt om den utdøende furuskog i Troldheimen. *Tidsskrift for Skogbruk 31*, 25-29

Olafsen, O. 1911: Skogen i Norge fra Oldtiden til begyndelsen af det 16de Aarhundrede. *Tidsskrift for Skogbruk 19*, 11-19.

Rabot, C. 1911: La recul du pin sylvestre dans les montagnes de la Suède. *La Géographie 23*, 270-276.

Renvall, A. 1912: Om orsakerna till depressionen av tallens skogsgräns. *Finska Forstföreningens Meddelanden 29*, 233-241.

Rekstad, J. 1903: On the greater height formerly of the forest line and snow line in Norway. *Norges Geologiske Undersökelses Aarbog 33/36*, 14-18.

Sernander, R. 1899: Studier öfver vegetationen i mellersta Skandinaviens fjälltrakter. *Bihang till Kungliga Svenska Vetenskaps-Akademiens Handlingar 24 Afd III, No. 11*, 1-56.

Vesterlund, O. 1900: En liten tallstudie på Rotivare. *Tidskrift för Skogshushållning 1900(4)*, 215-218.

Wibeck, E. 1929: Kan den nuvarande barrskogsgränsen förskjutas uppåt? *Skogen 14*, 401-404.

Zetterstedt, J.W. 1833: Resa genom Umeå Lappmarker i Västerbottens län. Örebro.

The appreciation of nature and landscape by tourism service providers and visitors in the Ore Mountains (Germany)

Olaf Bastian[1]*, Christian Stein[2], Gerd Lupp[3], Jan Behrens[4], Christina Renner[5], Karsten Grunewald[2]

1 State Capital of Dresden, Environmental Authority, P.O. Box 12 00 20, D-01001 Dresden, Germany

2 Leibniz Institute of Ecological Urban and Regional Development, Weberplatz 1, D-01217 Dresden, Germany

3 Chair for Strategic Landscape Planning and Management, TU Munich, Emil-Ramann-Str. 6, 85354 Freising, Germany

4 Technische Universität Dresden, Chair of Tourism Economics and Management, D-01062 Dresden, Germany

5 Alaunstr. 90, D-01099 Dresden, Germany

Abstract

The paper presents empirical studies on the appreciation of nature and landscape in the Eastern Ore Mountains (Saxony, Germany) by tourism service providers (TSP) and visitors. Attractive landscape and experience of nature are the most important reasons to visit this region and to spend leisure time there. Particularly mountain meadows, raised bogs and mixed forests are highly appreciated. Deforestation, industrial development and the decline of biodiversity would reduce attractiveness for visitors. We also assessed whether the tourism sector is prepared to contribute to the funding of nature conservation and landscape management. Use of general tax revenues is favoured, but other modes would also be accepted, e.g. a nature tax. Willingness to pay (WTP) is ranging between €0.75 and €1.36 per guest per night by TSP, or between €1.06 and €2.73 per day by visitors. With respect to landscape preference and WTP we found in some cases significant differences among visitors, depending on region of residence, age and education level. A major part of the annual costs for nature conservation and landscape could be covered by public funds (taxes), if the results of the WTP approach were understood as a sign of societal demand and a call to action.

Keywords:
Landscape attractiveness; landscape management; nature tax; stated preference valuation; ecosystem services; willingness to pay

*Corresponding author. Email: Olaf.Bastian@web.de

1 Introduction

The concept of ecosystem services comprises both the supply of services, which is based on the structures, processes and potentials of ecosystems and landscapes, and the demand for these services by particular beneficiaries, stakeholder groups or society as a whole. To date, the ranking of services by beneficiaries has received only limited attention (Lamarque et al. 2011; Seppelt et al. 2012). Hence, there is a need for additional research on the demand for and supply of ecosystem services, rather than for a focus on supply alone (Termorshuizen & Opdam 2009; García-Nieto et al. 2013). Also Honey-Rosés & Pendleton (2013) note that "better information about ecological processes or abstract valuations will not spur better decision-making", and that "despite a large and growing body of studies on the supply of ecosystem services, ecosystem services thinking and information is still not being used by the decision makers who need it."

While there is a general broad social consensus for the permanent preservation and development of valuable landscapes with their ecosystems, and numerous studies on willingness to provide the financial means to do this exist (e.g. Spangenberg & Settele 2010; Grunewald et al. 2014), it is difficult to achieve consensus on the specific development direction in an area (Lupp et al. 2014) – e.g. in a tourist region – and the preferred or the most favorable management options. Demand driven analyses of the public may support politicians and planners to design management policies and to set priorities in regard where to invest public money to maintain landscapes and protect biodiversity and gain acceptance (e.g. Casado-Arznaga et al. 2013).

Nature conservation, especially the maintenance of human-influenced, biodiversity-rich, so-called semi-natural ecosystems, e.g. the flowering mountain meadows which may shape the character of many cultural landscapes or tourist regions attracting especially nature-oriented tourists need considerable financial means. Such analyses should be also seen against the background of the further development of existing and the establishment of new funding models involving tourism to mobilize financial sources for landscape management (Heuchele et al. 2014).

The social expenditures and costs for landscape management society is ready to bear, represent an indicator of the economic valuation of ecosystems, since the existence of some cultural landscape ecosystems is closely linked to permanent human intervention in the form of specific land use or management measures (Farber et al. 2006; Grunewald et al. 2014).

Nature and natural landscapes are key elements of numerous recreation and tourism forms. They range from active nature experiences (e.g. wildlife watching) to more hedonistic activities with natural features as backdrop (e.g. skiing, hunting or fishing, combined culture/nature trips) (Strasdas 2006; Arnegger et al. 2010; Coghlan & Buckley 2013). Nature-based tourism is "marked by a desire to spend time in places and spaces (...) that provide meaningful, transformative, spontaneous and extraordinary experiences" (Curtin 2013). Ecotourism can be seen as a special subset of nature(-based) tourism conserving the environment and improving the well-being of local people (TIES 1990), and thus as a special manifestation of sustainable tourism.

A nationwide representative survey of nature awareness in Germany reveals that a vast majority of people places high value on nature in terms of quality of life, health and recreation (BMU & BfN 2014). For more than half of all Germans, "experiencing nature" is a key motivation for holidays (FUR 2013). Domestic tourists in Germany regard beautiful nature and a good price-performance ratio as most important aspects for their personal happiness during their holidays; this is particularly distinctive in the age group of 65 years or more (SfZ 2014). Three-quarters of German holidaymakers feel psychologically deterred by obstructed landscapes that have lost their pristine character (Kösterke & von Laßberg 2005). Therefore, opportunities to experience nature often represent decisive factors within the selection process of holiday destinations (Kösterke & von Laßberg 2005; Pröbstl-Haider & Haider 2014).

Due to their attractive landscapes and wildlife as well as their cultural attractions mountainous regions around the world are important destinations for recreation and tourism (Fredman & Lindberg 2008; Beza 2010; Pickering & Barros 2013). In Germany, the Alps and several low mountain ranges offer various opportunities for outdoor activities like hiking, trekking, mountain biking or skiing (BMWi 2013).

This paper presents questionnaire-based empirical studies on the appreciation of nature, semi-natural and natural ecosystems, landscape elements, landscape qualities and ecosystem services in the Eastern Ore Mountains in the German state of Saxony, by different stakeholder groups: tourism service providers (TSP) and visitors (tourists with overnight stays and same-day visitors). Also the role of landscape changes on the visitor behaviour is addressed. Moreover, it examines the question as to whether these groups are prepared to participate in the costs for nature conservation and landscape management in the region, and which funding models for nature conservation are favoured. It is also investigated whether appreciation of nature and the landscape, and the willingness to pay (WTP), correlate with socio-demographic data (e.g. place of residence, age, level of education, lifestyle).

2 Methods

2.1. General remarks

The supply side of ecosystem services (ecosystem properties and potentials – Bastian et al. 2012) can be analyzed with classical (landscape) ecological methods. It has been carried out already for a wide spectrum of provisioning, regulation and socio-cultural services in the Ore Mountains on the example of Natura 2000 sites (Bastian et al. 2010; Bastian 2013) and is not in the focus here. For the assessment of the demand side, we designed standardized questionnaires, one for TSP (see chapter 2.3) and one for visitors (outdoor recreationists) (see chapter 2.4).

The study deploys the contingent valuation method in order to assess the willingness to pay of TSP and visitors for conservation purposes. Stated preference analyses like the contingent valuation method or choice experiments do not only reveal the amount that people would be prepared to pay, but also the conditions or developments in the environment which they desire, or want to avoid. On the basis of the results of such assessments, it is then possible to determine the actions and the limits required to avoid deteriorations (Grunewald et al. 2012).

Thus, the contingent valuation method is frequently applied to determine the appreciation of visitors for particular qualities of nature and landscape (Lee & Han 2002; Tisdell 2006), referring to recreational and other non-material benefits of ecosystem services like aesthetic pleasure, spiritual values or educational purposes (Butler & Oluoch-Kosura 2006). This method can also be used to estimate non-use values in monetary terms like existence values of natural attributes (for example wildlife species) or bequest values, reflecting individual attitudes towards conservation efforts for future generations (Tisdell 2006; Mayer & Job 2014).

Moreover, numerous scientific studies employ the contingent valuation method and other stated preference techniques to assess WTP with regard to user fees in relation to tourism and outdoor recreation activities in protected areas or other (non-protected) public natural areas (Mmopelwa et al. 2007; Reynisdottir et al. 2008; Rosenberger et al. 2012; Wang & Jia 2012; Mayer 2014). Concerning the legitimacy of imposing entrance and other user fees on nature tourists a "user pays" principle aiming at contributions of tourists applies. This opposes a "public good" view that advocates public funding of nature conservation by general taxes (Reynisdottir et al. 2008). An alternative tourism-oriented approach is the imposition of special taxes dedicated to the management and conservation of natural areas (e.g. room tax, "nature tax") (Laarman & Gregersen 1996; Degenhardt et al. 1998; Oom do Valle et al. 2012; Renner et al. 2012).

In recent years, a growing number of contingent studies have been undertaken world-wide, e.g. Degenhardt et al. (1998); Elsasser et al. (2009);

Meyerhoff et al. (2010); Tacconi (2012). Barrio & Loureiro (2010) presented a comprehensive review of contingent studies, and the pros and cons of such methods was discussed by Chee (2004) but also by Riera et al. (2012) who also developed a set of good practice guidelines for the non-market valuation (of forests), elaborating on stated and revealed preference methodologies.

2.2.The study area

The study area (Fig. 1) is situated in the German part of the Eastern Ore Mountains (the district Sächsische Schweiz-Osterzgebirge, in southern Saxony), in the rural municipalities of Altenberg (population: 8500), Hermsdorf (900), and Bad Gottleuba-Berggießhübel (6000). The average annual income in the district in 2012 was € 18,332 per person (DESTATIS 2014).

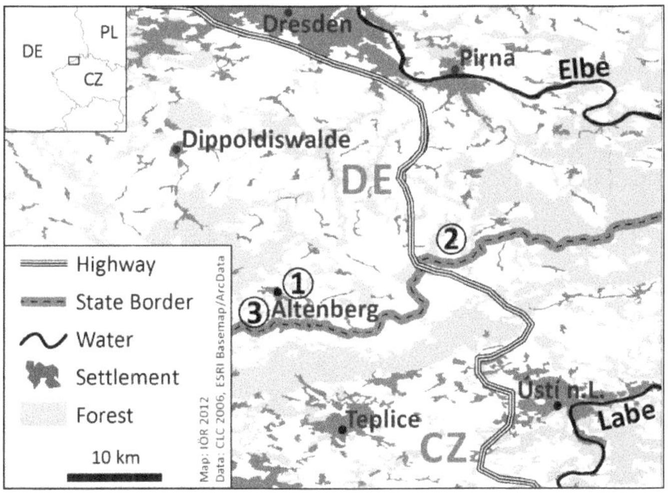

Figure 1: The Eastern Ore Mountains with the study sites *Geisingberg* meadows (1), *Oelsen* meadows (2) and the *Georgenfeld* raised bog (3)

The Ore Mountains (German *Erzgebirge*), a traditional cultural landscape of European significance particularly marked by ore mining, have the shape of a slanted desk of approx. 150 km in length, formed by tectonic forces. Acid rocks such as gneiss, phyllite and granite are typical for this mountain range. The harsh climate has given rise to such characteristic ecosystems as raised bogs and mountain meadows, with such anthropogenic features as stone walls, cairns and hedgerows (Fig. 2 and 3). Often, mixed mountain forests and streams are still in a state close to nature. The Ore Mountains ridge, which rises an average of 800–1000 m above sea level, also forms the border between Germany and the Czech Republic. Particularly this ridge zone features many valuable natural areas, and thus has a high attractiveness for nature-related tourism (Bastian et al. 2010; Bastian 2013). The land use pattern consists of forests (especially on the ridge), grassland, arable fields and settlements (villages, small towns).

2.3. Survey of tourism service providers (TSP)

In a first step, we analyzed TSP in the study area. Data banks provided by the communities like tourism catalogues were used to identify TSP. For the study, all TSP having e-mail addresses were selected. One disadvantage of this method was, that only those TSP being listed in an accommodation directory or with an own website and having e-mail were within the survey group. Thus, especially small size TSP and small private overnight stay providers like holiday apartments are probably underrepresented in this sample. The perception of the TSP in the study area was examined using an online survey conducted between April and May 2012. We first established an address database with the aid of an internet search in accommodation directories.

The survey was performed with the online questionnaire *"oFb-der onlineFragebogen"* on www.soscisurvey.de. The questionnaire was available only via a serial number, which was sent together with an e-mail to all the addresses in the database. This procedure ensured that the participants replied to the questionnaire only once, and that only the addressed target group was able to participate. The questionnaire was sent by e-mail to 223 recipients. After two weeks' time, we sent out a reminder to all recipients who had not yet returned the questionnaire.

2.4. Survey of visitors

For the survey of outdoor recreationists (see Blamey 2001; in the following we prefer the shorter term "visitor") in the area, we used similar questions (Table 1), slightly adapted for this target group. Here, we used on-site interviews with passers-by at well attended natural sites in typical ecosystems/nature reserves for the Ore Mountains: mountain meadows, which were in bloom during the survey period, at the mountain *Geisingberg* (Fig. 2), close to the town of Altenberg, easily accessible by public transportation; near the village of Oelsen, a remote scenic outlook accessible only by car; and the *Georgenfeld* raised bog, the only nature reserve in the Ore Mountains, where visitors have to pay an entrance fee to walk along the nature trail (Fig. 3).

Before the final survey we had conducted pre-tests to adjust the questionnaire to ensure that it is well understood and correctly worded, and that (only) the

Figure 2: Typical landscape of the Eastern Ore Mountains near the Geisingberg mountain with mountain meadows and stone walls (photo O. Bastian)

Figure 3: The *Georgenfeld* raised bog – view from the nature trail (photo O. Bastian)

most important questions are included. The survey was performed between the end of May and the end of June 2012, mainly on weekends and during the Whitsun holiday, so as to reach many visitors. Questioning started at 10:00 a.m. and ended around 5:00 p.m. Passers-by were approached at random and asked to participate in the survey. We expected this approach would provide us with more answers and more reliable information than by sending letters or displaying questionnaires at hotels or other accommodation places.

2.4. Structure of the questionnaires

The questionnaires comprised between 13 and 20 questions: The introductory part asked for the reason and the character of the visit. The two main parts dealt with the appreciation of nature and the landscape (elements), and with the importance and funding of nature conservation and landscape management (Table 1). Finally, socio-demographic data was collected (only in the questionnaires for visitors). The questionnaires were designed not to require more than 15 to 20 minutes to be answered. We felt it wise to ask partly different questions to tourism service providers (TSP) and visitors. In general, the first group is more competent in tourism as a whole, while individual visitors express their opinions from other points of view.

We applied different forms of questions: mainly closed questions, some semi-open, and a few open ended questions (e.g. Degenhardt et al. 1998, Atteslander 2010). In some cases, we asked persons to rate items on a Likert scale from 1 to 5 (e.g., 1 = not important/not attractive ... 5 = very important/ very attractive). Multiple answers were allowed in most cases.

Table 1: Main content of questionnaires for the survey of tourism service providers (TSP) and visitors in the Eastern Ore Mountains

Question	Survey group	
	TSP	**Visitors**
What are the main motivations for visiting the Eastern Ore Mountains?	x	x
What are the favoured leisure activities exercised here?	x	
Which landscape elements are found especially attractive?	x	x
In which manner landscape changes influence visiting behaviour?	x	x
Which importance is attached to particular ecosystem services?		x
What is the role of protected areas for the tourist attractiveness of the region?	x	
Do you know any protected area in the region and the Natura 2000 system?	x	
Do you think landscape management is necessary, and if yes, for what reasons?	x	x
Who should be responsible for funding nature conservation (and landscape management)?	x	x
Do you think that tourism should contribute to nature conservation by financial means?	x	
Would you be ready to support nature conservation by own financial means?	x	x
What is the height of payments you would accept?	x	x
Do you think that the introduction of a nature tax in the Eastern Ore Mountains is justified and feasible?	x	x
Who should receive these payments?	x	

Questions concerning the monetary value (WTP) we formulated differently. On the one hand, we asked for the maximum amount the respondent is prepared to pay (open question). On the other, we wanted to know whether the respondent was prepared to pay a predefined amount of money (closed question: agreement or disagreement).

The tourism sector profits from the management and restoration of habitats and ecosystems, because landscape attractiveness is enhanced. Therefore, we inquired whether and to which extent the tourism sector would be prepared and able to contribute to the funding of such management, and whether the costs of nature conservation and landscape management are justified from the point of view of tourism. We presented a filter question to the TSP concerning their willingness to provide financial contributions. If they answered "no", only one additional question followed regarding the reasons of the denial, and the survey was concluded. In order to estimate the amount of a possible financial contribution of the tourism sector or to quantify the WTP, the TSP were asked to measure the amount of taxes per guests and night (in the form of an open question).

To detect possible discrepancies between different users, sub-groups were defined and compared using statistical tests (χ^2-, t-test): Residents of the district vs. external visitors; older (> 40) vs. younger persons (up to 40), and persons with higher education (university graduates) vs. those with lower education (secondary school, vocational training, etc). The latter two differentiations also enable application of a simple lifestyle group concept, according to the German sociologist Schulze (1997). Recent studies indicate that value orientation (Müller & Job 2009; Lupp, et al. 2011) or lifestyles strongly affect opinions and attitudes towards the environment, and seem to provide a better differentiation and understanding than do such purely socio-demographic characteristics as age or gender (UBA, 2009).

3 Results

3.1. Level and profiles of the respondents

A total of 72 recipients among the tourism service providers (TSP) completed the questionnaire; the response rate was 32%. This response rate is in line with the average response rate for questionnaires at the SoSci Panel (20%) and other internet-based surveys (Cook et al. 2000). Of the TSP surveyed, 74% owned holiday accommodations; the others worked in restaurants, trade and such other tourism-related services as retailing, tour operating, or cultural facilities.

Table 2: Selected characteristics of surveyed visitors in the Eastern Ore Mountains

Gender	Male:	145	**Duration of stay**	Day trip	252
	Female:	157		Overnight stay	56
Age	<18:	1	**Origin**	Altenberg:	17
	18-30:	18		District of Sächsische Schweiz-Osterzgebirge:	75
	31-40:	28		Saxony:	179
	41-50:	72		Germany:	29
	51-60:	59		Czech Republic:	2
	>60:	125			
Household income	≤1000€:	47	**Education**	No qualification:	0
	≤2000€:	92		Vocational training:	75
	≤3000€:	43		A Level:	15
	≤4000€:	15		Secondary school:	14
	>4000€:	4		Foreman:	20
	Not specified:	81		University or college degree:	169

Of the 308 surveyed visitors (passers-by), 75 came from the district of Sächsische Schweiz-Osterzgebirge the study area is belonging to, and 210 came from outside but mostly from Saxony. By age, 256 (85%) interviewees were older than 40, many of them older than 60. Most of the respondents visit the Eastern Ore Mountains and its natural highlights frequently. Of the respondents, 82% were on a day trip; the others stayed four days on average (Table 2).

3.2. Appreciation of nature, landscape and ecosystem services

The TSP emphasized – besides the good price-performance ratio and the opportunity for sports (skiing) – the particularly attractive nature and landscape as the main reasons for tourists to visit the region, with 86% stating that their guests preferred to hike in nature. One third of the respondents mentioned that their guests liked to observe nature (plants, animals).

For the majority of surveyed visitors (221 = 71.8 %), experiencing nature was an important motivation for their visit (Fig. 4). Pair-wise comparisons (visitors from the district/from outside, younger/older people, lower/higher education) showed no significant differences.

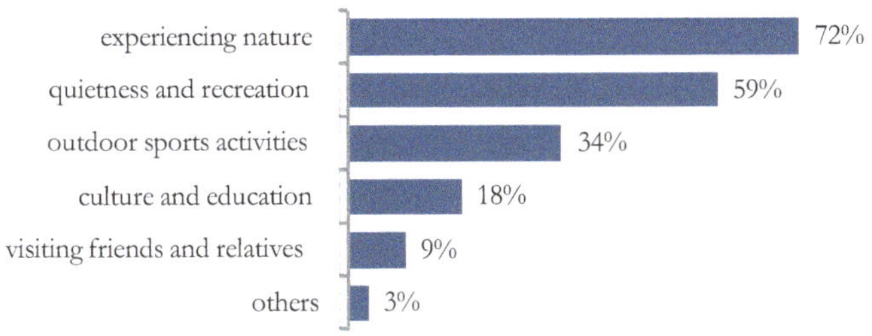

Figure 4: Motivation for recreation in the Eastern Ore Mountains (multiple responses), visitor survey (n=308)

Most TSP (60%) attributed the **highest attractiveness of landscape elements** for tourists to the mountain meadows and mixed mountain forests, and to a slightly lesser degree, to running waters, raised bogs, stone walls/hedgerows and spruce forests. Particularly the management of mountain meadows was seen as important for tourism (average value of 4.5 on the Likert scale). Management of hedges and stone walls (value 4.0) and restructuring spruce forests to mixed stands (3.9) were also seen as important for tourism. Species protection received a slightly lower value (3.7).

The visitors surveyed attached the highest priority to maintaining mountain meadows (4.7), raised bogs (4.5), watercourses (4.4), mixed forests (4.3) and hedgerows (4.2). It is interesting that they valuate the importance of landscape elements higher than the TSP do (Fig. 5). The largest differences concern raised bogs and hedgerows/stone walls.

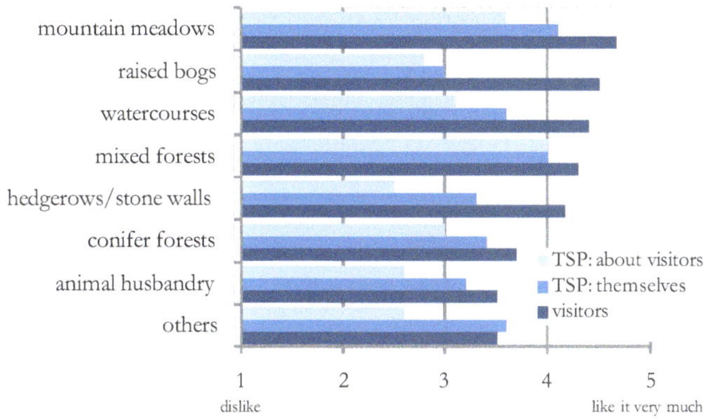

Figure 5: Valuation of different landscape elements in the Eastern Ore Mountains by visitors (visitors n=308; TSP n=66 average values; 5: like it very much; 1: dislike)

While there are almost no significant differences between both groups of visitors (residents of the district/visitors from outside the district) in terms of ranking landscape elements, there are such differences with regard to age classes and educational levels (pairwise comparisons) for mountain meadows, with the assignment of a ranking of 5 (like it very much) as follows: older vs. younger: 82%/60%, and of a 1 (dislike): 12%/26%; for hedgerows and stone walls: older vs. younger: ranking of 5: 53%/31%; of 1: 23%/36%. Thus, mountain meadows and hedgerows/stone walls are more appreciated by older persons. People with lower education ranked conifer forests higher than did more highly educated persons (only weak significance), and they gave a higher priority to the maintenance of animal pasturing (cp. the higher appreciation of the less educated group for animal habitats – in this section, below).

More than the half of TSP (53%) believed that **protected areas** would enhance the attractiveness of the region and would contribute to the development of tourism; 21% stated that protected areas would not be attractive for tourists. 71% of TSP were able to mention at least one protected area in the Eastern Ore Mountains. The by far most frequently mentioned protected areas were the nature reserves *Geisingberg* (32 mentions) and *Georgenfeld* raised bog (31), which are among the most important and typical protected areas in the region. Only 14% of TSP knew the term Natura 2000. They associated it with "conservation of flora and fauna", the "maintenance of the natural heritage and biodiversity in Europe", or "a network of protected areas"; however, two of them also associated it with restrictions, strict regulations and negative effects on the tourism sector.

The visitors stated that the provision of habitats for animal and plant species would be very important and should be secured on a permanent basis. They also placed high value on the landscape beauty and the recreational values. This is in accord with their reasons for travelling to the Ore Mountains. Compared with other **ecosystem services**, the production of food and other commodities was regarded as less important (Fig. 6).

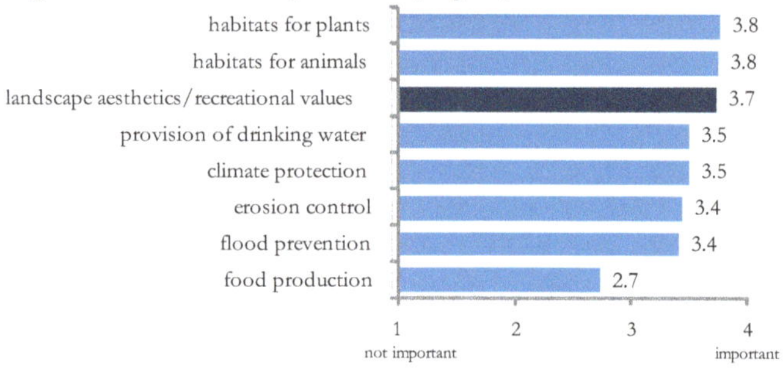

Fig. 6. The appreciation of various ecosystem services by visitors surveyed in the Eastern Ore Mountains (average values; large numbers values = high ranks)

There were no significant differences between the two groups of visitors (except with respect to animal habitats, which were more frequently mentioned as important by visitors from outside the district – weak significance – 10% level). Landscape aesthetic values, flood prevention (highest ranking: 53%/32%), and drinking water (60%/44%) were significantly more important for older than for younger persons. People with lower education levels significantly valued food production (highest ranking: 28%/14%) and habitats for animals (88%/72%) higher than persons of higher education levels (cp. similar results for lower education levels in the Müritz National Park, for natural features that are often associated with animals; Lupp, et al. 2011).

3.3. Attitudes towards ecosystem and landscape changes

More than the half of the TSP feared that deforestation (70%), wind turbines (60%), the decline of biodiversity and the loss of mountain meadows (50%) would reduce touristic attractiveness. Other unfavourable impacts mentioned included:

- The loss of panoramic views and the landscape character due to excessive afforestation of open areas
- Conflicts of interests between forestry and natural development, but also unfavourable recreational use of forests, e.g. snowmobiles and quad bikes
- General increase in motor traffic, also for leisure purposes
- Flood protection dams in beautiful valleys
- Air pollution from industry in the neighbourhood
- Too many protected areas that encumber the movement of tourists
- Elimination/limitation on skiing, hiking and mountain-biking routes
- Unmaintained, polluted rivers and river banks

The visitors stated that especially the loss of biodiversity and more technical infrastructure (wind turbines, industry, housing) would affect their choice of travel destination (Fig. 7).

With respect to wind turbines, visitors living in the district felt less deterred than visitors from outside the district (no impact 35%/25%, would choose

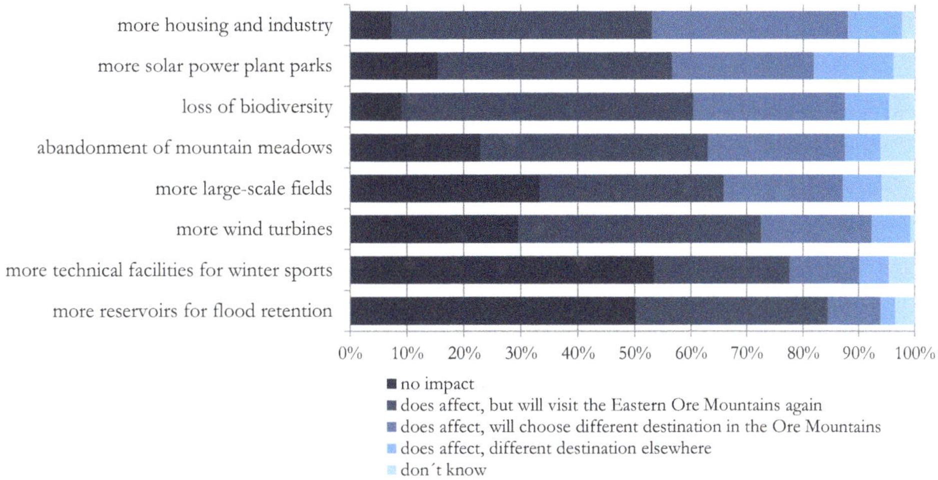

Figure 7: Impact of landscape changes on choice of destination for tourism/ recreation (external visitors' view, n=288)

other destinations in the Ore Mountains 16%/22% or elsewhere 9%/7%), but there are no significant differences between older and younger persons. Younger people felt less deterred by solar power plants (weak significance – 10% level). A significantly higher acceptance of water reservoirs and winter sports facilities among older persons could be observed. Visitors from outside the district were more sensitive to biodiversity losses than residents of the district, and younger people saw it as a greater problem than older persons (Fig. 8).

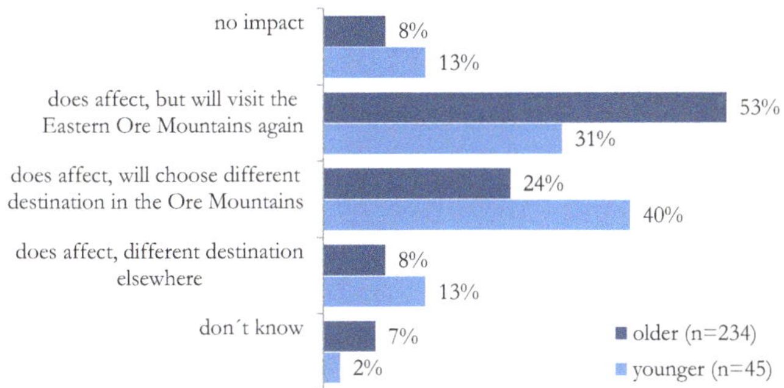

Figure 8: Assessment of biodiversity losses by people of different age groups and recreation patterns (younger people = up to 40 years)

3.4 Willingness to pay (WTP) for nature conservation and landscape management

More than two thirds of the responding TSP felt no responsibility for funding nature conservation and landscape management, stating that it is the task of the federal or state governments, or of landowners. Only 21% expressed the opinion that TSP and/or tourists should contribute. Tourists, they noted, had to pay the visitors' tax in any case, and the financial situation of the TSP precluded payment for nature conservation. Most TSP would prefer funding from general tax revenues (Table 3).

Of the visitors, 86% argued that the state or federal government should be responsible for nature conservation and landscape management (Fig. 9).

Table 3: Reasons of tourism service providers (TSP) for refusing payment for nature conservation

Reasons	Complaints	Share
Additional funding should come from donations	3	6 %
Additional funding should come from taxes	32	63 %
Tourism does not benefit from nature conservation	3	6 %
Demands of nature conservation for more funding are not justified	6	12 %
Other reasons	14	27 %
No response	6	12 %

Here, too, most considered general tax revenues as the preferred instrument (Fig. 10).

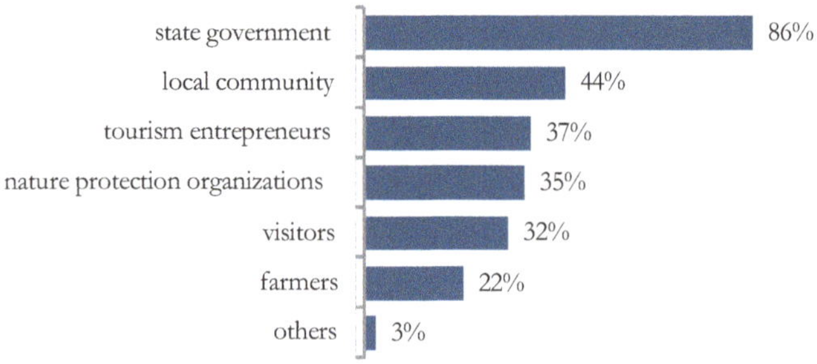

Figure 9: Responsibility for funding nature conservation in the view of visitors surveyed in the Eastern Ore Mountains (n=308)

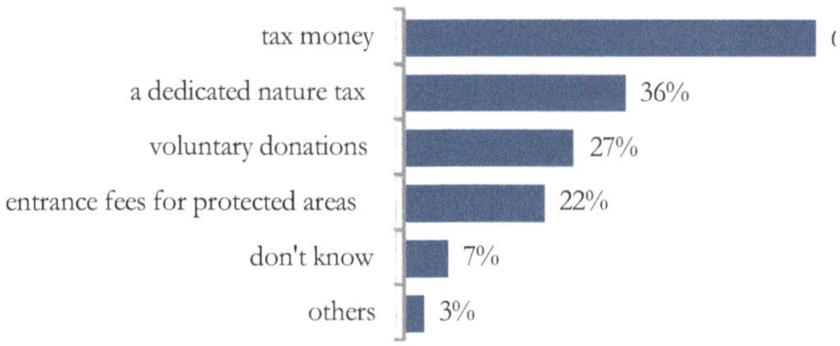

Figure 10: Types of payments for nature conservation accepted by visitors (n=308)

Younger people, more than older ones, saw a dedicated nature tax as an option. 68% of all visitors called for more and extra funding via such a tax, donations or entrance fees (Fig. 11).

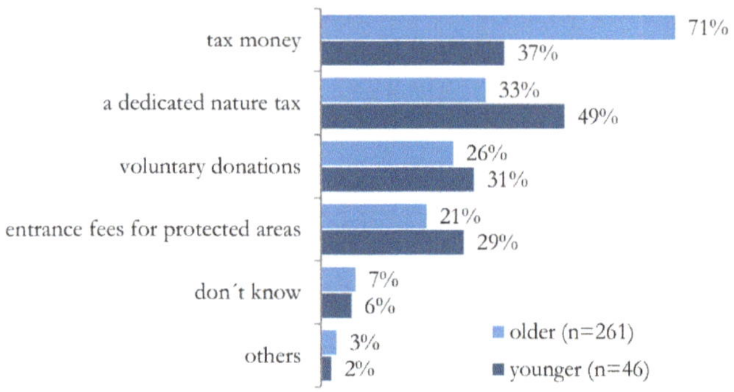

Figure 11: Differences between older and younger visitors in terms of accepted payment types for nature conservation (younger people = up to 40 years)

Of those visitors not willing to pay, more than two thirds (71%) argued that nature conservation should be funded from general taxes, and 22% believed that they could not afford more money. The last group can be considered real payment deniers, as they assign other goods a higher value than they do to nature. For the first group, this is not certain.

In order to estimate the **amount of a possible financial contribution** of the tourism sector or to quantify the WTP, the TSP were asked to measure the amount of taxes per guests and night (in the form of an open question); nine TSP replied, and the average of their estimates, which ranged between €0.25 - €3.00, was €1.36.

Less than half of the respondents (42% of TSP surveyed) endorsed an increase in the visitors' tax to the amount of a nature tax in the Ore Mountains; 37% wanted no increase in the visitors' tax. Attitudes on this issue were quite divided. The proponents of the nature tax called for amounts ranging between €0.20 and €3.00, with an average of €0.75 per person per night (n=14). Two thirds of those TSP who would be willing to bear some of the management costs believed that this money should best be paid to the local or regional landscape management associations, while 50% considered the municipalities and farmers to be suitable recipients; environmental organizations were also mentioned.

The survey of visitors resulted in a WTP amount of between €1.06 and €2.73 per visitor per day, or between €5.03 and €18.91 per resident of the district per year (see Table 4). The lower values represent the arithmetic mean of all respondents, including those not willing to pay (entered into the calculation with a WTP of €0). The higher values exclude the payment deniers (84). The five respondents who expressed WTP, but did not call for any particular amount, were treated as having an average WTP equal to that of those who participated (methodological notes: Meyerhoff et al. 2010).

The question as to the acceptance of the current entrance fee (€1.80) to the *Georgenfeld* raised bog nature reserve resulted in even higher WTP amounts: the 261 external visitors surveyed considered it appropriate or even too low. Only 5% did not accept the existing fee. The acceptance quota of the entrance fee to the bog was dependent on the site where the survey was carried out. It was higher among the visitors to the bog than among those to the mountain meadows.

Both groups of visitors felt a responsibility to better support nature conservation. Respondents living in the district stated a significantly higher willingness to pay (significant on the 5% level) €7.59 per year, on average – than respondents from outside the district expected from them (€3.51, on average). The reverse was also true: residents of the district expected "guests" to pay an average of €0.71 per day, while the latter actually stated a willingness to pay €1.20, on average (5% level). The overall WTP of younger respondents was higher than that of older ones (highly significant – 1% level; but no significance concerning the *Georgenfeld* raised bog entrance fee). With respect to education levels, no significant differences could be found. We could establish no clear correlation between income levels and WTP of visitors (statistically checked by regression analysis and Spearman rank correlation coefficient).

Table 4: WTP of visitors for nature conservation and landscape management (in the Eastern Ore Mountains generally and for the *Georgenfeld* raised bog)

	Suggested amount per day for visitors from outside the district	Suggested amount per resident (of the district) and year[1]	Proposed entrance fee for the *Georgenfeld* raised bog (current fee: €1.80)[2]
All interviewees (n=308)			
Mean	€1.06	€5.03	€1.95
Quote "Willing to pay"	€2.73	€18.91	€3.30
Residents of the district (n=92)			
Mean	€0.71	€7.59	€2.03
Quote "Willing to pay"	€1.98	€22.52	€2.42
Visitors from outside the district (n=216)			
Mean	€1.20	€3.51	€1.91
Quote "Willing to pay"	€2.32	€16.48	€2.24
Older (>40 years, n=261)			
Mean	€0.89	€3.76	€1.98
Quote "Willing to pay"	€2.11	€17.54	€2.32
Younger (≤ 40 years, n=47)			
Mean	€1.88	€9.27	€1.66
Quote "Willing to pay"	€2.71	€22.70	€2.14
Lower Education level (n=88)			
Mean	€0.95	€4.43	€1.84
Quote "Willing to pay"	€2.15	€14.98	€2.22
Higher Education level (n=204)			
Mean	€1.15	€5.18	€2.06
Quote "Willing to pay"	€2.32	€21.13	€2.31

[1] Five interviewees gave no amount, we assumed the average value for them.

[2] Those willing to pay but giving no amount → assumption of €1.80

4 Discussion

4.1. Appreciation of nature and landscape

The results of the surveys showed that the attractiveness of the Eastern Ore Mountain landscape is very important for the tourism sector. It is seen as the main reason to spend holidays or leisure time there. This agrees with the results of similar studies, by Grunewald et al. (2012) in the Western Ore Mountains and Lupp & Konold (2008) in the Müritz Region (northern Germany), where most tourists cited beauty, harmony and naturalness of the landscape as key travel motivations. Our interviewees mentioned experiencing nature as a key reason for visiting the Ore Mountains more often than the TSP assumed for their guests (see also Fig. 5 – the higher appreciation of landscape elements by visitors than by TSP). The reason might be that the survey was conducted on-site near the nature reserves, where mostly nature-loving visitors can be found, while the TSP had a wider spectrum of tourists in mind. This is in line with the impression that the visitors regretted a possible loss of biodiversity more than the TSP did.

Most TSP in the Eastern Ore Mountains argued that the loss of landscape elements, particularly mountain meadows, would have negative consequences for tourism, and that conservation and management measures are very important, too, from a touristic point of view. This high regard for mountain meadows could be also due to the fact that great efforts are presently being made to protect and restore mountain meadows in the region through the large-scale conservation project "Mountain Meadows in the Eastern Ore Mountains" (Hachmöller et al. 2001).

The fact that no more than 71% of TSP could name a protected area, and that the term "Natura 2000" was familiar to a mere 14% of them revealed knowledge deficits in terms of nature conservation highlights. Of the only 20 surveyed TSP in the Western Ore Mountains, 18 (90%) could name protected areas in that region, and 11 (55%) viewed them positively (Grunewald et al. 2012). With respect to knowledge of the term biodiversity, only 42% of respondents in a nationwide survey (BMU 2012) were familiar with the term. These facts indicate despite a general high appreciation of "nature", there is a need for more publicity work aimed at broadening the scope of nature conservation awareness especially considering the protected areas and their needs for respective management to maintain their character, which are also important key cornerstones of the tourism in this area. This is very important, in particular for TSP, as they play a multiplier role for the many tourists visiting the region. Appropriate education and awareness-raising initiatives, especially by the tourism associations, should be enhanced.

Impacts of interventions in nature (e.g. reforestation, extension of agricultural activities or construction of wind power plants) on landscape perceptions may vary between different social groups (Hunziker et al. 2008). Moreover, studies on the impact of particular landscape changes on the tourism attractiveness of destinations come to different conclusions. In regard to the installation of wind turbines, some authors point to possible shifts in the destination choice of tourists (Bundesverband deutsche Mittelgebirge/ Centouris 2012; Grunewald et al. 2012), whereas others do not believe in major negative impacts on landscape perceptions and experiences (Frantal & Kunc 2011). Not surprisingly, the sensitivity to the placement of turbines rises in accordance with higher aesthetic quality levels of landscapes (Molnarová et al. 2012).

The surveys in the Western Ore Mountains (Grunewald et al. 2012) confirmed the findings that deforestation and wind turbines were seen as impairing the scenery significantly; there, 82% (deforestation) and 73% (wind turbines), respectively, of summer tourists responded accordingly, while the figures in the present Eastern Ore Mountains survey were 82% and 74%, respectively (see Fig. 4). Highly visible landscape changes were mentioned more frequently than qualitative impacts on such sensitive ecosystems as raised bogs, or the decline of biodiversity in general. These clear statements of aesthetic preferences should be seen as a warning signal from the tourism sector to limit the expansion of wind power use, and to keep sensible areas, particularly the ridge zone free of wind turbines.

The spoiling of the landscape would, according to the Western Ore Mountains study (Grunewald et al. 2012), also change the travel behaviour of the tourists surveyed there: 28% of winter tourists and 23% of summer tourists would definitely change their travel destinations, and 34%/28% would probably do so. Only 24% of winter tourists and 39% of summer tourists would remain faithful to the region, while 13%/9% are still undecided. Depending on the particular impact, between 10% and 36% of visitors in the Eastern Ore Mountains would change the destination of travel and recreation.

Without knowing the concept of ecosystem services explicitly, Germans in general appreciate nature due to its manifold services for humans (BMU 2012): The majority demands that nature be used in such a manner that it remains available in full for future generations (93%), that the diversity of plants, animals and their habitats is secured (93%), and that the beauty and peculiarity of the landscape is maintained (92%). More than half of all Germans can imagine becoming active for nature conservation, or declared that they are already active (18%). This is a remarkably considerable segment of the population. Persons over the age of 65 years and well-educated people display an above average level of activity (BMU 2012). The great majority see the conservation of biological diversity as an important task of the society. About one half of the surveyed people feel personally responsible for ensuring this goal. Among the responding visitors in the Eastern Ore Mountains survey, 68% would be ready to contribute in monetary terms! This is in line with the nation-wide survey (BMU 2012): 65%!

We could find some differences in the answers depending on origin (residents of the district/ visitors from outside), age and education level. Petrosillo et al. (2007) showed a high dependence of tourists' perception of recreational environment and management on gender, education level, and place of residence, and stressed the importance of better identification of visitors' profile, to achieve better management of tourism development.

4.2. Funding of nature conservation and landscape management

An attractive landscape is a public good; nobody can be excluded. However, conservation and management costs must be covered. Tourism providers benefit from the increased attractiveness of a landscape without bearing any of the associated costs. This is a typical "free rider problem". The nature tax may be a payment model to solve this problem at least partially. All guests are obliged to contribute to an attractive landscape (Huhtala & Pouta 2008). However, not all users of the improved state of nature and landscape are reached by this model, e.g. day-trippers, local residents and their guests (Degenhardt et al. 1998). As most TSP see not the tourism sector, but rather the general public as being responsible for nature conservation and landscape management, they are opposed to the nature tax. Some of them argued that after establishing a nature tax, the state might shrink its responsibility in times of tight budgets. The financial means available for nature conservation would depend on the tourism development of the region. But even in regions less important for tourism, funds are needed. Moreover, nature conservation should not depend on tourism, since nature is not protected for touristic reasons alone. Such funding models are also perceived as an inadequate tool because of the well-known and very common right to visit nature free of charge. Also, many other WTP studies have ascertained that financing through general tax revenues is the most popular model (Degenhardt et al. 1998).

The height of WTP (or nature tax) determined in the Eastern Ore Mountains (€0.75 tourists' nature tax or €1.36 as payment of TSP per day and tourist, or between €1.06 and €2.73 as average visitor's WTP per day) is similar to the Western Ore Mountains (€1-2) (Grunewald et al. 2012). This is in line with the results from Ireland (Yadav & O'Neill 2013), were the tourists believed that the TSP should pay almost twice as much as their contribution and just 22% of the TSP stated that tourists should make any contribution (Eastern Ore Mountains: 21%).

Numerous critics, however, have pointed out that the method is fraught with technical and conceptual problems. Thus, the survey design and the description and framing of what is to be valued is critical to the reliability of the method, and the information a survey provides as well as the order in which questions are asked substantially influences WTP (Riera et al. 2012). The results may be influenced by the specific location of the survey, the composition and characteristics of the reference group, particularly their level of income and education, prior knowledge, preconceived opinions, level of understanding, and psychological aspects. Also various socio-economic variables, like the GDP or the income per capita, have an influence on WTP estimates (Barrio & Loureiro 2010).

Jones-Walters & Mulder (2009) argue that the fundamental objections with respect to WTP valuations should not be dismissed, not only due to the discrepancy between the stated preferences and the actual behaviour of respondents, but also with regard to their knowledge and rationality (Jessel et al. 2009). WTP is not based on actual market behaviour (Chee 2004). An overestimation, i.e. the amount by which the actual WTP exceeds the theoretical WTP obtained in surveys, by a factor of 2.5 is possible. On the other hand, stated preferences and actual payments are often not adequate to the real benefits and values of the natural environment. WTP analyses fail to capture all benefits and beneficiaries. Nature conservation provides not only recreation benefits, but also existence values, which benefit all people, not only the tourists and residents seeking recreation, or the TSP of the tourist resorts (Schweppe-Kraft 2009).

Much of the attention devoted to ecosystem services valuations incl. WTP approaches stems from the hope that it will result in better decisions about the environment. This requires that valuation is actually used for decisions, and that it has a significant impact on these decisions (Laurans & Mermet 2014). Such valuations have the potential to inform policy decisions by highlighting the benefits of sustainable ecosystem management. Unfortunately, as already stated above, the techniques used for valuation suffer from serious limitations. Ecosystem management problems are often complex, multi-faceted, socially contentious and fraught with uncertainty. For policy makers ecosystem services valuation is not the most important factor, but it provides a set of tools with which to make better and more informed decisions. As Chee (2004) emphasises, the welfare economic approach to decision-making is too narrowly focussed and needs to be supplemented by more comprehensive approaches like substantive stakeholder participation with opportunities for social learning, value formation, problem exploration, risk assessment, analysis of uncertainty, broad-based debate and reconciliation of interests.

One could argue that from the fact that in our study the interviews were carried out at nature reserves methodological weaknesses may result, because the survey over-represents nature-oriented people, but not the opinions and behaviour of the "average" total population visiting or living in the region. But: As a nation-wide survey on nature consciousness (BMU 2012) shows, there is a huge demand of high environmental standards and a great relevance for environmental protection in almost all groups of society. About three quarters of the total population in Germany expressed the view that nature is very important for them. Hence, the methodical bias of our survey near nature reserves might be relativised.

The funds spent for landscape management are an expression of the economic value of landscape and nature. There is a broad political and social consensus for protecting environment, biodiversity and valuable landscapes (revealed public preferences). National and international laws underpin such goals and regulations as the European network Natura 2000. The measures for nature conservation and landscape management entail costs, which can be seen as a measure of the public's WTP to achieve defined goals (UBA 2007). In this respect it would be interesting to determine whether this WTP really covers the level of funding needed or actually spent for landscape management.

4.3. Willingness to pay and landscape management costs

But what about the real costs of landscape management? In Saxony 420€/ha are needed annually for the management of the protected biotope type "mountain meadow" (Grunewald & Syrbe 2012; Grunewald et al. 2014). In 2010, €163,380 were used for the regular management of 389 ha of mountain meadows in the district of Sächsische Schweiz-Osterzgebirge (€420 x 389 ha). For all valuable biotope types of the district which need management measures, a total of €1,119,678 were calculated (Grunewald & Syrbe 2012).

We have noticed that under different assumptions, the WTP analyses among TSP resulted in monetary amounts varying between €0.75 and €1.36 in the study area (see sect. 3.4). If the TSP of Altenberg and Hermsdorf had collected €1.36 per guest per night in 2011, €565,000 would have been available for nature conservation and landscape management (415,844 overnight stays without campsites – http://www.statistik.sachsen.de/appsl1/Gemeindetabelle). Based on the proposed nature tax of €0.75, €312.000 would have been available. The costs of the large-scale conservation project "Mountain Meadows of the Eastern Ore Mountains" (see sect. 4.1), which is located within the municipal borders of Altenberg, were €600,000 annually from 2000 to 2008, and are €567,000 annually from 2010 to 2015. These costs include biotope management, acquisition of land, and such transfer costs as administration and public relations. Thus, the majority of the annual costs for nature conservation and landscape management could be covered by public funds, primarily tax revenues, if the results of the WTP approach were to be used as orientation for the demand (stated preference) of society and a measure of justified redistributions of tax-payers' money for nature conservation and landscape management.

Grunewald et al. (2012) determined a WTP of €170,000 per year for nature conservation and landscape management in their study area in the Western Ore Mountains (€4.50 per tourist arrival). By contrast, only one tenth of this amount was actually spent in 2009! It is not easy to calculate a total amount from the WTP data obtained from the survey among tourists. Total tourist frequentation of all relevant biotopes or nature reserves per year in the area is not known. For the Georgenfeld raised bog the number of tickets sold is a measure for this particular site: In 2012, a total of 13,949 persons and 462 families with four and more members visited the bog. The sum of the fees paid was €24,040 (there were different entrance fees for adults, adults with visitor's cards, children, children with visitor's cards, and school classes). The fees together with the sale of drinks, ice-cream and postcards covered the costs for the staff of the nature reserve.

The Schellerhau botanical garden, which is specialized in the mountain flora, was visited by 8,500 paying visitors in 2012 (average entrance fee: €1.72), and provided 3,123 free entries. If we assume theoretically the same fee for the last group, the total sum would have been €20,124. These two items represent only a very small part of all nature sights in the study area. However, for legal and practical reasons, it is impossible and not reasonable to charge entrance fees for all of them.

It is interesting that the actual WTP for public environmental goods is – as a rule – much higher than the costs of the provision and maintenance of these goods (Schweppe-Kraft & Grunewald 2014). Hampicke et al. (1991) found that the number the Germans prepared to pay for biodiversity in their country would be more than twice the estimated costs of measures needed to preserve all ecosystems essential for the conservation of species in Germany. Recent studies have shown an even higher WTP (Schweppe-Kraft 2009; Meyerhoff et al. 2010). Each hectare of protected areas (nature reserves) would benefit to the tune of €1000 per year. The cost-benefit-ratio is positive, which means that sustaining biodiversity is economical, even if only existence and aesthetic motives are considered. For Natura 2000 in Europe, about €63/ha/yr are necessary; however, these costs are by far exceeded by the benefits, e.g. from carbon sequestration and tourism, even if not all ecosystem services are included in the calculation (Kettunen et al. 2009).

The question is: Why are adequate nature conservation programmes not implemented in practice? The reasons may be the general mistrust of contingent valuations and the lack of hard facts in the area of biodiversity (Schweppe-Kraft 2009).

The issue of the tourism sector's contribution to nature conservation and landscape management should be kept under review. In further investigations larger samples might allow for better differentiation among respondents, and for the sampling of larger groups for statistical analyses of lifestyles. It would also be interesting to reveal any differences in appreciation of nature and the landscape, and in WTP, between the German and Czech sides of the Ore Mountains, respectively.

6 Conclusions

Traditional landscape planning concepts are oriented mainly towards the properties and potentials of ecosystems and the landscape, i.e. the supply side. The concept of ecosystem services is focussed more on the demand side, i.e. on the values and benefits as well as the providers and beneficiaries of services. At the interface of nature conservation and tourism, stated preference analyses are useful to reveal the needs, wishes and motivations of stakeholders. The survey carried out in the Eastern Ore Mountains has revealed the clear expression by tourism service providers (TSP) and tourists that they regard the attractive landscape as the main reason to visit this region, and that leisure activities in nature have by far the highest priority for visitors.

Attractive sceneries are considered being very important, especially such landscape elements as mountain meadows and mountain mixed forests. Deforestation, wind turbines, biodiversity loss and decline of mountain meadows reduce the attractiveness for outdoor recreation. Thus, nature conservation, landscape management and sustainable use are seen as very important, from the point of view of tourism as well. The lack of knowledge of nature conservation issues (protected areas, Natura 2000) draws attention to the need for information and education, particularly in view of the fact that TSP are important multipliers in the region.

Although the majority of TSP and visitors see the state and the public as responsible for funding nature conservation and landscape management, and tax revenue as the most appropriate financial source, a remarkably significant segment is prepared to contribute to covering such expenses. The amounts ascertained by the WTP analyses would be sufficient to cover most of the costs of nature conservation and landscape management in the study area. Since especially rural and peripheral regions benefit from nature-based tourism (Bell et al. 2007), politicians could see this as a call to implement these measures, which also support and secure tourism and jobs in this rural region.

Economic valuation, including contingent valuation, is just one aspect within the ecosystem service concept, and not necessarily the adequate and optimum indicator or instrument for resource management (Seppelt et al. 2012). Especially for goods without real markets (e.g. biodiversity, beauty of landscape), it should be applied only as an accompanying factor, not as the only basis to justify decisions or to choose options for action, particularly when weighing long-term aspects against short-term benefits (e.g., the construction of wind turbines in sensitive areas, or the underfunding of biotope management). Despite all problems and limitations, stated preference analyses belong to the most widely accepted and used methods for expressing the demand side in ecosystem services valuations (Riera et al. 2012).

They reveal not only the amount that people would be prepared to pay, but also the conditions or developments in the environment which they desire, or want to avoid. Valuations and surveys as performed in this study can help to set priorities for landscape management and for optimizing open space planning, in particular with regard for the preferences of people for certain landscape elements.

On the basis of such assessments, it is then possible to determine the actions and the limits required to avoid deteriorations (Grunewald et al. 2012). In order to establish successful environmental policies for the maintenance of biodiversity, ecosystem services and landscape qualities, and for the sustainability of an area, the users must be involved, and their opinions and preferences as well as their attitudes towards its conservation must be properly understood (Martín-López et al. 2007; Radford & James 2013). It is also useful to develop suitable funding models to balance the distribution of benefits and costs of nature conservation and landscape management, e.g. by involving beneficiaries (Macke & Schweppe-

Kraft 2011). Nevertheless, there is an urgent need for further studies and enhanced methodological approaches to bridge from the assessment of nature and landscape to practical decision-making, particularly in the area of landscape management.

Acknowledgements

The study was conducted in the framework of the projects "Grünes Netzwerk Erzgebirge" (Ore Mountains Green Network) and "Mehrwert Natur Ost-Erzgebirge" (Eastern Ore Mountains Ecosystem Services), and supported by the European Union, the State of Saxony and the Czech Republic. The students of the Dresden University of Technology, Chair of Tourism Economics and Management, M. Arjes, D.C. Bestfleisch, N. Jacob, P. Kleber, and M. Milde carried out a major part of the on-site surveys. Ms. A. Zimmermann (Gemeinnützige GmbH Naturbewahrung Osterzgebirge) provided visitor statistics of the *Georgenfeld* raised bog and the *Schellerhau* botanical garden. We thank Phil Hill, Berlin, for polishing the language.

References

Arnegger, J.; Woltering, M. & H. Job 2010. Toward a product-based typology for nature-based tourism: a conceptual framework. Journal of Sustainable Tourism 18, 915-928.

Atteslander, P. 2010. Methoden der empirischen Sozialforschung. *(Methods of empirical sociological research)*. Erich Schmidt: Berlin – New York. 10th ed.

Barrio, M. & M.L. Loureiro 2010. A meta-analysis of contingent valuation forest studies. Ecological Economics 69, 1023-1030.

Bastian, O. 2013. The role of biodiversity in supporting ecosystem services in Natura 2000 sites. Ecological Indicators 24, 13-22.

Bastian O.; Haase D. & K. Grunewald 2012. Ecosystem properties, potentials and services – The EPPS conceptual framework and an urban application example. Ecological Indicators 21, 7-16.

Bastian, O.; Neruda, M.; Filipová, L.; Machová, I. & M. Leibenath 2010. Natura 2000 sites as an asset for rural development: the German-Czech Ore Mountains Green Network Project. J. Landscape Ecol. 3, 41-58.

Bell, S.; Tyrväinen, L.; Sievänen, T.; Pröbstl, U. & M. Simpson 2007. Outdoor Recreation and Nature Tourism: A European Perspective. Living Reviews in Landscape Research 1(2). http://www.livingreviews.org/lrlr-2007-2. (Date: 29.10.2014).

Beza, B.B. 2010. The aesthetic value of a mountain landscape: A study of the Mt. Everest Trek. Landscape and Urban Planning 97, 306-317.

Blamey, R.K. 2001. Principles in ecoturism. In: Weaver, D.B. (ed.). The encyclopedia of ecotourism. CABI Publishing: Oxon – New York; pp. 5-22.

BMWi 2013. Tourismusperspektiven in ländlichen Räumen: Handlungsempfehlungen zur Förderung des Tourismus in ländlichen Räumen. German Federal Ministry of Economics and Technology *(Tourism perspectives in rural areas: recommendations for improving tourism in rural areas);* Berlin.

BMU 2012. Naturbewusstsein 2011. Bevölkerungsumfrage zu Natur und biologischer Vielfalt. (*Natural awareness 2011. Population survey about nature and biological diversity*). German Federal Ministry for the Environment, Nature Conservation, Building and Nuclear Safety, Berlin & ECOLOG-Institute; Hannover.

BMU & BfN 2014. Naturbewusstsein 2013. Bevölkerungsumfrage zu Natur und biologischer Vielfalt. (*Natural awareness 2013. Population survey about nature and biological diversity*). German Federal Ministry for the Environment, Nature Conservation, Building and Nuclear Safety, Berlin & German Federal Agency for Nature Conservation, Bonn.

Bundesverband deutsche Mittelgebirge/Centouris 2012. Akzeptanz von Windenergieanlagen in deutschen Mittelgebirgen (*Acceptance of wind turbines in German mow mountain ranges*). Passau.

Butler, C.D. & W. Oluoch-Kosura 2006. Linking future ecosystem services and future human well-being. Ecology and Society 111:30. http://www.ecologyandsociety.org/vol11/iss1/art30 (Date: 4.10.2014).

Casado-Arzuaga, I.; Madariaga, I. & M. Onaindia 2013. Perception, demand, and user contribution to ecosystem services in the Bilbao Metropolitan Greenbelt. J. Environm. Management 129, 33-43.

Chee, Y.E. 2004. An ecological perspective on the valuation of ecosystem services. Biological Conservation 120, 549-565.

Coghlan, A. & R. Buckley 2013. Nature-based tourism. In: Holden, A. & D.A. Fennell (eds.): The Routledge Handbook of Tourism and the Environment. Routledge; London et al.; pp. 334-344.

Cook, C., Heath, F., Thompson, R.I. 2000. A meta-analysis of response rates in web- or internet-based surveys. Educational and Psychological Measurement 60, 821-836.

Curtin, S. 2013. The intrinsic motivations and psychological benefits of eco- and wildlife tourism experiences. In: Ballantyne, R. & J. Packer (eds.): International Handbook on Ecotourism. Edward Elgar; Cheltenham – Northampton; pp. 203-216.

Degenhardt, S.; Hampicke, U.; Holm-Müller, K.; Jaedicke, W. & C. Pfeiffer 1998. Zahlungsbereitschaft für Naturschutzprogramme (*Willingness to pay for nature conservation programmes*). Angewandte Landschaftsökologie 25. Bonn; Federal Agency for Nature Conservation.

DESTATIS 2014. Einkommen der privaten Haushalte in den kreisfreien Städten und Landkreisen der Bundesrepublik Deutschland 2000 bis 2012, Reihe 2, Band 3. (*Income of households in the districts of the Federal Republic of Germany from 2000 to 2012*) https://www.destatis.de/DE/Publikationen/Thematisch/VolkswirtschaftlicheGesamtrechnung/VGRderLaender/VGR_KreisergebnisseBand3.html (Date: 3.12.2014)

Elsasser, P.; Meyerhoff, J.; Montagné, C. & A. Stenger 2009. A bibliography and database on forest valuation studies from Austria, France, Germany and Switzerland – A possible base for a concerted European approach. Journal of Forest Economy 15, 93-107.

Farber, S.; Costanza, R.; Childers, D.L.; Erickson, J.; Gross, K.; Grove, M.; Hopkinson, C.S.; Kahn, J.; Pincetl, S.; Troy, A.; Warren, P. & M. Wilson 2006. Linking ecology and economics for ecosystem management. Bioscience 56, 117-129.

Frantál, B. & J. Kunc 2011. Wind turbines in tourism landscapes: Czech Experience. Annals of Tourism Research 38, 499-519.

Fredman, P. & K. Lindberg 2008. Special Focus: Mountain tourism – Introduction. Tourism Economics 14, 245-247.

FUR 2013. Reiseanalyse 2013 (*Travel analysis 2013*). Kiel; Forschungsgemeinschaft Urlaub und Reisen.

García-Nieto, A.P.; García-Llorente, M.; Iniesta-Arandia, I. & B. Martín-López. 2013. Mapping forest ecosystem services: From providing units to beneficiaries. Ecosystem Services 4, 126-138.

Grunewald, K. & R.-U. Syrbe 2012. Landschaftspflegestrategie Sachsen 2020 – Bilanz und Aufgabenschwerpunkte. (*Landscape management strategy, Saxony 2020 – Accounting and key tasks)*. State Agency for the Environment, Agriculture and Geology of Saxony. Series, No. 20, Dresden.

Grunewald, K.; Syrbe, R.-U. & O. Bastian 2014. Landscape management accounting as a tool for indicating the need of action for ecosystem maintenance and restoration – Exemplified for Saxony. Ecological Indicators 37, 241-251.

Grunewald, K.; Syrbe, R.-U. & C. Renner 2012. Analyse der ästhetischen und monetären Wertschätzung der Landschaft am Erzgebirgskamm durch Touristen (*Aesthetic and monetary appreciation of Ore Mountains scenery by tourists*). GEOÖKO 33, 34-65.

Hachmöller, B.; Menzer, H.; Kafurke, B. & B. König 2001. Naturschutzgroßprojekt Bergwiesen im Osterzgebirge. (*Large nature conservation project "Mountain Meadows in the Eastern Ore Mountains"*). Natur und Landschaft 76, 442-453.

Hampicke, U.; Horlitz, T.; Kiemstedt, H.; Tampe, K.; Timp, D. & M. Walters 1991. Kosten und Wertschätzung des Arten- und Biotopschutzes. (*Costs and appreciation of the conservation of species and biotopes*), Reports 3/91. German Federal Environment Agency, Berlin.

Heuchele, L.; Renner, C.; Syrbe, R.-U.; Lupp, G. & W. Konold 2014. Nachhaltige Entwicklung von Tourismusregionen im Kontext von Klimawandel und biologischer Vielfalt. *(Sustainable development of tourism regions in the context of climate change and biodiversity)*. Culterra 64, University Freiburg.

Honey-Rosés, L. & L.H. Pendleton 2013. A demand driven research agenda for ecosystem services. Ecosystem Services 5, 160-162.

Huhtala, A. & E. Pouta 2008. User fees, equity and the benefits of public outdoor recreation services. Journal of Forest Economics 14, 117-132.

Hunziker, M.; Felber, P.; Gehring, K.; Buchecker, M.; Bauer, N. & F. Kienast 2008. Evaluation of landscape change by different social groups. Mountain Research and Development 28, 140-147.

Jessel, B.; Tschimpke, O. & M. Walser 2009. Produktivkraft Natur. (*Nature as a productive factor*). Hoffmann und Campe: Hamburg.

Jones-Walters, L. & I. Mulder 2009.Valuing nature: The economics of biodiversity. Journal of Nature Conservation 17, 245-247.

Kettunen, M.; Bassi, S.; Gantioler, S. & P. ten Brink. 2009. Assessing socio-economic benefits of Natura 2000 – a toolkit for practitioners (September Edition). Output of the European Commission project Financing Natura 2000: Cost estimate and benefits of Natura 2000. Institute for European Environmental Policy (IEEP), Brussels.

Kösterke, A. & D. von Laßberg. 2005. Urlaubsreisen und Umwelt: eine Untersuchung über die Ansprechbarkeit der Bundesbürger auf Natur- und Umweltaspekte in Zusammenhang mit Urlaubsreisen. (*Holiday trips and environment: an investigation into the responsiveness of German citizens to aspects of nature and environment related with holiday trips.*) Ammerland; Studienkreis für Tourismus und Entwicklung.

Laarman, J.G. & H.M.Gregersen 1996. Pricing policy in nature-based tourism. Tourism Management 17, 247-254.

Lamarque, P.; Tappeiner, U.; Turner, C.; Steinbacher, M.; Bardgett, R.D.; Szukics, U.; Schermer, M. & S. Lavorel 2011. Stakeholder perceptions of grassland ecosystem services in relation to knowledge on soil fertility and biodiversity. Regional Environm. Change 11, 791-804.

Laurans, Y. & L. Mermet 2014. Ecosystem services economic valuation, decision-support system or advocacy? Ecosystem Services 7, 98-105.

Lee, C.-K. & S.-Y. Han 2002. Estimating the use and preservation values of national parks' tourism resources using a contingent valuation method. Tourism Management 23, 531-540.

Lupp, G. & W. Konold 2008. Landscape preferences and perception of both residents and tourists. A case study in Müritz National Park (Germany). In: Siegrist, D.; Clivaz, C.; Hunziker, M. & S. Iten (eds.). Visitor Management in Nature-based Tourism – Strategies and Success Factors for Parks and Recreational Areas. HSR College Rapperswil, Switzerland; pp. 47-58.

Lupp, G.; Höchtl, F. & W. Wende 2011. "Wilderness" – A designation for Central European landscapes? Land Use Policy 28, 594-603.

Lupp, G.; Syrbe, R.-U.; Heuchele, L.; Renner, C.; Konold, W. & D. Siegrist 2014. Partizipative Szenarien als Erfolgsmodell für eine integrierte Entwicklung? Strategieentwicklung im Kontext von Tourismus, erneuerbarer Energie, Sicherung der biologischen Vielfalt und Klimaanpassung in Großschutzgebieten. (*Participative Scenarios as Recipe for Success for Integrated Development? Strategies in the context of tourism, renewable energies, conservation of biodiversity, and climatic adaptation on large-scale conservation areas)*. Naturschutz und Landschaftsplanung. Naturschutz und Landschaftsplanung 46, 336-344

Macke, S. & B. Schweppe-Kraft 2011. Ökonomisches Denken für den Naturschutz – ein Plädoyer für gemeinsame Argumente (*Economic thinking for nature conservation – a call for common arguments*). Natur und Landschaft, 86, 146-147.

Martin-López, B.; Montes, C. & J. Benayas 2007. Influence of user characteristics on valuation of ecosystem services in Donana Natural Protected Area (south-western Spain). Environmental Conservation 34, 215-224.

Mayer, M. 2014. Can nature-based tourism benefits compensate for the costs of national parks? A Study of the Bavarian Forest National Park, Germany. Journal of Sustainable Tourism 22, 561-583.

Mayer, M. & H. Job 2014. The economics of protected areas – a European perspective. Zeitschrift für Wirtschaftsgeographie 58, 73-97.

Meyerhoff, J.; Angeli, D. & V. Hartje 2010. Valuing the benefits of implementing a national strategy on biological diversity. The case study of Germany. Environ Science & Policy 23, 109-119.

Mmopelwa, G.; Kgathi, D.L. & L. Molefhe. 2007. Tourists' perceptions and their willingness to pay for park fees: a case study of self-drive tourists and clients for mobile tour operators in Moremi Game Reserve, Botswana. Tourism Management 28, 1044-1056.

Molnarová, K.; Sklenicka, P.; Stiborek, J.; Svobodová, K.; Šalek, M. & E. Brabec 2012. Visual preferences for wind turbines: Location, numbers and respondent characteristics. Applied Energy 92, 269-278.

Müller, M. & H. Job 2009. Managing natural disturbance in protected areas: Tourists' attitude towards the bark beetle in a German national park. Biological Conservation 142, 375-383.

Oom do Valle, P.; Pintassilgo, P.; Matias, A. & F. André 2012. Tourist attitudes towards an accommodation tax earmarked for environmental protection: A survey in the Algarve. Tourism Management 33, 1408-1416.

Petrosillo, I.; Zurlini, G.; Corliano, M.E.; Zaccarelli, N. & M. Dadamo 2007. Tourist perception of recreational environment and management in a marine protected area. Landscape and Urban Planning 79, 29-37.

Pickering, C. & A. Barros 2013. Mountain environments and tourism. In: Holden, A. & D.A. Fennell (eds.): The Routledge Handbook of Tourism and the Environment. Routledge; London et al.; pp. 183-191.

Pröbstl-Haider, U. & W. Haider 2014. The role of protected areas in destination choice in the European Alps. Zeitschrift für Wirtschaftsgeographie 58, 144-163.

Radford, K.G. & P. James 2013. Changes in the value of ecosystem services along a rural–urban gradient: A case study of Greater Manchester, UK. Landscape and Urban Planning 109, 117-127.

Renner, C.; Lupp, G.; Stein, C.; Siegrist, D. & O. Bastian 2012. Maintaining high biodiversity and landscape diversity for and through tourism – approaches for co-financing models. In: Fredman, P. et al. (eds.): Proc 6th Int Conf "Monitoring and management of visitors in recreational and protected areas. Outdoor recreation in change – Current knowledge and future challenges, Stockholm", Sweden, 21.-24.8. 2012. (Forskningsprogrammet Friluftsliv i förändring Rapport 19), pp. 188-189

Reynisdottir, M.; Song, H. & J. Agrusa 2008. Willingness to pay entrance fees to natural attractions: An Icelandic case study. Tourism Management 29, 1076-1083.

Riera, P.; Signorello, G.; Thiene, M.; Mahieu, P.-A.; Navrud, S.; Kaval, P.; Rulleau, B.; Mavsar, R.;

Madureira, L.; Meyerhoff, J.; Elsasser, P.; Notarol, S.; De Salvo, M.; Giergiczny, M. & S. Dragoi 2012. Non-market valuation of forest goods and services: Good practice guidelines. Journal of Forest Economics 18, 259-270.

Rosenberger, R.S.; Needham, M.D.; Morzillo, A.T. & C. Moehrke 2012. Attitudes, willingness to pay, and stated values for recreation use fees at an urban proximate forest. Journal of Forest Economics 18, 271-281.

Schulze, G. 1997. Die Erlebnisgesellschaft: Kultursoziologie der Gegenwart. (*The adventure society: cultural sociology in the present*). Campus: Frankfurt – New York. 7th ed.

Schweppe-Kraft, B. 2009. Natural capital in Germany – State and valuation; with special reference to biodiversity. In: Döring, R. (ed.). Sustainability, natural capital and nature conservation. Marburg; 24 pp.

Schweppe-Kraft, B. & K. Grunewald (2014). Approaches to the economic valuation of natural assets. In: Grunewald, K. & O. Bastian (eds.). Ecosystem Services: Concept, Methods and Case Studies (pp. 90-109). Heidelberg, Berlin: Springer.

Seppelt, R.; Fath, B.; Burkhard, B.; Fisher, J.L.; Grêt-Regamey, A.; Lautenbach, S.; Pert, P.; Hotes, S.; Spangenberg, J.; Verburg, P.H. & A.P.E. van Oudenhoven 2012. Form follows function? Proposing a blueprint for ecosystem service assessments based on reviews and case studies. Ecological Indicators 21, 145-154.

SfZ 2014. Tourismusanalyse 2014. (*Tourism analysis 2014*). Stiftung für Zukunftsfragen; Hamburg.

Spangenberg, J.H. & J. Settele 2010. Precisely incorrect?Monetising the value of ecosystem services. Ecological Complexity 7, 327-337.

Strasdas, W. 2006. The global market for nature-based tourism. In: Job, H. & J. Li (eds.): Natural heritage, ecotourism and sustainable development (pp. 55–64). Lassleben; Kallmünz.

Tacconi, L. 2012. Redefining payments for environmental services. Ecological Economics 73, 29-36.

Termorshuizen, J.W. & P. Opdam 2009. Landscape services as a bridge between landscape ecology and sustainable development. Landscape Ecology 24, 1037-1052.

TIES 1990. About ecotourism: Definitions & principles. The International Ecotourism Society. https://www.ecotourism.org/what-is-ecotourism (Date: 4.10.2014).

Tisdell, C. 2006. Valuation of tourism´s natural resource. In: Dwyer, L. & P. Forsyth (eds.): International Handbook on the Economics of Tourism. Edward Elgar; Cheltenham – Northhampton; pp. 359-378.

UBA 2007. Ökonomische Bewertung von Umweltschäden. (*Economic assessment of environmental damages*). German Federal Environment Agency. Dessau.

UBA 2009. Umweltbewusstsein und Umweltverhalten der sozialen Milieus in Deutschland. (*Environmental awareness and environmental behavior of social milieus in Germany*). German Federal Environment Agency. Dessau.

Wang, P. & J. Jia 2012. Tourists' willingness to pay for biodiversity conservation and environment protection, Dalai Lake protected area: implications for entrance fee and sustainable management. Ocean & Coastal Management 62, 24-33.

Yadav, L.P. & S. O'Neill 2013. Is there agreement between beneficiaries on who should bear the costs of conserving farm landscapes? Tourism Management 39, 62-70.

8

Urban-Rural Gradients of Ecosystem Services and the Linkages with Socioeconomics

Ying Hou[1, 2]*, Felix Müller[1], Bo Li[2], Franziska Kroll[3]

[1] Christian Albrechts University of Kiel, Institute for Natural Resource Conservation, Olshausenstr. 75, D 24118 Kiel, Germany

[2] Beijing Normal University, College of Resource Science and Technology, Beijing, China 100875, Beijing, China

[3] Regierungspräsidium Darmstadt, Abteilung Arbeitsschutz und Umwelt Frankfurt, Gutleutstraße 114, D 60327 Frankfurt am Main, Germany

Abstract

The principle of urban-rural gradients can reveal the spatial variations of ecosystem services and socioeconomic dimensions. The interrelations between ecosystem services and socioeconomics have scarcely been considered in the context of urban-rural areas. We investigated the spatial gradients and the mutual linkages of several ecosystem services and socioeconomic variables in the urban-rural areas of Leipzig, Germany, and Kunming, China. The results showed some regularities in the spatial patterns of ecosystem services and socioeconomic dimensions in both study areas. Habitat quality and f-evapotranspiration of Leipzig and habitat quality of Kunming demonstrated apparent increasing trends along all gradient patterns. However, the other ecosystem services presented divergent spatial variabilities in different gradient patterns. Road density, urban fabric and population density showed identical declining trends in both case study areas. Differently, household size, housing area as well as unemployment rate in Leipzig presented inconsistent spatial dynamics with considerable fluctuations. With regard to the gradient interrelations, road density, urban fabric and population density were strongly correlated with most ecosystem services in both case study areas. In contrast, the gradients of household size, housing area and unemployment rate of Leipzig showed inconsistent correlations with the ecosystem services gradients. Our study provides evidence to the argument that typical patterns of ecosystem service gradients do not exist in different urban-rural areas.

Keywords:
human-environmental systems, spatial dynamics, gradient patterns, quantitative, comparison, uncertainties

*Corresponding author. Email: yhou@ecology.uni-kiel.de

1 Introduction

1.1. The role of ecosystem services in tackling the issue of human-environment interactions

As a multidisciplinary principle, the concept of ecosystem services can be used to tackle the issue of human-environment interactions by linking natural sciences with social sciences. Ecosystem services cannot be tenable without human perception and use as it is defined as the benefits people acquire from natural system (Costanza et al., 1997; TEEB, 2010). The relationship between ecosystem services and human well-being can be either synergies or trade-offs, depending on which types of ecosystem services and human well-beings and what scales are considered (Duraiappah, 2011; Haines-Young and Potschin, 2010; MA, 2005; Raudsepp-Hearne et al., 2010). Human activities can act as drivers for the changes of ecosystem services. These drivers primarily involve demography, economy growth, social policies, culture, religion as well as science and technology (MA, 2005; Nelson et al., 2005). Drivers induce pressures, which directly affect ecological integrity and furthermore, influence ecosystems' generation of services (Hou et al., 2014) . Pressures often relate to land use changes, agriculture production activities and industrial activities (Alcamo et al., 2005; Bastian, 2013; Nelson et al., 2005). Apart from the impacts caused by drivers and pressures, the response, which relates to social political influences, e.g., governmental regulations, can play a positive role in preserving ecological integrity and enhancing the supply of ecosystem services (Couch and Karecha, 2006; Nuissl and Rink, 2005; Schwarz et al., 2011). Owing to the encouraging role in addressing human-environment interactions, ecosystem services approach is promising in assisting decision-making processes of local and regional environmental management (de Groot et al., 2010; Polasky et al., 2011; TEEB, 2010).

1.2. Ecosystem services and socioeconomic gradients in urban-rural areas

Urban-rural areas have received increasing concerns from ecosystem service investigators due to the accompanying intensive human-environment interactions (Kroll et al., 2012; Radford and James, 2013). Among the associated surveys, the ones performed in European urban-rural areas constitute a primary part (e.g., Larondelle and Haase, 2013). These investigations involve food provisioning, regulating services such as local and global climate regulation, pollination as well as recreation (Haase et al., 2012; Larondelle and Haase, 2013; Lautenbach et al., 2011; Radford and James, 2013; Schwarz et al., 2011; Strohbach and Haase, 2012). A German research group has conducted comprehensive surveys of ecosystem services in Leipzig urban-rural gradient in eastern Germany. These surveys mainly relate to the issues such as ecosystem services synergies and trade-offs (Haase et al., 2012; Larondelle and Haase, 2013), ecosystem services value transfer (Strohbach and Haase, 2012), ecosystem services' historical changes as well as the sensitivity and robustness of the evaluation results (Lautenbach et al., 2011).

The principle of urban-rural gradients was initially proposed to delineate the features of urbanization (McDonnell and Pickett, 1990). These characteristics primarily involve land cover structures, the biota of urban-rural area and socioeconomic dimensions (Hahs and McDonnell, 2006; McDonnell and Pickett, 1990). Based on the assessments of ecosystem services, some scientists further sketched urban-rural gradients of ecosystem services. The results concerning different provisioning, regulating and cultural services paint such a picture: Some services, such as food and water supply (Kroll et al., 2012), aesthetic services as well as spiritual services (Radford and James, 2013), present an increasing trend along the urban-rural gradient in the associated study areas. Others such as above ground carbon storage or sequestration (Larondelle and Haase, 2013; Radford and James, 2013; Strohbach and Haase, 2012), energy supply (Kroll et al., 2012), water flow regulation, pollination potential as well as recreation (Radford and James, 2013), show fluctuations from city center to fringe. Noticeably, although apparent ecosystem services gradients have been found in different urban-rural areas, typical models of ecosystem services gradients are evidenced not to be existing in different cities (Larondelle and Haase, 2013).

The issue of socioeconomic gradients in urban-rural areas has gained a growing popularity, not only in social science community, but also in some interdisciplinary subjects including environmental science (Dow, 2000; McDonnell and Hahs, 2008). Kroll and Kabisch (2012) and Kroll and Müller (2011) depicted the spatial gradients of human dimensions in urban-rural landscapes in several agglomerations in Europe (see Piorr et al., 2011). These human aspects primarily refer to population density, urban fabric, dwelling area, sealed surface, employment state, mean age as well as household size. The findings suggest the suitability of the gradient approach to delineate and understand the structural differences between urban areas and the surrounding environment. Another study performed in Australia reveals the demographic and employment gradients in urban-rural gradients in Melbourne (Hahs and McDonnell, 2006). In addition to the gradients of objective socioeconomic states, the gradient of subjective well-being expressed as "happiness" was explored in the US. This survey presents a rising trend of happiness from large central cities to small towns or the rural periphery (Berry and Okulicz-Kozaryn, 2011).

The literature investigation shows that ecosystem services and socioeconomic gradients in urban-rural areas have been drawing increasing concerns from interdisciplinary scientists as these two approaches can uncover human-environmental interactions in urban-rural space. Whereas, studies on these two issues are still limited. Furthermore, we found little literature investigating the linkages between the two issues. These shortages diminish the ability of scientists to understand how human and environmental systems are interplaying in urban-rural areas.

Consequently, our research aimed at bridging the revealed gap by spatially explicitly and quantitatively characterizing urban-rural gradients of ecosystem services and socioeconomics and exploring their mutual linkages. Accordingly, we propose the following three research questions:

1) What are the spatial and quantitative characteristics of ecosystem service and socioeconomic gradients in the investigated urban-rural areas?

2) What are the similarities and differences of ecosystem service and socioeconomic gradients between different gradient patterns of distinct areas?

3) Are there linkages between ecosystem service and socioeconomic gradients in urban-rural areas?

In order to answer these questions, we performed empirical studies in the urban-rural areas of two cities, Leipzig, Germany, and Kunming, China. The former is a representative of the cities in the east of Germany witnessing their reurbanization, while the latter can be understood as a proxy of the cities in the southwest of China experiencing rapid urbanization. These two research areas are both facing intensive human-environmental interactions.

2 Research areas

2.1. Leipzig urban-rural area

The research area is situated in Eastern Germany, covering Leipzig city and the adjacent municipalities. It is a circular area with the center of Leipzig city as the centroid and a radius of 25 kilometers. The primary land covers of the entire research area are artificial areas, agricultural land, forests as well as semi-natural areas (Figure 1). Located in a temperate transition zone, Leipzig has an oceanic-continental climate with an average annual temperature and precipitation of 8.8 °C and 511 mm, respectively (Leipzig Amt für Statistik und Wahlen, 2009). Leipzig covers large areas of commercial and industrial land, and is surrounded by intensively cultivated land. Eight percent of Leipzig is covered by green space and seven percent by forest (Leipzig Amt für Statistik und Wahlen, 2009).

Leipzig has a population of approximately 520,000 in 63 urban districts covering an area of 297 km^2 (Leipzig Amt für Statistik und Wahlen, 2009). The densest population can be found in the districts with prefabricated multi-story housing estates. Surrounding the city are the municipalities of low density incorporated into Leipzig since 1999.

Leipzig's population dropped by 12% between 1990 and 2000 due to suburbanization and out-migration, but grew by 2.7% between 2000 and 2006 (Saxon Office of Statistics, 2010). Historically, Leipzig was a center of commerce and trade. Lignite mining and chemical industry constituted the two primary economic sectors from the 19th century to the German reunification in 1990. After the German reunification, the economy of Leipzig recessed facing national and international competition. Throughout these years new developments tentatively emerged in the service sector and in fields such as education, transportation and tourism (Haase et al., 2012).

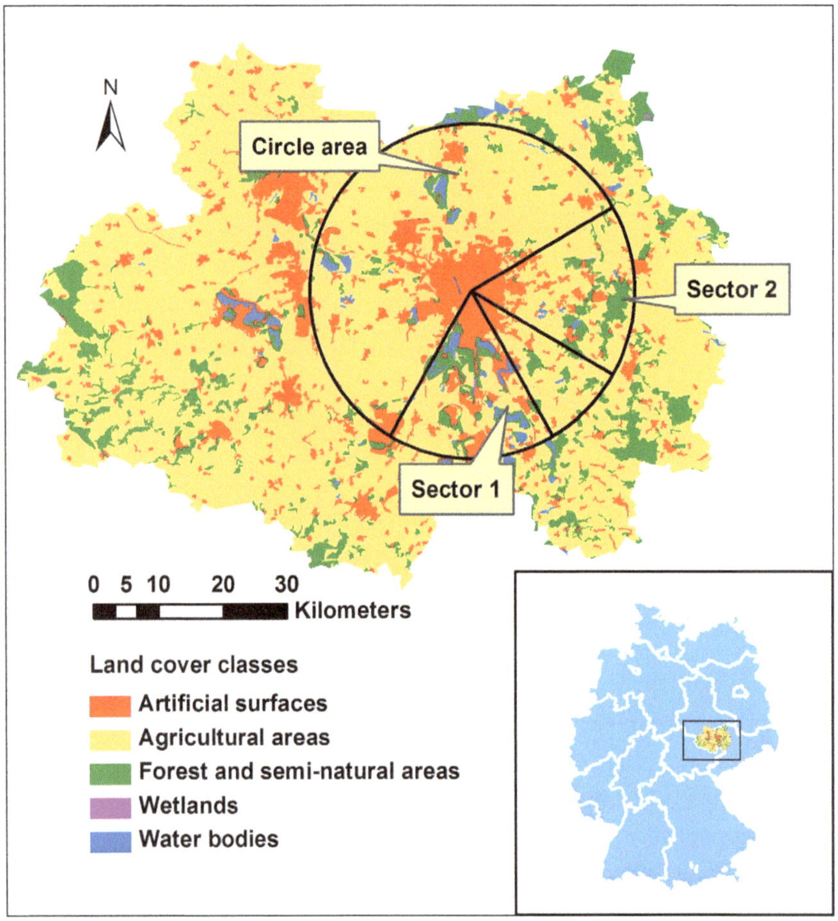

Figure 1: Land cover map of the Leipzig urban-rural area for 2006. The five primary classes are aggregated from Corine land cover classes (German Federal Environmental Agency, 2014). The circle area, sector 1 and sector 2 are the investigation areas with the city center of Leipzig in the centers and radiuses of 25 kilometers for gradient analysis

2.2. Kunming urban-rural area

Kunming is the capital of Yunnan province in Southwestern China (Figure 2). This research area consists of six districts within the Kunming region: Wuhua, Panlong, Guandu, Xishan, Chenggong and Jinning. Kunming is surrounded by mountains in the north, east and west and connected with the Dianchi Lake in the southern direction. It is situated at an elevation of around 1890m on the Yunnan-Guizhou Plateau. Kunming has a subtropical, mild highland climate throughout the whole year with humid summers and dry winters. The mean annual precipitation and mean annual temperature are 1000 mm and 15 °C, respectively (Statistics Bureau of Kunming, 2011). The entire study area covers 4076.30 km^2 including Kunming city with an area of 212 km^2. The primary land covers of this area are urban and rural residential areas, cropland in the southeast of the city and in the east and south of the Dianchi Lake, water bodies virtually consisting of the Dianchi Lake as well as forests mainly covering the peripheral mountains (Figure 2).

The research area has a population of 3,870,000 with 3,000,000 belonging to Kunming city (Statistics Bureau of Kunming, 2011). Kunming has experienced rapid economic growth and urbanization during the last two decades. The gross domestic product (GDP) increased from 16.9 billion RMB Yuan in 1992 to 63.6 billion in 2000. From 2000 to 2010, the GDP more than tripled, reaching 212 billion RMB Yuan (Statistics Bureau of Kunming, 2011). The urbanization process of Kunming city had caused a great loss of green space in the suburban areas and an expansion of urbanized area from 184.4 km^2 in 1992 to 275 km^2 in 2010 (National Bureau of Statistics of China, 2011).

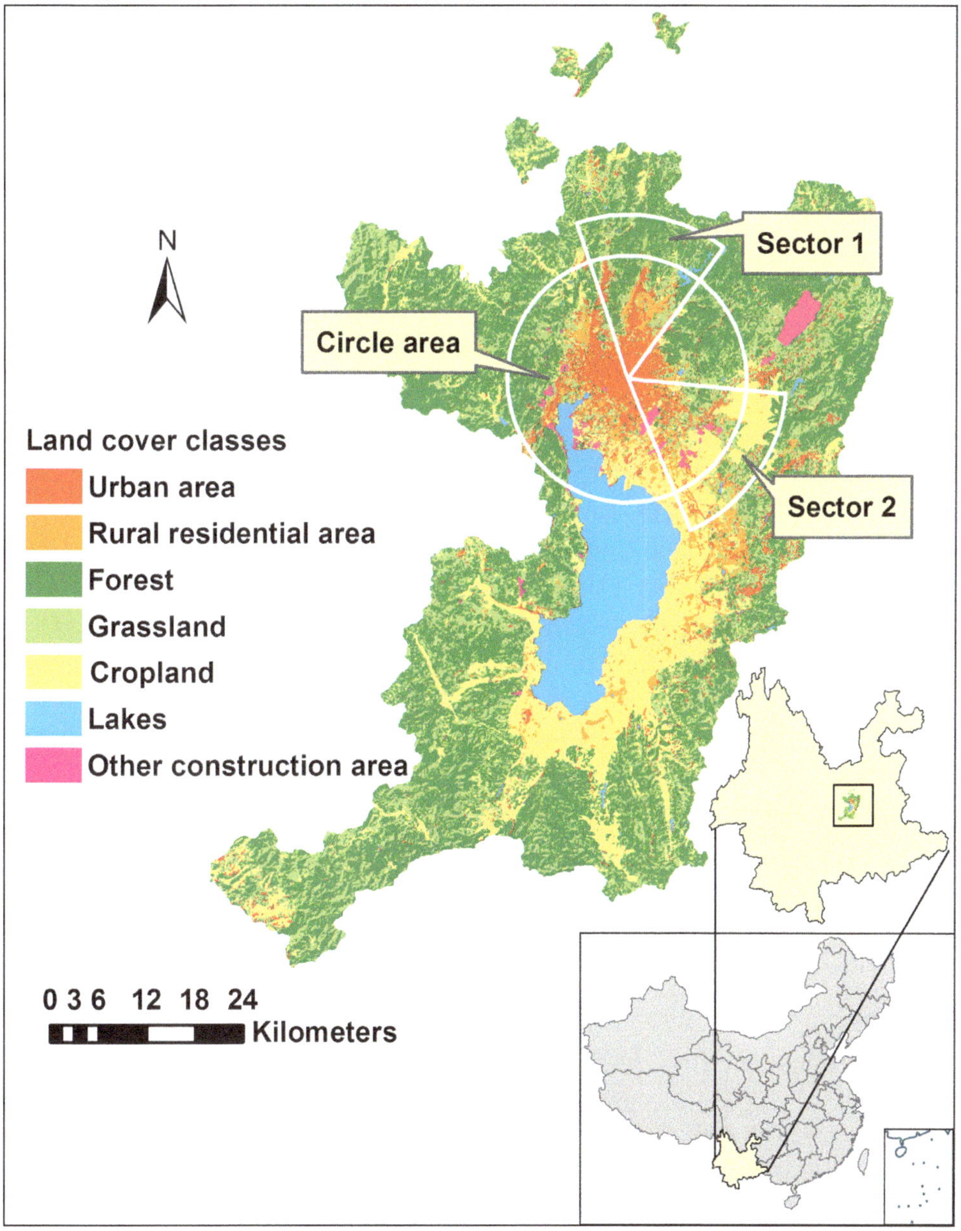

Figure 2: Land cover map of the Kunming urban-rural area for 2010. The seven primary classes have been aggregated from the classes of the National Land Classification (applicable in transitional period) released by the Ministry of Land and Resources of China in 2002 . The circle area with a radius of 15 kilometers and sectors 1 and 2 with radiuses of 20 kilometers are the areas for gradient analysis

3 Methods and data sources

3.1. Data sources

In this research we mainly used biophysical data (land cover, climate, soil, etc.) and socioeconomic statistical and spatial data (see Table 4 and Table 5 in the appendix for the data, time span, scales and sources). The data were derived from associated regional or national governmental agencies through online or offline accessibilities, online open access data sources of scientific institutions and published work.

3.2. Urban-rural gradient analysis

The gradient approach is proven to be promising in revealing the spatial variations of environmental states, ecosystem processes and functions as well as socioeconomic characteristics in urban-rural areas (Kroll and Müller, 2011; McDonnell et al., 1997; McDonnell and Hahs, 2008; McDonnell and Pickett, 1990; Pouyat and McDonnel, 1991). These delineations can be either qualitative or quantitative (Kroll and Müller, 2011; McDonnell and Pickett, 1990). Moreover, the changes can be explicitly visualized by combining GIS methods with the urban-rural gradient approach (Kroll et al., 2012; Larondelle and Haase, 2013). So far, two mainstreaming gradient patterns have been developed and applied. One is concentric ring gradients dispersing from the city center to rural outskirts (Kroll et al., 2012); the other one is transect gradient directionally crossing the city center and extending to rural areas with its two ends (Hahs and McDonnell, 2006; Zhou and Wang, 2011).

In this study, we used the concentric ring gradient method reported by Kroll et al. (2012) to spatially and quantitatively describe the distributions and variations of ecosystem services and socioeconomics of the research areas. Additionally, we modified this type of gradient and developed concentric sector ring gradients which can detailedly sketch specific sectors of the total circle area. For the Leipzig urban-rural area we created three concentric buffers to investigate gradients. One is a 25-km concentric ring area with 1-km intervals spreading out from the Leipzig city center. The other two are two concentric sector ring areas with the same radius and intervals dispersing from the city center to the south and east, respectively (Figure 1). For the Kunming urban-rural area also three concentric gradients with 1-km intervals were created including a circle area and two sector areas. All starting from the Kunming city center, the circle area extends 15 km and the two sectors disperse to the north and southeast with radiuses of 20 km (Figure 2). When selecting the respective two concentric sector ring areas of the two study areas, we considered that the two sectors can capture the areas with apparently different land use structure and distribution compared to the rest of the entire circle area.

Land covers have significant impact on ecosystem service supplies and the urban-rural distributions of socioeconomic elements (Burkhard et al., 2014; Jenerette et al., 2007; Luck et al., 2009). In this study land cover gradients were quantified and shown by area proportions of different land cover classes in each rings of all the gradient areas of the two study areas. With these results, we can to some extent explain the gradient characteristics of ecosystem services and socioeconomics.

In this study we used GIS techniques to delineate the features of several ecosystem services and socioeconomic dimensions across the urban-rural space in the two case study areas. Quantitative changes of the gradients were further shown with the average values of each ecosystem service and socioeconomic indicators for each concentric ring area. The findings in the forms of maps and line charts are given in the results section. Considering the uncertainty propagation with spatial aggregation and value averaging (Hou et al., 2013), we quantified the uncertainties in the calculations of selected indicators. The findings shown in the discussion section can provide insights into the magnitudes and variation trends of uncertainties along the urban-rural gradients of the two research areas.

3.3. *Indicator selections, calculations and gradient creations*

Ecosystem services indicators can be proxies of different ecosystem service types. Calculating the indicators or measuring the values is a mainstreaming approach to quantify ecosystem services (Müller and Burkhard, 2012). Appropriate ecosystem service indicators should be quantifiable, sensitive to land cover changes, spatially and temporally explicit and scalable (Burkhard et al., 2012). The ecosystem service indicators chosen in this study are highly sensitive to land cover changes, so their spatial variation in the considerably dynamic urban-rural environments can be efficiently captured. Additionally, we based the selections of the indicators on their typicality in the two study areas, the comparability between the two research areas and data availability. Thereafter, three regulating and two provisioning service indicators, which are habitat quality, above ground carbon storage, f-evapotranspiration, food supply and energy supply were selected for Leipzig urban-rural area. For Kunming urban-rural area, two regulating service and one provisioning service indicators were selected. They are habitat quality, net primary productivity (NPP) and food productivity. The indicators were calculated by robust methods from published literature, or based on empirical or statistical data (see Table 1 for the detailed information on the indicators).

Demonstrating the intensity and extensity of socioeconomic activities requires appropriate indicators which are sensitive to socioeconomic

Table 1: Overview of the quantified ecosystem service an socioeconomic indicators and the calculation methods

Study areas	Indicators	Units	Quantification methods	Data sources and references
Leipzig ecosystem services	Habitat quality	relative value	InVEST biodiversity model	Tallis et al., 2013
	Above ground carbon storage	MgC/ha	Look-up table with above ground carbon storage per land cover type	Larondelle and Haase, 2013
	f-evapotranspiration	relative value	Look-up table with f-evapotranspiration value per land cover type	Larondelle and Haase, 2013
	Food supply	GJ/ha	Regression model for land cover and soil types (yield)	Kroll et al., 2012
	Energy supply	GJ/ha	Assignation of energy supplies of different energy types to associated land cover classes	Kroll et al., 2012
Leipzig socioeconomics	Road density	km/km^2	Road length divided by area of the associated ring	See data sources in Table 4 in the appendix
	Urban fabric	%	Urban fabric area divided by total area of the associated ring	See data sources in Table 4 in the appendix
	Population density	persons/km^2	Leipzig municipality and city district population statistical data	See data sources in Table 4 in the appendix
	Household size	persons/household	Leipzig municipality and city district population statistical data	See data sources in Table 4 in the appendix
	Housing area	m^2/person	Leipzig municipality and city district population statistical data	See data sources in Table 4 in the appendix
	Unemployment rate	%	Leipzig municipality and city district population statistical data	See data sources in Table 4 in the appendix
Kunming ecosystem services	Habitat quality	relative value	InVEST biodiversity model	Tallis et al., 2013
	NPP	gC/m^2/yr	MODIS GPP/NPP Project (MOD17)	Numerical Terradynamic Simulation Group, 2013; Lin et al., 2008; Sun et al., 2011; Chinnaiah and Madhu, 2010
	Food productivity	yuan/km^2	Assignation of food outputs of different food type to associated land cover classes	See data sources in Table **5** in the appendix
Kunming socioeconomics	Road density	km/km^2	Road length divided by area of the associated ring	See data sources in Table **5** in the appendix
	Urban fabric	%	Urban fabric area divided by total area of the associated ring	See data sources in Table **5** in the appendix
	Population density	persons/km^2	Kunming districts and counties population statistical data	See data sources in Table **5** in the appendix

spatial and temporal dynamics. In this study we chose indicators involving the three primary human dimensions of urban-rural system: artificial environment, socio-demography and economy (Kroll and Müller, 2011). The calculations of the indicators were based on simple formulas with statistical or spatial data (Table 1).

I. Ecosystem service indicators

Habitat quality

Habitat quality is a proxy of land's biodiversity support capability. It involves the capability of the ecosystem to provide conditions appropriate for individual and population persistence (Tallis et al., 2013). Habitat quality modeling can be an acceptable solution to reflect the overall potential of regional biodiversity. One novel and promising modeling tool is the InVEST biodiversity model, which combines information on LULC and threats to biodiversity to produce habitat quality maps (Tallis et al., 2013). For the details of the model, see the InVEST 2.5.5 User's Guide (Tallis et al., 2013).

This study employed the InVEST biodiversity model to quantify and visualize the habitat qualities of the two study areas. For the Leipzig urban-rural area, we used the Corine land cover spatial data of 2006 as the current land cover map and considered cropland, urban area, roads and railways as biodiversity threats. For the Kunming urban-rural area, a land cover map rasterized from a 1:100,000 shapefile map of 2010 was used and cropland, urban and rural residential area, construction area as well as roads were assigned as threats. The two biophysical tables of each study area were adapted from the sample data of the InVEST biodiversity model with changes based on the land cover classifications of the associated study area. The derived habitat quality raster maps were converted to polypoint shapefiles in order to calculate the average values of the concentric rings of the circle and sector areas. Based on this, the habitat quality gradients were generated in the form of line charts. The modeling results are relative values between 0 and 1. Areas with higher values on the habitat quality maps enjoy better habitat quality compared to the rest of landscape. It is important to note that the habitat quality values of Leipzig and Kunming are not comparable as the model parameter inputs (e.g., the biophysical tables) were different from each other.

Above ground carbon storage

Being an important part of the global carbon cycle, terrestrial carbon storage consists of four layers: above ground biomass, deadwood and litters, below ground biomass and soil carbon storage. It is difficult to completely count the carbon storage of the four layers in terrestrial ecosystems. Whereas, measuring and estimating above ground carbon storage is feasible, and the results are credible (Haase et al., 2012; Strohbach and Haase, 2012). In this study the indicator "above ground carbon storage" was chosen to represent urban-rural ecosystems' global climate regulation services. As above ground biomass is sensitive to land cover types, estimating above ground carbon storage of different land cover classes can reveal the spatial dynamics of contributions from urban-rural ecosystems to mitigate climate change.

In this study, empirical values from research proceeded in Leipzig city were used to calculate this indicator (Larondelle and Haase, 2013; Strohbach and Haase, 2012). Above ground carbon storage values of different Corine land cover classes were derived by transferring the measured above ground carbon stock of the trees in the urban-rural gradient of Leipzig (Larondelle and Haase, 2013). By linking the empirical values to the associated land covers, we mapped the spatial distributions of the above ground carbon storage of the Leipzig urban-rural area. Furthermore, the quantitative changes of this service along the urban-rural gradient were characterized by plotting the average values of each concentric ring of the circle and the two sector areas.

Net primary production (NPP)

NPP represents the net carbon uptake from the atmosphere into biosphere. It is widely used to evaluate the patterns, processes and dynamics of carbon cycling in ecosystems, assess ecosystem status and monitor environmental changes (Ingraham and Foster, 2008; Melillo et al., 1993; Zhao and Zhou, 2005). Given the role NPP plays in carbon cycling, this study uses it to indicate the global climate regulation service of Kunming urban-rural area.

We distinguished the NPP quantification into three methods based on different data sources and existing research. For vegetation NPP, the MOD17A3 NPP product produced by the Numerical Terradynamic Simulation Group of the University of Montana was used (see Table 5 in the appendix for data sources). The NPP value of urban area was transferred from Lin et al. (2008). The NPP of the Dianchi Lake was assessed with reference to Sun et al. (2011) and Chinnaiah and Madhu (2010).

The MOD17A3 NPP product is generated from NASA MODIS data by the Numerical Terradynamic Simulation Group. It provides continuous estimates of NPP across the earth's entire vegetated land surface. For this research we used the annual global NPP of 2006 and extracted the Kunming urban-rural area from the global map. Lin et al. (2008) estimated the NPP of the urban area of Shanghai, China based on urban forest inventory data of 2006 and remote sensing. Considering the similarities of the climates between Kunming and Shanghai, we transferred the NPP value of Shanghai city from Lin et al. (2008) to the Kunming urban area. Sun et al. (2011) measured the surface water gross primary production (GPP) of the north, middle and south of the Dianchi Lake from the winter of 2009 to the spring of 2010. Chinnaiah and Madhu (2010) estimated the NPP/GPP ratio from October 2009 to September 2010 for the Darmasagar Lake in India. This study adapted the findings of these two studies to estimate the NPP values for the parts of the Dianchi Lake that belong to the six districts of the research area. For the calculation, the GPP values of spring from Sun et al. (2011) were used to calculate the annual average GPP of the Dianchi Lake. The mean NPP/GPP ratio from Chinnaiah and Madhu (2010) was used to convert GPP to NPP.

By joining the NPP derived from the above described methods using the union tool of ArcGIS 10, we produced the NPP map for the entire Kunming urban-rural area. Furthermore, the map was resampled to grids of 200 m × 200 m and intersected with the concentric rings of the circle and two sector areas. Thereafter, the Kunming urban-rural gradients of NPP was generated by averaging the values in each concentric ring and shown by line chart in the result section of this article.

f-evapotranspiration

Evapotranspiration plays a significant role in regulating local temperature, humidity and precipitation. Therefore, it is usually employed to indicate ecosystem's capacity to regulate local climate (Burkhard et al., 2009; Kandziora et al., 2013; Schwarz et al., 2011).

In this study "f-evapotranspiration" was used as a local climate regulation service indicator. Referring to latent heat flows, f-evapotranspiration is an approximation for the evapotranspiration potential of a land cover class (Schwarz et al., 2011). According to Schwarz et al. (2011), f-evapotransipiration is calculated with the following formula:

$$f(i) = \frac{{}_{\max}evapotranspiration(i)}{ET_{(0)}}$$

Here ${}_{max}$*evapotranspiration(i)* is the maximal evapotranspiration of land cover type *i*, and $ET_{(0)}$ is the reference evapotranspiration of the grass of 12 mm high in the local climate.

The f-evapotranspiration values of different Corine land cover classes were derived from a lookup table after Larondelle and Haase (2013). These values were transferred from the calculation and estimation (Schwarz et al., 2011) in the Leipzig urban-rural area. By linking the empirical values to the associated land covers, we showed the distribution of f-evapotranspiration across Leipzig urban-rural area. The quantitative dynamics of this service along the gradient were further presented by a line chart with the average values of each concentric ring of the circle and two sector areas.

Food supply and food productivity

As one of the basic ecosystem goods, food is a fundamental material input of urban systems and crucial for the survival of the urban population (Jansson and Polasky, 2010). In this study we employed food supply and food productivity to indicate the food provisioning services of Leipzig and Kunming urban-rural areas, respectively. For the details of the calculation of food supply, see Kroll et al. (2012).

The food productivity of the Kunming urban-rural area was computed in RMB yuan/km^2 for each land cover type of the six districts. Statistical data of 2010 on crop types and outputs (cereals, oil crops, cane, tobacco, vegetables fruits, nuts, spices and Chinese herbal medicine), sheep and caws, poultries as well as aquaculture products of different districts were used to calculate the average values of food output per km^2 (food productivity) of cropland, forest, grassland, rural residential area and lakes of the associated district (for data sources see Table 5 in the appendix). Pig production was not considered as pig farms cannot clearly be classified into a certain land cover type.

Based on the above described calculations, we linked the values of food supply and food productivity to the associated land cover types of the spatial land cover data of Leipzig and Kunming with ArcGIS, respectively. Accordingly, the maps of food provision of the Leipzig and Kunming urban-rural areas were produced. Moreover, food provision values were averaged in each concentric ring of the circle and sector areas of Leipzig and Kunming.

Energy supply

Energy is another important good natural system provides for human well-being. Although some energy sources are not regarded to be strongly linked with ecosystem processes and functions (e.g., wind energy and solar energy), they can have synergies or tradeoffs with some ecosystem services, and be considerably influential in the decision making procedures of environmental management (Burkhard et al., 2014). This study calculated the energy supply of Leipzig urban-rural area for 2007 in the unit of GJ/ha and delineated its spatial distributions and gradients (for data sources, see Table 4 in the appendix). For the details of the respective quantification methods, see Kroll et al. (2012).

II. Socioeconomic indicators

Road density and urban fabric

To calculate the road density of the Leipzig urban-rural area, we used road network maps of 2006 for the entire Germany and the Leipzig city district level and a railway network map of 2008 for the Leipzig city district level (see Table 4 in the appendix for data sources). For Kunming, a road network map of 2007 at the China national level was used (see Table 5 in the appendix for data sources). Then we intersected these maps with the maps of the circle and sector areas of concentric rings of Leipzig and Kunming. Based on this, we calculated the road densities of each concentric ring by dividing road length by area of the corresponding concentric ring. Urban fabric was computed in the unit of percent by dividing the urban fabric area by the total area of each concentric ring. This method was applied both to the circle and sector areas of Leipzig and Kunming. To accomplish this calculation, urban fabric maps were extracted from the land cover maps of Leipzig for 2006 and Kunming for 2010 (see Table 4 and Table 5 in the appendix for data sources).

Population density, household size, housing area and unemployment rate

Socio-demographic data from local and regional statistic offices were used to estimate the values of these four indicators of the Leipzig urban-rural area, (see Table 4 in the appendix for data sources). The calculation methods were developed and described by Kroll and Kabisch (2012).

To estimate the population density of the Kunming urban-rural area, we used the population and land cover data of the six districts constituting the research area (for data sources see Table 5 in the appendix). Urban population and rural population were assigned to urban area and rural residential area, respectively. The other five land cover classes, forest cropland, grassland, lakes and construction were assumed not to be populated. Then we calculated the population densities of the urban and rural residential areas of the six districts. Afterward a population density map was produced by joining the results with the land cover map in ArcGIS. Furthermore, we converted the map to a grid map with the cell size of 200 m ×200 m, intersected it this the concentric ring and concentric sector ring maps and computed the average values of each concentric ring. Finally, the population gradient line chart of Kunming urban-rural area was derived.

3.4. Comparisons of gradient patterns and case study areas

In this study we compared the three gradients of the concentric ring and concentric sector ring gradient patterns in order to examine their appropriateness. The comparisons were performed with the land cover structure, each ecosystem service and socioeconomic item investigated in this study for Leipzig and Kunming urban-rural areas. To achieve this purpose, we presented the three gradients in the same line chart to reveal the mutual similarities and differences. These comparisons can on the one hand reveal the spatial heterogeneities of the compared objects and on the other hand, address the uncertainties of the concentric ring gradient pattern, which are generalized from the heterogeneous patches in the ring. Furthermore, we quantified and presented the uncertainties of the three gradients with four indicators in the discussions. The indicators are above ground carbon storage and housing area of Leipzig and habitat quality and population density of Kunming.

The two study areas of this study may exhibit different spatial dynamics and urban-rural gradients with ecosystem services and socioeconomic dimensions. These variations can be caused by their distinct physical conditions, urbanization processes and socio-political systems. To examine the applicability of the concentric ring approach in these two areas, we compared their ecosystem service and socioeconomic gradients and described the similarities and differences in the results section.

3.5. Analysis of ecosystem service-socioeconomics gradient linkages

Ecosystem service supply and demand in urban-rural systems are reported to be closely linked with regional human activities (Burkhard et al., 2012; Hou et al., 2014; Kroll et al., 2012). Accordingly, we assume that ecosystem service and socioeconomic gradients should correlate in the urban-rural contexts of the two research areas.

In this study we firstly observed the qualitative linkages between the investigated ecosystem services and socioeconomic items by comparing their gradients shown by the line charts. Furthermore, we quantified the correlations between ecosystem service and socioeconomic gradients applying Spearman correlation analyses between each pair of ecosystem service and socioeconomic indicators for the three concentric ring areas of the two study areas.

4 Results

4.1. Land cover gradients in Leipzig and Kunming urban-rural areas

The artificial surfaces proportions showed overall declining trends along the urban-rural gradients in the circle and the two sectors areas in Leipzig (Figure 3). Contrarily, the proportions of agriculture areas presented overall growing trends along the gradients in these three gradient areas. The gradients of these two land cover classes were less smooth in the two sector areas, especially in sector 2, than in the circle area. The proportions of the other three land cover classes: forest and semi-natural areas, wetlands and water bodies, presented apparent irregular changes from the city center to 25 km away from the city center in all the three gradient areas of Leipzig.

In Kunming urban-rural area, the urban area proportions showed apparent declining trends along the gradients in the circle and the two sector areas (Figure 4). In contrast, the forest shares kept increasing along the gradients in the circle and sector 1 areas. In sector 2, it fluctuated at low levels along the gradient. The proportions of cropland showed different trends between the circle and the two sector areas. In the circle area, it generally kept increasing along the gradient; in sector 1, it fluctuated slightly at low levels; in sector 2, cropland share showed a first soaring then plummeting trend. The proportions of the other three land cover types: grassland, rural residential area and lakes, showed irregular changes along the urban-rural gradient in all the three gradient areas.

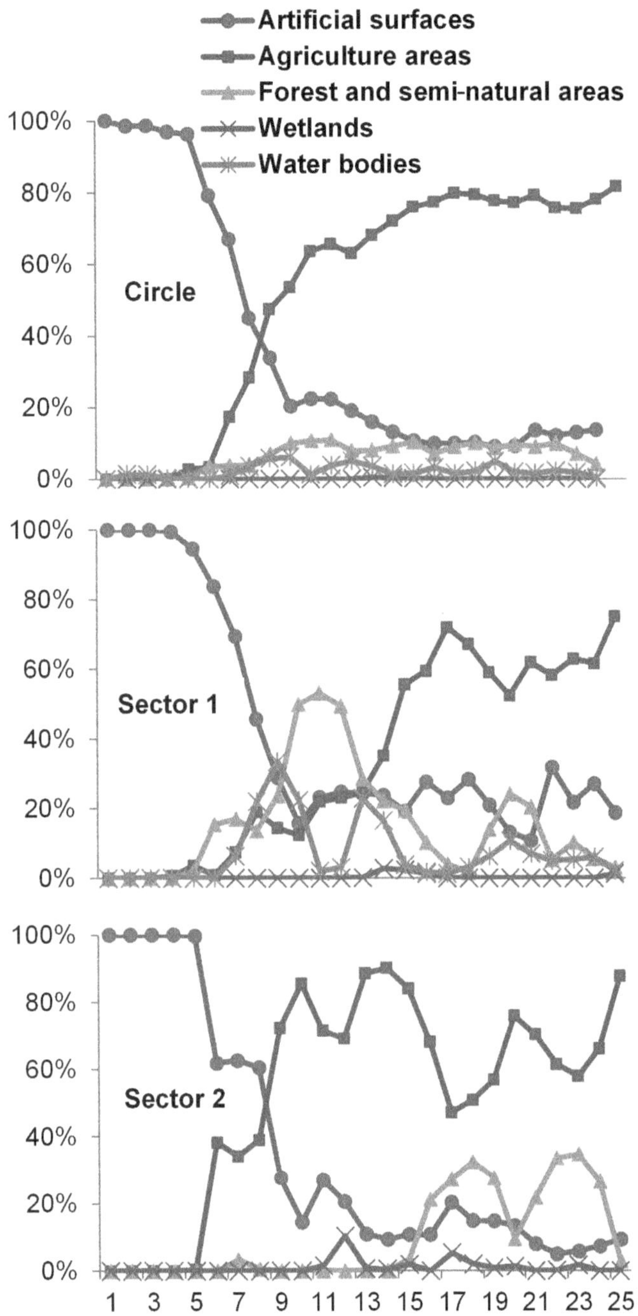

Figure 3: Land cover gradients along the Leipzig urban-rural area. The X-axis is the distance in kilometers from the city center of Leipzig. The Y-axis is the area proportions of different land cover classes in the total area of the associated concentric ring or concentric sector ring

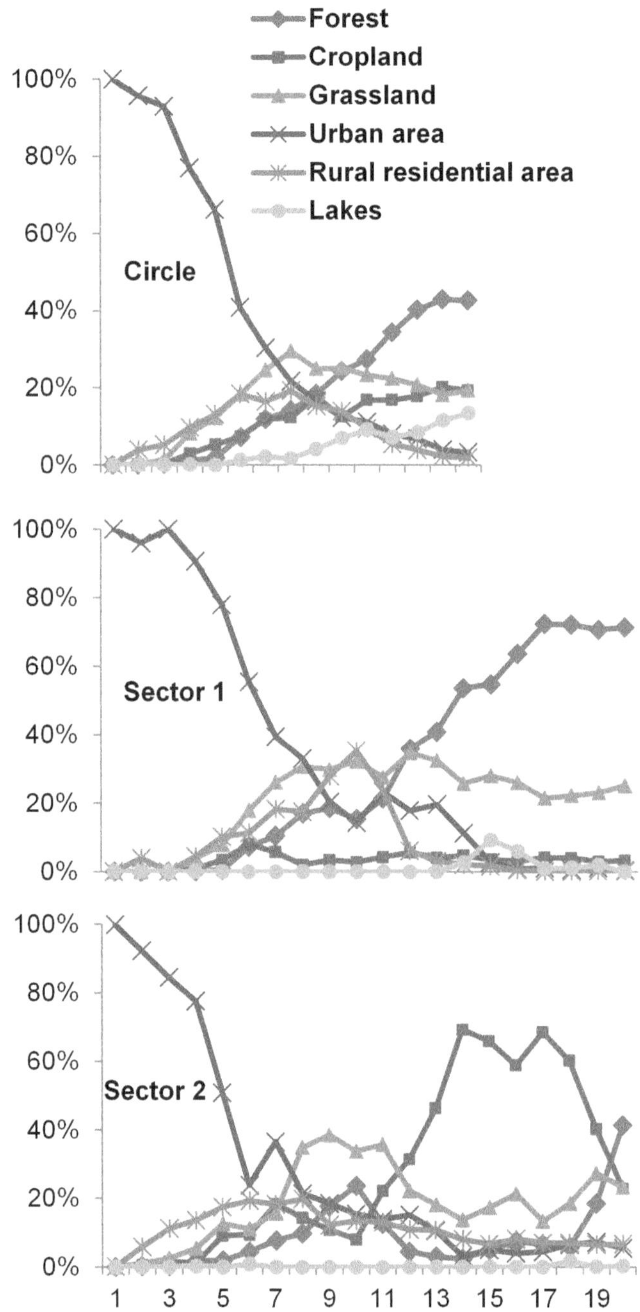

Figure 4: Land cover gradients along the Kunming urban-rural area. The X-axis is the distance in kilometers from the city center of Kunming. The Y-axis is the area proportions of different land cover classes in the total area of the associated concentric ring or concentric sector ring

4.2. Ecosystem service and socioeconomic gradients in Leipzig urban-rural area

All the five ecosystem services generally demonstrated an increase along the urban-rural gradient except above ground carbon storage (Figure 5). This exception showed a reversed trend that the urban area roughly had higher above ground carbon storage than the rural area. Having said this, the urban-rural spatial variations of the five ecosystem services were not even in different parts of the circle area. For example, sector 1 revealed highly irregular distributions of all the five ecosystem services and sector 2 showed special irregularities with above carbon storage, food supply and energy supply.

The five ecosystem services presented gradient characteristics in different degrees in Leipzig urban-rural area (Figure 6). Habitat quality obviously kept increasing from the city center to the concentric ring 25 kilometers away in all the three investigated areas (circle, sector 1 and sector 2). It showed a smoother rising trend in the circle area than in sector 1 and sector 2. Above ground carbon storage generally declined along the urban-rural gradient of Leipzig. It is noteworthy that this service showed two sudden surges in the regions around 17 and 23 kilometers away from the city center in sector 2.

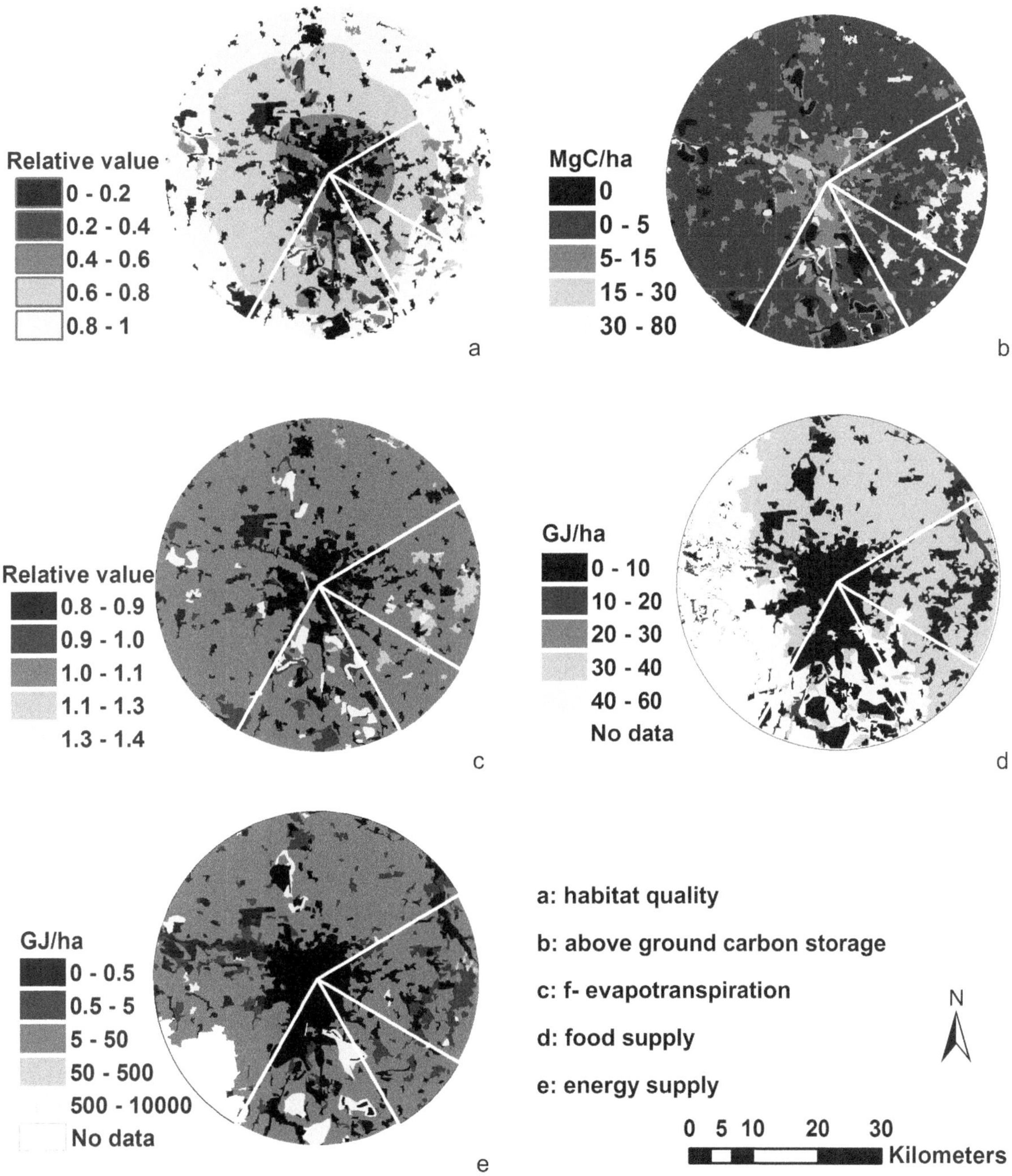

Figure 5: Spatial distributions of the five ecosystem services in the circle and sector areas of the Leipzig urban-rural gradient. a, b, c items are for 2006 and d, e items are for 2007 (d and e are adapted from Kroll et al., 2012)

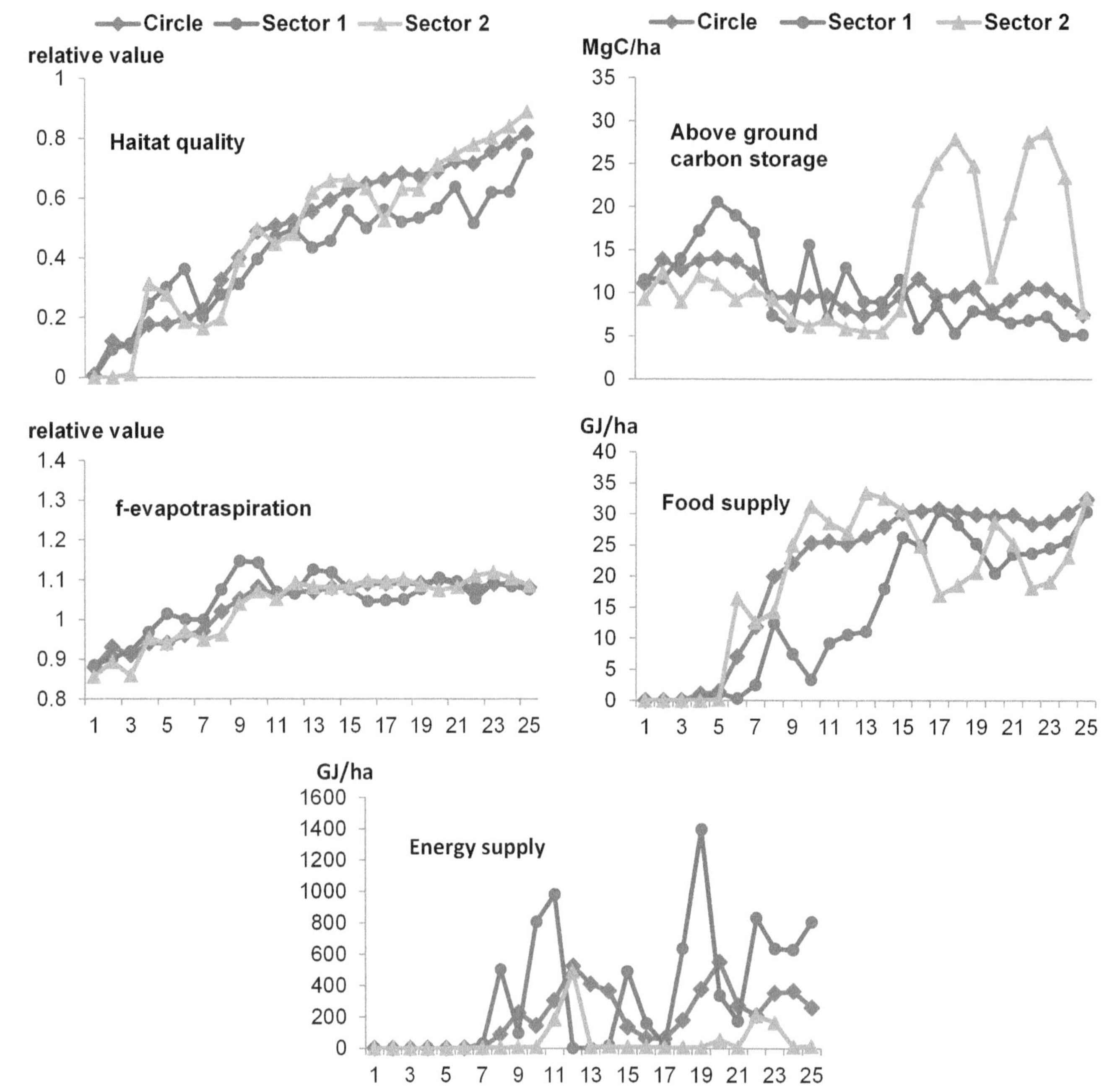

Figure 6: Gradients of the five ecosystem services in the circle and sector areas of the Leipzig urban-rural area. The X-axis is the distance in kilometers from the city center of Leipzig. The Y-axis is the average ecosystem service value of the associated concentric ring or concentric sector ring of the circle or sectors (food supply and energy supply are adapted from Kroll et al., 2012)

The characteristic of the f-evapotranspiration gradient was similar to that of habitat quality, revealing an apparent value increase along the urban-rural gradient. Although food supply and energy supply appeared with higher values in the rural area than in the urban area, they enjoyed considerably irregular gradients than the other three services. Noteworthily, the services in the circle areas changed (increased or decreased) more smoothly than in the two sector areas for all the five ecosystem services.

All the socioeconomic gradients of the Leipzig urban-rural area were much more regular than the ecosystem service gradients (Figure 7). Among the

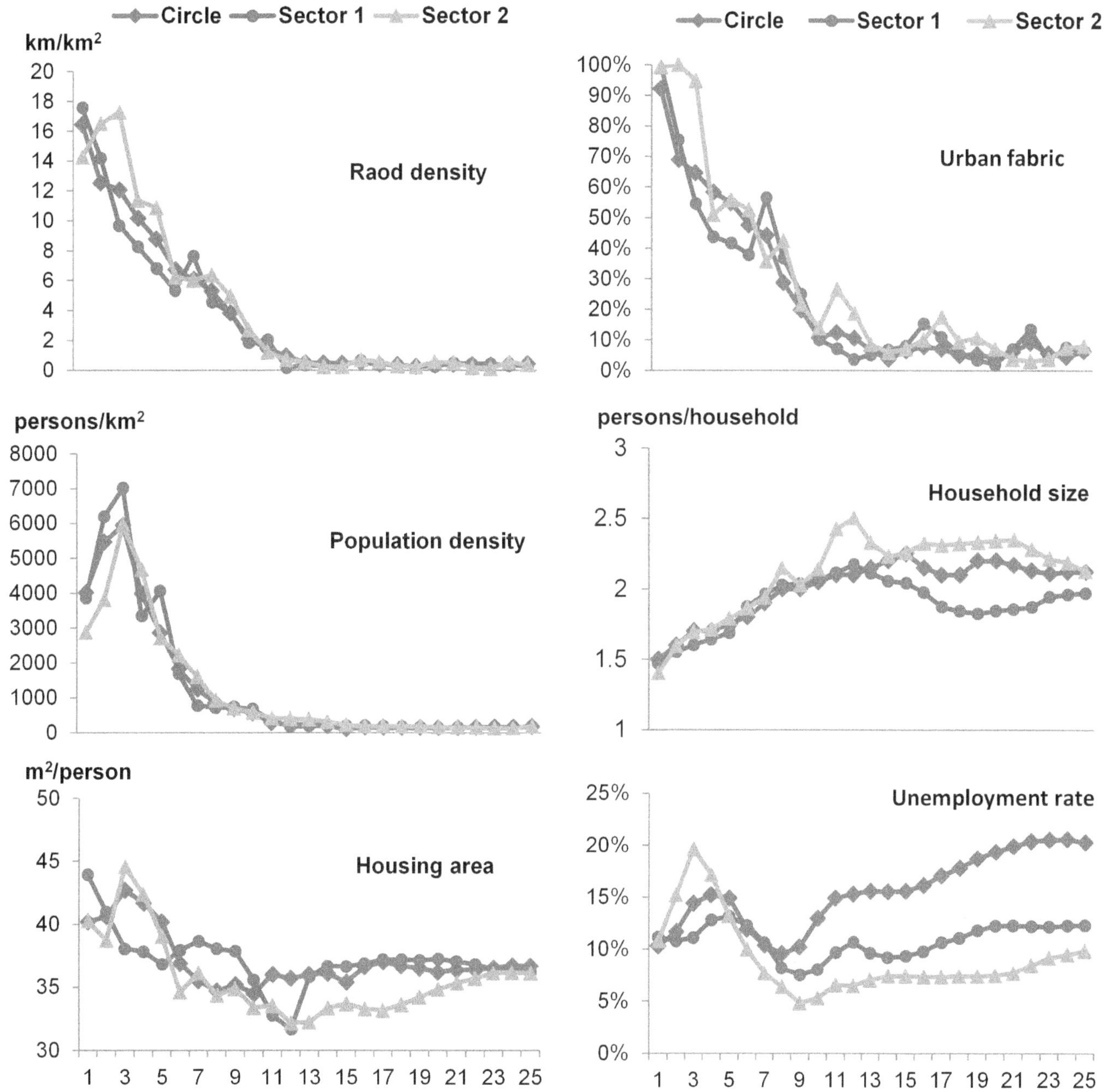

Figure 7: Gradients of the six socioeconomic items in the circle and sector areas of the Leipzig urban-rural area. The X-axis is the distance in kilometers from the city center of Leipzig. The Y-axis is the average value of the socioeconomic indicators in the associated concentric ring or concentric sector ring of the circle or sectors. Road density and urban fabric are for 2006 and the other four indicators are for 2005 (after Kroll and Kabisch, 2012)

socioeconomic dimensions, road density and urban fabric showed generally declining trends along the gradient. Differently, population density, household size, housing area as well as unemployment rate appeared notable turning points in some concentric rings. Among them, population density reached the peak 3 kilometers away from the Leipzig city center and then kept decreasing. Household size grew to the peaks around 13 kilometers away from the city center and either remained stable or slightly decreased in the circle or the two sector areas. Housing area declined to the minimum around 12 kilometers away from the city center and then turned growing slowly. Unemployment rate experienced firstly increasing and then decreasing and finally rising trends in all the three areas.

4.3. Ecosystem service and socioeconomic gradients in Kunming urban-rural area

The spatial distributions of the three ecosystem services presented consistency to some extent (Figure 8). Habitat quality appeared with an apparently increasing trend from the city center to the northern and western directions. Along the eastern and southern directions, the spatial changes of habitat quality were less regular than the other two directions. The NPP gradient was not as fine as the gradient of habitat quality, whereas, apparent spatial changes of the NPP gradient, which were similar to that of habitat quality, can be identified. With regard to food productivity, cropland area had considerably higher values than the other areas. The southern part of the Dianchi Lake had moderate values higher than the northern part of the lake and most of the forest, urban and grassland areas. The five branch-shaped areas can be even more clearly seen. However, different from habitat quality and NPP, these five branches areas had higher values than the surrounding areas.

The three ecosystem services had apparent gradient characteristics in the three gradient areas (the circle and two sectors, Figure 9). Habitat quality virtually kept increasing along the urban-rural gradient in all the three areas. Among them the growing line in the circle area was smoother than the two sectors'. NPP in the circle area monotonically rose from about 300 $gC/m^2/yr$ in the city center to nearly 900 $gC/m^2/yr$ in the concentric ring 15 kilometers away from the city center. In the two sectors, NPP appeared roughly growing trends. The food productivity gradients showed noticeable divergences in the three areas. In the circle area, food productivity appeared with a generally growing trend with trivial fluctuations. In sector 1 it remained below 100×10^4 yuan/km^2 all along the gradient; in sector 2, it fluctuated significantly along the urban-rural gradient.

The three socioeconomic dimensions also showed noticeable gradient features (Figure 8 and 9). All the three socioeconomic dimensions showed apparent high densities in the urban area and reductions in the areas spreading out from the city center. In addition, they extended to the southeast of the urban area and were distributed between the Dianchi Lake and the mountainous area in the east in moderate densities. Notably decreasing trends of the three socioeconomic indicators' values can be found along the urban-rural gradients in the circle and the 2 sector areas. Road density monotonically kept reducing along the urban-rural gradient in all the three investigated areas. Urban fabric quickly dropped from 100% at the city center to nearly 0% at the end of the gradient. Moreover, population density showed identical gradients in the three areas decreasing from 20,000 persons/km^2 to nearly zero. The differences from urban fabric and population density to road density were the several slight fluctuations in sector 2.

4.4. Gradient comparisons between Leipzig and Kunming

Notable similarities and differences emerge when ecosystem service gradients are compared between the urban-rural areas of Leipzig and Kunming. With regard to habitat quality, both research areas enjoyed marked increases along the gradients in the circle and sector areas. The most significant difference was the much smoother changing trends of habitat quality in the three investigated areas of Kunming compared to that of Leipzig. Another comparison can be found between above ground carbon storage of Leipzig urban-rural area and NPP of Kunming. Although these two indicators both relate to ecosystem's production function based on photosynthesis and can in some degree represent global climate regulation service, they demonstrated different gradient characteristics. As described in the preceding sections, above ground carbon storage roughly declined along the urban-rural gradient in the circle and sector 1 areas of Leipzig, despite of some fluctuations. Contrarily, NPP showed noticeably growing trends along the urban-rural gradients in the circle and sector 1 areas of Kunming. Food provisioning services of the two research areas showed similar variations along some gradient patterns and different spatial dynamics along others. In the circle area of Leipzig, food supply appeared with a firstly increasing and then stable trend. This phenomenon was similar to that of food productivity in the circle area of Kunming. However, food productivity in sector 1 of Kunming

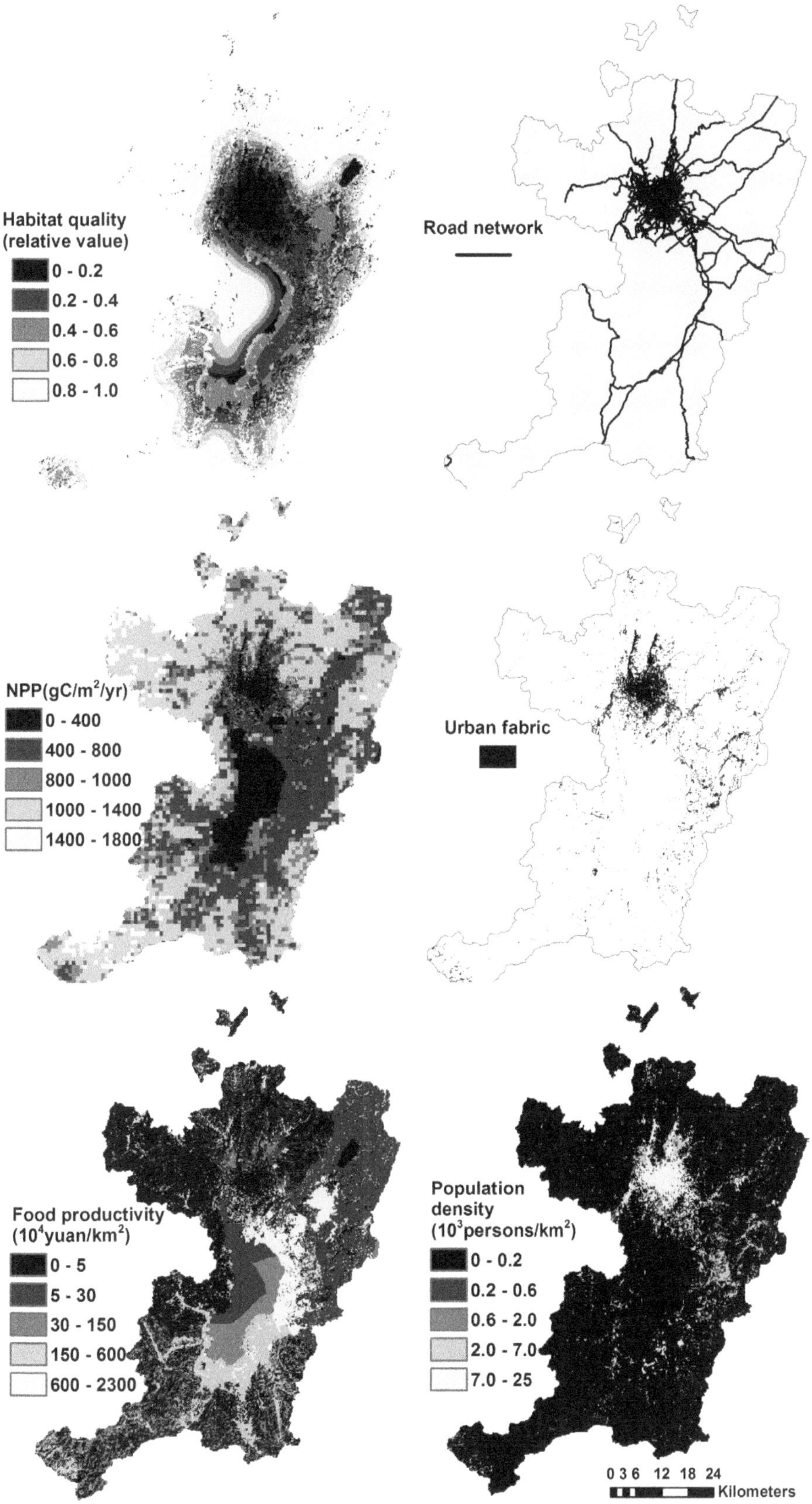

Figure 8: Spatial distributions of the three ecosystem services (left column) and the three socioeconomic items (right column) in the Kunming urban-rural area. NPP data are for 2006, road network relates to 2007 and the other four item are for 2010

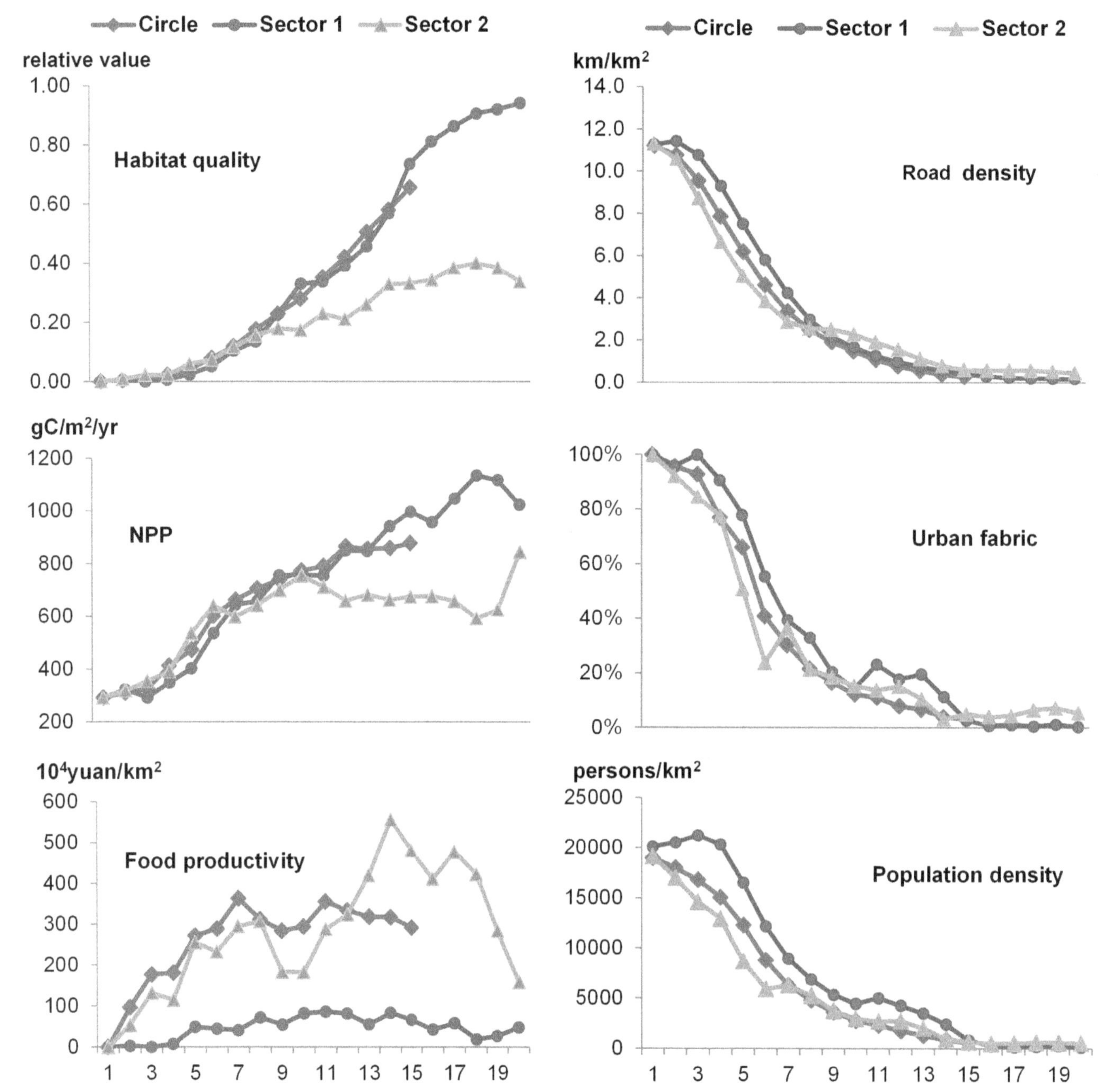

Figure 9: Gradients of the three ecosystem services (left column) and the three socioeconomic dimensions (right column) in the circle and sector areas of the Kunming urban-rural area. The X-axis is the distance in kilometers from the city center of Kunming. The Y-axis is the average value of the indicators in the associated concentric ring or concentric sector ring of the circle or sectors

remained stable at a low level along the gradient. This characteristic was considerably different from food supply along the two sector gradients of Leipzig, which can be characterized by marked fluctuations.

With respect to the socioeconomic gradients, the two research areas provided significant consistencies with the three identical indicators. The urban - rural gradients of road density and urban fabric of Kunming were apparently negative and considerably similar to those of Leipzig, respectively. These similarities occurred in all the three investigated areas. The most significant difference was that Kunming showed smoother trends of the indicator values than Leipzig. Population density also showed similarities between the two research areas by generally keeping dropping in the three analyzed gradient areas. The distinctions primarily existed in the first few concentric ring areas. In Leipzig, population density turned to drop quickly from the peak to zero from the third concentric ring in all the three gradient patterns. Contrarily, in the three analyzed areas of Kunming, population densities did not show apparent turning points and the declining velocities were lower than that of Leipzig.

4.5. Linkages between ecosystem service and socio-economic gradients

Leipzig

In Leipzig urban-rural area, some ecosystem service gradients showed similarities to that of socioeconomics; whereas, some others showed noticeable differences (Figure 6 and 7). The gradients of habitat quality, f-evapotranspiration, food supply and energy supply were significantly negatively correlated with the gradients of road density, urban fabric and population density in the circle and the two sector areas with $P<0.01$ (Table 2). The sole exception was the weaker correlation between energy supply and road density in sector 1, which was still significant at the 0.05 level. The gradient of above ground carbon storage was significantly positively correlated with the gradients of road density, urban fabric and population density in the circle and sector 1 areas. However, in sector 2, the linkages were much looser.

For the other three socioeconomic gradients, household size, housing area and unemployment rate, their correlations with the ecosystem service gradients were not as regular as road density, urban fabric or population density. For example, the correlations of household size with habitat quality, food supply and energy supply were not significant in sector 1. Furthermore, it was only negatively correlated with above ground carbon storage in the circle area. However, for its linkage with f-evapotranspiration, highly significant correlations were found in all the three areas. With regard to housing area and unemployment rate, their correlations with the five ecosystem services showed even more divergences. For instance, housing area was significantly negatively correlated with habitat quality solely in sector 1 and positively correlated with above ground carbon storage solely in the circle area. Whereas, its correlation with energy supply showed significance in all the three gradient areas. Regarding unemployment rate, positive significant correlations were solely found with habitat quality, f-evapotranspiration, food supply and energy supply in the circle area. In contrast, it had no significant correlations with the five ecosystem services in sector 1. Even more unexpectedly, it was negatively correlated with food supply and energy supply in sector 2.

Kunming

Ecosystem service-socioeconomic linkages can also be identified along the urban-rural gradient of Kunming (Figure 9). The marked increases of habitat quality in the three analyzed areas were noticeably contrary to the declining trends of road density, urban fabric and population density in the three areas. The NPP gradient in the circle area also showed opposite trend to the gradients of these three socioeconomic dimensions in the circle areas. The correlations between ecosystem service and socioeconomic gradients of Kunming presented more consistency than that of Leipzig (Table 3). Habitat quality and NPP both showed strongly negative correlations with road density, urban fabric and population density at the 0.01 significance level in the circle and the two sector areas.

Unexpected finding refers to the linkages of food productivity gradients with those of road density, urban fabric and population density in sector 1. Unlike the circle and sector 2 areas, where food productivity gradients were significantly negatively correlated with the three socioeconomic gradients, sector 1 did not exhibit close correlations between this service and the three socioeconomic dimensions.

Table 2: Spearman correlations between the ecosystem service and socioeconomic gradients of the circle and sector areas in the Leipzig urban-rural area. The sample values are the mean values of the indicators in the associated rings with all sample sizes of 25 (25 concentric rings in the circle area and 25 concentric sector rings in both of the two

		Habitat quality			Above ground carbon storage			f-evapotranspiration			Food supply			Energy supply		
		Circle	Sector 1	Sector 2	Circle	Sector 1	Sector 2	Circle	Sector 1	Sector 2	Circle	Sector 1	Sector 2	Circle	Sector 1	Sector 2
Road density	Circle	-0.927**			0.502*			-0.928**			-0.845**			-0.726**		
	Sector 1		-0.769**			0.425*			-0.640**			-0.744**			-0.472*	
	Sector 2			-0.876**			-0.244			-0.849**			-0.622**			-0.724**
Urban fabric	Circle	-0.879**			0.675**			-0.901**			-0.825**			-0.797**		
	Sector 1		-0.771**			0.454*			-0.698**			-0.651**			-0.548**	
	Sector 2			-0.940**			-0.205			-0.800**			-0.668**			-0.735**
Population density	Circle	-0.860**			0.513**			-0.919**			-0.897**			-0.616**		
	Sector 1		-0.878**			0.649**			-0.566**			-0.780**			-0.631**	
	Sector 2			-0.937**			-0.432*			-0.885**			-0.547**			-0.713**
Household size	Circle	0.791**			-0.631**			0.809**			0.792**			0.756**		
	Sector 1		0.225			-0.181			0.588**			0.253			0.300	
	Sector 2			0.567**			0.065			0.652**			0.634**			0.688**
Housing area	Circle	-0.197			0.574**			-0.242			-0.185			-0.528**		
	Sector 1		-0.563**			0.232			-0.555**			-0.426*			-0.421*	
	Sector 2			-0.259			0.282			-0.477*			-0.655**			-0.516**
Unem-ployment rate	Circle	0.903**			-0.351			0.800**			0.777**			0.552**		
	Sector 1		0.246			0.041			-0.327			.000			-0.157	
	Sector 2			-0.152			0.325			-0.320			-0.591**			-0.453*

Table 3: Spearman correlations between the ecosystem services and socioeconomic gradients of the circle and sector areas in Kunming urban-rural area. The sample values are the mean values of the indicators in the associated rings. The sample sizes are 15 for the circle area and 20 for the sectors (15 concentric rings in the circle area and 20 concentric sector rings in both of the two sectors). ** P< 0.01, * P<0.05

		Habitat quality			NPP			Food productivity		
		Circle	Sector 1	Sector 2	Circle	Sector 1	Sector 2	Circle	Sector 1	Sector 2
Road density	Circle	-1.000**			-0.989**			-0.736**		
	Sector 1		-0.995**			-0.977**			-0.375	
	Sector 2			-0.971**			-0.629**			-0.642**
Urban fabric	Circle	-1.000**			-0.989**			-0.736**		
	Sector 1		-0.976**			-0.973**			-0.384	
	Sector 2			-0.928**			-0.657**			-0.789**
Population density	Circle	-1.000**			-0.989**			-0.736**		
	Sector 1		-0.982**			-0.970**			-0.398	
	Sector 2			-0.972**			-0.642**			-0.687**

5 Discussion

5.1. Relationships between ecosystem service and socioeconomic gradients

Leipzig

As described in the results section, the gradients of habitat quality, f-evapotranspiration, food supply and energy supply had significantly negative correlations with the gradients of road density, urban fabric and population density in the circle and the two sector areas. These findings are logical as the share of artificial surfaces in the total area generally kept declining and the vegetation cover proportion (mainly including forest, grassland and agricultural areas) appeared with an increasing trend along the gradient (Figure 3). In other words, it is the land cover variabilities in urban-rural areas that build the linkages between ecosystem services and socioeconomics. The close connections between ecosystem service supplies and land cover patterns in urban-rural areas have been evidenced by an extensive body of literature (Burkhard et al., 2012; Haase et al., 2012; Kroll et al., 2012; Larondelle and Haase, 2013). High habitat quality (or biodiversity) and evapotranspiration are usually due to intensive natural vegetation (Burkhard et al., 2009; Schwarz et al., 2011; Tallis et al., 2013). Food supply is undoubtedly attributed to agricultural land and energy, as estimated in this research, is mainly provided by peripheral areas of cities. Although not as much as ecosystem services, the spatial characteristics of socioeconomics have been investigated in the land cover context of urban-rural areas by some studies (Hope et al., 2003; Luck et al., 2009). With regard to the direct relationships between ecosystem services and socioeconomic elements, we found limited literature. Among these studies, one provided by Buyantuyev and Wu (2010) revealed the negative correlation between daytime surface temperature and median family income in the Phoenix metropolitan region. Another source from the same area additionally reported a positive relationship between mean temperature and population density (Jenerette et al., 2007). This finding can to some extent prove the correlation between f-evapotranspiration and population density in our research because evapotranspiration tends to go in an opposite direction of temperature in urban-rural gradients (Adler and Tanner, 2013).

Looking into the possible reasons for the irregular correlations of the gradients of household size, housing area and unemployment rate with the ecosystem service gradients, we assumed that the uneven spatial variations of household size, housing area and unemployment rate had done major contributions (see the irregular gradients of these three indicators in Figure 7). These irregular spatial dynamics should originate from more complex socioeconomic determinants of these three elements compared to road density, urban fabric and population density.

Kunming

As found in this research, habitat quality and NPP had significantly negative correlations with road density, urban fabric and population density in the circle and the two sector areas. These findings agree with the gradient characteristics of the five indicators. In other words, the three socioeconomic dimensions presented apparent decreasing trends and the two ecosystem services showed opposing trends along the gradient in all the three areas (Figure 9). The determinants of these phenomena should be the spatial dynamic characteristics of land covers as the values of all these five indicators are strongly associated with the land cover status (Burkhard et al., 2009; Hawbaker et al., 2005; Helmut Haberl, 2001; Kroll and Kabisch, 2012). For the spatial variations of land covers, the gradients showed noticeable declines of urban area shares and general increases of forest and grassland shares in all the three areas (Figure 4). The phenomenon that food productivity didn't show close correlations with the three socioeconomic dimensions in sector 1 can be easily explained by the ignorable proportion of cropland along the gradient of sector 1 (Figure 4, also see the land cover map in Figure 2).

5.2. Comparisons between gradient patterns and case study areas

As described in the results section, some ecosystem services showed similar spatial variations in the three gradient types (the circle and the two sectors areas) in the Leipzig and the Kunming urban-rural areas. Others, such as food supply, above ground carbon storage and energy supply of Leipzig and food productivity of Kunming presented different gradient patterns (Figure 6 and 9). These variation divergences are understandable when looking into the land cover characteristics of the different gradient patterns. For example, forest and semi-natural areas intensively emerged in the southeast of sector 2 area of Leipzig (Figure 1), causing the sharp increase and the two peaks of above ground carbon storage after the 15th concentric sector ring (Figure 6). The differences of the gradient patterns of ecosystem services in different areas imply that although the concentric ring gradient pattern can comprehensively capture the spatial variation information along the urban-rural gradient, it can be unsuitable in some cases. For instance, it is problematic when the land cover structure does not exhibit the same pattern in all directions from the city center as shown by the two study areas in this research. In the regions with unique land cover patterns (e.g., sector 1 and 2 of Leipzig and sector 2 of Kunming in this study), refined gradient patterns that can more accurately capture the details of the whole research area will be more preferable. These gradient patterns include the concentric sector ring gradient proposed in this study and the transect gradient used in some other research (Banzhaf et al., 2009; Hahs and McDonnell, 2006; Luck and Wu, 2002). Having said this, as the concentric sector ring and the transect gradients rely on directions, they can solely treat the chosen directions and lose the information of the regions in other directions. Considering the advantages and disadvantages of different gradient patterns, a combination of them can be a promising solution to both comprehensively and detailedly investigate ecosystem service gradients.

Compared to ecosystem services, the investigated socioeconomic dimensions presented much more consistency between the three gradient areas of both Leipzig and Kunming. The major disagreements referred to household size, housing area and unemployment rate of Leipzig (Figure 7). As discussed earlier, these three aspects are affected by more complex socioeconomic reasons compared to road density, urban fabric and population density, which are primarily based on urbanization levels (Hahs and McDonnell, 2006). Accordingly, we assume that the concentric ring gradient is appropriate to reveal the spatial dynamics of road density, urban fabric and population density in urban-rural gradients. Whereas, gradient patterns that can capture details (e.g., concentric sector ring and transect) are additionally required to treat more complex socioeconomic dimensions, such as household size, housing area and unemployment rate.

Comparing different study areas, we found limited ecosystem services presenting similar gradient patterns in urban-rural area. Different regions tend to exhibit divergent spatial variation characters of ecosystem services from the city center to outskirts. For example, with respect to f-evapotranspiration and above ground carbon storage, our findings in Leipzig are inconsistent with those from Larondelle and Haase (2013) in several other European cities. Their results also showed divergent ecosystem service gradient patterns in urban-rural areas of different regions. According, we can induce that although ecosystem service gradients in certain patterns can be found in urban-rural areas, typical gradient patterns suiting different regions are scarcely exiting. In other words, the gradient patterns of ecosystem services found in our study cannot be generalized to other cities. The possible reason, as we mentioned before, are the distinct land cover structures and distributions of different urban-rural regions.

With respect to socioeconomic dimensions, identical urban-rural gradients have been found between the two study areas in our study. Additionally, similar gradient patterns are identified in other studies. For example, Leipzig and Kunming have identical urban fabric gradients patterns (Figure 7 and 9), which are similar to the ones of Manchester, Berlin, Hamburg as well as Munich delineated by Kroll and Müller (2011) and Kroll and Kabisch (2012). With regard to population density, consistency of urban-rural gradient patterns exit between Leipzig (this study) and Berlin, Hamburg and Munich (Kroll and Müller, 2011) and between Kunming (this study) and Warsaw (Kroll and Müller, 2011). Regarding the three more intricate socioeconomic dimensions, household size, housing area and unemployment rate, urban-rural gradient patterns primarily present disagreements between different regions (e.g., household size between Leipzig in this study and Hamburg and Munich in the research of Kroll and Kabisch (2012) and housing area between Leipzig and Manchester and Hamburg presented by Kroll and Müller (2011)). To sum up, we can tentatively assume the existence of typical gradient patterns of some socioeconomic dimensions (e.g., urban fabric discussed above) in different urban-rural areas.

5.3. Uncertainty analysis

An ecosystem service assessment is always accompanied by uncertainties originating from the complexity of the ecological system, respondents' preferences and technical problems (Hou et al., 2013; Elmqvist et al., 2010; Scolozzi et al., 2011). In this research, the creation of gradients relied on averaging the values of heterogeneous ecosystem services and socioeconomic dimensions in each ring area, leading to uncertainties with the produced gradient. These uncertainties may cause bias in assessment (Li and Wu, 2006), such as unreliable characterization of gradients and doubtable revelations of relationships between ecosystem services and socioeconomics. Therefore, it is imperative to obtain the information of uncertainties with the results. Considering the large number of ecosystem services and socioeconomic dimensions investigated in this study, we solely chose four of them as examples to quantify the uncertainties of gradients in the three gradient areas (the circle and the two sectors). The analyzed indicators are above ground carbon storage and housing area of Leipzig and habitat quality and population density of Kunming. Referring to the mainstreaming methods showing uncertainties with value averaging in ecosystem service studies (Larondelle and Haase, 2013; Yanai et al., 2010; Zhu and Carreiro, 2004), we used standard deviation (SD) to show the value ranges of the four indicators in each ring of the circle area and the two sectors. Accordingly, uncertainty gradients of the four examined indicators were created (Figure 10).

In general, the established uncertainty gradients did not exhibit typical patterns. The magnitudes and changes of uncertainties were not depended on the magnitudes of the mean values or the distance from the city center. Taking the above ground carbon storage of Leipzig as an example, in sector 1, the greatest uncertainty (greatest SD value) appeared at the 12th ring, where the mean value was not the highest among all the rings (Figure 10, b). In sector 2, the uncertainties remained at low levels before the 15th ring and soared between the distance of 15 km and 24 km from the city center (Figure 10, c). Looking into the map of ecosystem services spatial distribution and the land cover map (Figure 1 and 5), we can attribute the considerable uncertainties in these rings to the divergences of the above carbon storage values and the heterogeneous land covers in these areas. For the other three indicators, it is also convenient to observe the uncertainty gradient patterns, compare the uncertainty magnitudes between different rings and find out the areas with most or least uncertainties. After testing this uncertainty gradient method, we assume that this method is appropriate to quantitatively capture the uncertainties in exploring ecosystem services and socioeconomic gradients in urban-rural areas.

6. Conclusions

In this article we draw on the urban-rural gradient principle to offer a methodology to investigate the linkages between ecosystem services and socioeconomics in urban-rural areas. Interesting findings were derived by performing empirical studies with the developed methods in two case urban-rural areas respectively in Germany and China. With these results, we can now answer the three research questions asked at the beginning of this article as follows:

1) The investigated ecosystem services and socio-economic dimensions of the two study areas presented regularities of spatial patterns in urban-rural gradients in different degrees. Some of them showed either generally increasing or decreasing trends along urban-rural gradients, others exhibited considerable fluctuations.

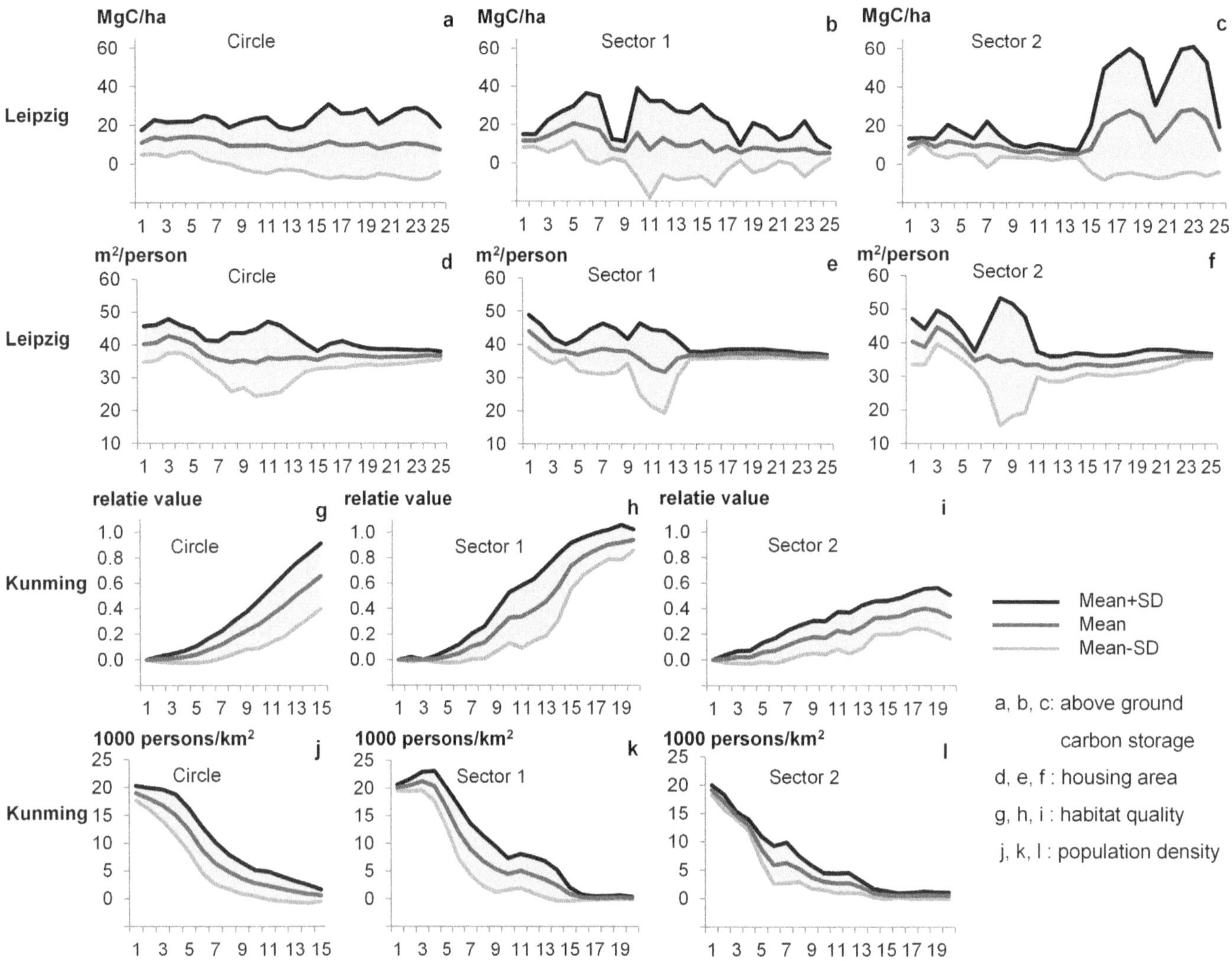

Figure 10: Quantified uncertainties of ecosystem services and socioeconomic gradient calculations for the Leipzig and Kunming urban-rural areas. The X-axis is the distance in kilometers from the city center. Above ground carbon storage and housing area were investigated for Leipzig. Habitat quality and population density were analyzed for Kunming. Uncertainties were quantified by the ranges between the values of mean plus standard deviation (SD), mean values and the values of mean minus SD

2) Ecosystem service gradient characters of the circle and the two sector gradient areas presented more divergences than that of socioeconomic gradients in both the two study areas. The concentric ring gradient pattern can more comprehensively capture the overall spatial dynamics of ecosystem services and socioeconomics along the gradients. In contrary, the concentric sector ring gradient pattern can better reveal the details of especially concerned areas in specific directions from the city center. For the two study areas, urban-rural gradients of identical ecosystem services did not necessarily present consistent patterns and gradients of the same socioeconomic dimensions showed similar patterns.

3) Strong linkages between some ecosystem service and socioeconomic gradients were observed and statistically proven in both the two study areas. The socioeconomic dimensions more based on land covers had stronger correlations with ecosystem services. Moreover, the correlations of a certain pair of ecosystem service and socioeconomic gradients could be inconsistent in different gradient areas of either study area.

In addition, the results of our study suggest another two important messages. By integrating our findings and the published work, we further proved that a typical model of ecosystem services gradient does not exist for different cities. The other message tells the necessity to be aware of uncertainties and effectively incorporate uncertainty analysis into the investigations of ecosystem service and socioeconomic gradients.

Novel innovations in the field of ecosystem services are emerging regularly. In this exiting time we call for future research to determine whether our arguments for the interrelationships between ecosystem services and socioeconomics in urban-rural areas are supported by more empirical evidence. We hope this paper will help draw more research interests to elevate our understanding of human-environmental interactions in urban-rural systems.

Acknowledgements

This work was financially supported by the EU Integrated Project PLUREL (contract No. 036921) and the project "Regulation of Integrative Land Use in Urban-Rural Areas with Ecological Principles in Lake Catchments of the Yungui Plateau" funded by the Department of Land and Resources of Yunnan Province, China. We are grateful for the time and efforts from the anonymous reviewers on improving this paper.

References

Adler, F.R. & C.J. Tanner 2013. Urban Ecosystems: Ecological Principles for the Built Environment. Cambridge University Press, Cambridge.

Alcamo, J.; van Vuuren, D. & W. Cramer 2005. Changes in ecosystem services and their drivers across the scenarios. In: MA (eds.): Ecosystems and Human Well Being: Scenarios. Island Press; Washington, DC.

Banzhaf, E.; Grescho, V. & A. Kindler 2009. Monitoring urban to peri-urban development with integrated remote sensing and GIS information: a Leipzig, Germany case study. International Journal of Remote Sensing 30, 1675-1696.

Bastian, O. 2013. The role of biodiversity in supporting ecosystem services in Natura 2000 sites. Ecological Indicators 24, 12-22.

Berry, B. & A. Okulicz-Kozaryn 2011. An urban-rural happiness gradient. Urban Geography 32, 871-883.

Burkhard, B.; Kandziora, M.; Hou, Y. & F. Müller 2014. Ecosystem service potentials, flows and demands – Concepts for spatial localisation, indication and quantification. Landscape Online 34, 1-32.

Burkhard, B.; Kroll, F. & F. Müller 2009. Landscapes' capacities to provide ecosystem services – A concept for land-cover based assessments. Landscape Online 15, 1-22.

Burkhard, B.; Kroll, F.; Nedkov, S. & F. Müller 2012. Mapping ecosystem service supply, demand and budgets. Ecological Indicators 21, 17-29.

Buyantuyev, A. & J. Wu 2010. Urban heat islands and landscape heterogeneity: linking spatiotemporal variations in surface temperatures to land-cover and socioeconomic patterns. Landscape Ecology 25, 17-33.

Chinnaiah, B. & V. Madhu 2010. Primary productivity of Darmasagar lake in Adilabad, Andhra Pradesh, India. International Journal of Pharmacy & Life Sciences 1, 437-439.

Costanza, R.; D'Arge, R.; de Groot, R.; Farber, S.; Grasso, M.; Hannon, B.; Limburg, K.; Naeem, S.; O'Neill, R.V.; Paruelo, J.; Raskin, R.G.; Sutton, P. & M. van den Belt 1997. The value of the world's ecosystem services and natural capital. Nature 387, 253-260.

Couch, C. & J. Karecha 2006. Controlling urban sprawl: Some experiences from Liverpool. Cities 23, 353-363.

de Groot, R.S.; Alkemade, R.; Braat, L.; Hein, L. & L. Willemen 2010. Challenges in integrating the concept of ecosystem services and values in landscape planning, management and decision making. Ecological Complexity 7, 260-272.

Dow, K. 2000. Social dimensions of gradients in urban ecosystems. Urban Ecosystems 4, 255-275.

Duraiappah, A.K. 2011. Ecosystem services and human well-being: Do global findings make any sense? Bioscience 61, 7-8.

Elmqvist, T.; Maltby, E.; Barker, T.; Mortimer, M. & C. Perrings 2010. Biodiversity, ecosystems and ecosystem services. In: P. Kumar (ed.): The Economics of Ecosystems and Biodiversity: Ecologic and Economic Foundations. Earthcan; London & Washington, DC, pp. 41-111.

Fachagentur für Nach wachsende Rohstoffe e.V. 2014. http://www.fnr.de/ (Date: 22.1.2015).

Federal Agency for Cartography and Geodesy 2014. http://www.bkg.bund.de/DE/Home/homepage__node.html__nnn=true (Date: 22.1.2015).

Geospatial Information and Surveying, Saxony 2014. http://www.landesvermessung.sachsen.de/ (Date: 22.1.2015).

German Federal Environmental Agency 2014. German Aerospace Center 2010. CORINE Land Cover 2006. http://www.corine.dfd.dlr.de/intro_de.html (Date: 22.1.2015).

Gibbs H.K 2006. Olson's major world ecosystem complexes ranked by carbon in live vegetation: An updated database using the GLC2000 land cover product. http://cdiac.ornl.gov/epubs/ndp/ndp017/ndp017b.html (Date: 22.1.2015).

Haase, D.; Schwarz, N.; Strohbach, M.; Kroll, F. & R. Seppelt 2012. Synergies, Trade-offs, and losses of ecosystem services in urban regions: An integrated multiscale framework applied to the Leipzig-Halle region, Germany. Ecology and Society 17, 22.

Haberl, H.; Erb, K. & F. Krausmann 2001. Changes in ecosystem processes induced by land use: Human appropriation of aboveground NPP and its influence on standing crop in Austria. Global Biogeochemical Cycles 15, 929-942.

Hahs, A.K. & M.J. McDonnell 2006. Selecting independent measures to quantify Melbourne's urban-rural gradient. Landscape and Urban Planning 78, 435-448.

Haines-Young, R. & M. Potschin 2010. The links between biodiversity, ecosystem services and human well-being. In: D. Raffaelli & C. Frid (eds.): Ecosystem Ecology: A New Synthesis. BES Ecological Reviews Series. CUP; Cambridge, pp. 110-139.

Hawbaker, T.J.; Radeloff, V.C.; Hammer, R.B. & M.K. Clayton 2005. Road density and landscape pattern in relation to housing density, and ownership, land cover, and soils. Landscape Ecology 20, 609-625.

Hope, D.; Gries, C.; Zhu, W.X.; Fagan, W.F.; Redman, C.L.; Grimm, N.B.; Nelson, A.L.; Martin, C. & A. Kinzig 2003. Socioeconomics drive urban plant diversity. Proceedings of the National Academy of Sciences of the United States of America 100, 8788-8792.

Hou, Y.; Burkhard, B. & F. Müller 2013. Uncertainties in landscape analysis and ecosystem service assessment. Journal of Environmental Management 127, Supplement, S117-S131.

Hou, Y.; Zhou, S.; Burkhard, B. & F. Müller 2014. Socioeconomic influences on biodiversity, ecosystem services and human well-being: A quantitative application of the DPSIR model in Jiangsu, China. Science of the Total Environment 490, 1012-1028.

Ingraham, M.W. & S.G. Foster 2008. The value of ecosystem services provided by the U.S. National Wildlife Refuge System in the contiguous U.S. Ecological Economics 67, 608-618.

Institute for Energy and Environment 2007. Schlüsseldaten Klimagasemissionen. http://www.ufop.de/files/6813/3918/4231/Bericht_Klimagas.pdf (Date: 22.1.2015).

Jansson, A. & S. Polasky 2010. Quantifying biodiversity for building resilience for food security in urban landscapes: Getting down to business. Ecology and Society 15.

Jenerette, G.D.; Harlan, S.L.; Brazel, A.; Jones, N.; Larsen, L. & W.L. Stefanov 2007. Regional relationships between surface temperature, vegetation, and human settlement in a rapidly urbanizing ecosystem. Landscape Ecology 22, 353-365.

Kandziora, M.; Burkhard, B. & F. Müller 2013. Interactions of ecosystem properties, ecosystem integrity and ecosystem service indicators—A theoretical matrix exercise. Ecological Indicators 28, 54-78.

Kroll, F. & N. Kabisch 2012. The relation of diverging urban growth processes and demographic change along an urban-rural gradient. Population Space and Place 18, 260-276.

Kroll, F. & F. Müller 2011. Can the principle of self-organized gradients be applied for human systems. International Journal of Design & Nature and Ecodynamics 6, 90-108.

Kroll, F.; Müller, F.; Haase, D. & N. Fohrer 2012. Rural–urban gradient analysis of ecosystem services supply and demand dynamics. Land Use Policy 29, 521-535.

KTBL 2005. Faustzahlen für die Landwirtschaft. KTBL-Schriftenvertrieb im Landwirtschaftsverlag, Münster-Hiltrup.

Larondelle, N. & D. Haase 2013. Urban ecosystem services assessment along a rural–urban gradient: A cross-analysis of European cities. Ecological Indicators 29, 179-190.

Lautenbach, S.; Kugel, C.; Lausch, A. & R. Seppelt 2011. Analysis of historic changes in regional ecosystem service provisioning using land use data. Ecological Indicators 11, 676-687.

Leipzig Amt für Statistik und Wahlen 2009. Statistisches Jahrbuch. Stadt Leipzig, Amt für Statistik und Wahlen (in German).

Li, H. & J. Wu 2006. Uncertainty analysis in ecological studies: An overview. In: J. Wu; K.B. Jones; H. Li & O.L. Loucks (eds.): Scaling and Uncertainty Analysis in Ecology. Springer; Netherlands, pp. 45.

Lin, W.; Wang, C.; Zhao, M.; Huang, J.; Shi, R.; Liu, Y. & J. Gao 2008. Estimation urban forests NPP based on forest inventory data and remote sensing. Ecology and Environment 17, 766-770 (in Chinese).

Luck, G.W.; Smallbone, L.T. & R. O'Brien 2009. Socio-economics and vegetation change in urban ecosystems: patterns in space and time. Ecosystems 12, 604-620.

Luck, M. & J. Wu 2002. A gradient analysis of urban landscape pattern: a case study from the Phoenix metropolitan region, Arizona, USA. Landscape Ecology 17, 327-339.

MA 2005. Ecosystems and Human Well-being-Synthesis. Island Press, Washington DC.

McDonnell, M.J. & A.K. Hahs 2008. The use of gradient analysis studies in advancing our understanding of the ecology of urbanizing landscapes: current status and future directions. Landscape Ecology 23, 1143-1155.

McDonnell, M.J. & S.T.A. Pickett 1990. Ecosystem structure and function along urban-rural gradients: An unexploited opportunity for ecology. Ecology 71, 1232-1237.

McDonnell, M.J.; Pickett, S.T.A.; Groffman, P.; Bohlen, P.; Pouyat, R.; Zipperer, W.; Parmelee, R.; Carreiro, M. & K. Medley 1997. Ecosystem processes along an urban-to-rural gradient. Urban Ecosystems 1, 21-36.

Melillo, J.M.; McGuire, A.D.; Kicklighter, D.W.; Moore, B.; Vorosmarty, C.J. & A.L. Schloss 1993. Global climate change and terrestrial net primary production. Nature 363, 234-240.

Ministry of Agriculture and Environment Saxony-Anhalt 2007. Biomassepotenzialstudie 2007 für das Land Sachsen-Anhalt, Magdeburg.

Müller, F. & B. Burkhard 2012. The indicator side of ecosystem services. Ecosystem Services 1, 26-30.

National Bureau of Statistics of China 2011. China City Statistical Year Book 2011. China Statistic Press, Beijing (in Chinese).

Nelson et al. 2005. Drivers of change in ecosystem condition and services. In: MA (eds.): Ecosystems and Human Well Being: Scenarios. Island Press; Washington, DC, pp. 173-222.

Nuissl, H. & D. Rink 2005. The 'production' of urban sprawl in eastern Germany as a phenomenon of post-socialist transformation. Cities 22, 123-134.

Numerical Terradynamic Simulation Group 2013. MODIS GPP/NPP Project (MOD17). http://www.ntsg.umt.edu/project/mod17 (Date: 22.1.2015).

Piorr, A.; Ravetz, J. & Tosics, I. (eds.) 2011. Peri-urbanisation in Europe: Towards a European Policy to Sustain Urban-Rural Futures, University of Copenhagen/Academic Books Life Sciences. p144.

Pouyat, R.V. & M.J. McDonnel 1991. Heavy-metal accumulations in forest soils along an urban-rural gradient in southeastern new-york, UAS. Water Air and Soil Pollution 57-8, 797-807.

Radford, K.G. & P. James 2013. Changes in the value of ecosystem services along a rural-urban gradient: A case study of Greater Manchester, UK. Landscape and Urban Planning 109, 117-127.

Raudsepp-Hearne, C.; Peterson, G.D.; Tengo, M.; Bennett, E.M.; Holland, T.; Benessaiah, K.; MacDonald, G.K. & L. Pfeifer 2010. Untangling the environmentalist's paradox: Why is human well-being increasing as ecosystem services degrade? Bioscience 60, 576-589.

Saxon Office of Statistics 2010. Population Statistics, territorial status. www.statistik.sachsen.de (Date: 22.1.2015).

Saxon State Ministry of the Environment and Agriculture 2007. Energie für die Zukunft, Sachsens Potenzial an nachwachsenden Rohstoffen/Biomasse, Dresden.

Saxon Upper Mining Authority 2007. Jahresbericht 2007, Freiberg.

Schwarz, N.; Bauer, A. & D. Haase 2011. Assessing climate impacts of planning policies—An estimation for the urban region of Leipzig (Germany). Environmental Impact Assessment Review 31, 97-111.

Scolozzi, R.; Morrib, E. & R. Santolinib 2011. Delphi-based change assessment in ecosystem service values to support strategic spatial planning in talian landscapes. Ecological Indicators 21, 134-144.

Statistics Bureau of Kunming 2011. Kunming Statistical Year Book 2011. China Statistics Press, Beijing (in Chinese).

Strohbach, M.W. & D. Haase 2012. Above-ground carbon storage by urban trees in Leipzig, Germany: Analysis of patterns in a European city. Landscape and Urban Planning 104, 95-104.

Sun, L.; Chen, D.; Wang, Q.; Gao, Y. & D. Jin 2011. Spatio-temporal distribution of chlorophyll a concentration and primary productivity between winter in 2009 and spring in 2010 of Lake Dianchi. Journal of Shanghai Normal University (Natural Sciences) 40, 191-196 (in Chinese).

Tallis, H.T.; Ricketts, T.; Guerry, A.D.; Wood, S.A.; Sharp, R.; Nelson, E.; Ennaanay, D.; Wolny, S.; Olwero, N.; Vigerstol, K.; Pennington, D.; Mendoza, G.; Aukema, J.; Foster, J.; Forrest, J.; Cameron, D.; Arkema, K.; Lonsdorf, E.; Kennedy, C.; Verutes, G.; Kim, C.K.; Guannel, G.; Papenfus, M.; Toft, J.; Marsik, M.; Bernhardt, J.; Griffin, R.; Glowinski, K.; Chaumont, N.; Perelman, A. & M. Lacayo 2013. InVEST 2.5.5 User's Guide. The Natural Capital Project, Stanford.

TEEB 2010. The Economics of Ecosystems and Biodiversity: Ecological and Economic Foundations. Earthscan, London and Washington.

Yanai, R.D.; Battles, J.J.; Richardson, A.D.; Blodgett, C.A.; Wood, D.M. & E.B. Rastetter 2010. Estimating uncertainty in ecosystem budget calculations. Ecosystems 13, 239-248.

Zhao, M. & G. Zhou 2005. Estimation of biomass and net primary productivity of major planted forests in China based on forest inventory data. Forest Ecology and Management 207, 295-313.

Zhou, X. & Y. Wang 2011. Spatial-temporal dynamics of urban green space in response to rapid urbanization and greening policies. Landscape and Urban Planning 100, 268-277.

Zhu, W. & M.A. Carreiro 2004. Temporal and spatial variations in nitrogen transformations in deciduous forest ecosystems along an urban - rural gradient. Soil Biology & Biochemistry 36, 267-278.

Appendix

Table 4: Date uses, time, scales and sources for the study of Leipzig urban-rural area

Data	Time	Scale	Publisher/source
Population number	2005	Municipality, city district	Statistical Office of the Free State of Saxony; Statistical Office of Saxony-Anhalt
Administrative borders of municipalities and city district	2001	Municipality, city district	German Federal Statistical Office (GENESIS)
Corine Land Cover	2006	1:100,000	German Federal Environmental Agency (2014)
Soil map of Saxony including usable field capacity and actual root depth	2007	1:200,000	Saxon State Department of the Environment and Geology
Soil survey map Germany No. CC 4734 Leipzig	2009	1:200,000	German Agency for Geology and Natural Resources Hanover
Soil fertility map of Saxony	2007	1:200,000	Saxon State Department of the Environment and Geology
Evapotranspiration values linked to land cover	2011	Corine Land Cover classes	Larondelle and Haase (2013); Schwarz et al. (2011)
Above ground carbon storage values linked to land cover	2013	Corine Land Cover classes	Gibbs (2006); Larondelle and Haase (2013); Strohbach and Haase (2012)
Total area of bioenergy crops (differentiated after crop types)	2007	Federal State	Fachagentur für Nach wachsende Rohstoffe e.V. (2014); Department of agriculture, forestry and horticulture Saxony-Anhalt; Ministry of Agriculture and Environment Saxony-Anhalt (2007); Saxon State Ministry of the Environment and Agriculture (2007)
Energy yield per ha crop type	2007	No scale	Institute for Energy and Environment (2007)
Wind, water and solar energy plants with installed power	2007	Exact locations	Vattenfall Europe AG (2009), URL: http://www.vattenfall.com/en/index.htm (Date: 22.1.2015)
Lignite extraction	2007	Lignite extraction site	Saxon Upper Mining Authority (2007)
Crop yield	2007	District	Statistical Office of the Free State of Saxony; Statistical Office of Saxony-Anhalt
Fruit yield, livestock, game	2007	Federal State	Saxon State Ministry of the Environment and Agriculture; KTBL (2005)
Road network	2006	Germany, city district	Federal Agency for Cartography and Geodesy (2014); Geospatial Information and Surveying, Saxony (2014)
Railway network	2008	Leipzig city district	Geospatial Information and Surveying, Saxony (2014)
Urban fabric	2006	1:100,000	Corine land cover from German Federal Environmental Agency (2014)
Household size	2005	Municipality, city district	Statistical Office Leipzig, Statistical Office of the Free State of Saxony
Housing area	2005	Municipality, city district	Statistical Office Leipzig; Statistical Office of the Free State of Saxony
Unemployment	2005	Municipality, city district	Statistical Office Leipzig; Statistical Office of the Free State of Saxony

Table 5: Date uses, time, scales and sources for the study of Kunming urban-rural area

Data	Time	Scale	Publisher/source
Administrative borders of districts and counties	2000	1:4,000,000, China	Data Sharing Infrastructure of Earth System Science, http://www.geodata.cn/Portal/?isCookieChecked=true (Date: 22.1.2015)
Land cover map	2010	1:100,000, Kunming prefecture-level city	Produced by the interpretation of 30-meter Landsat thematic mapper (TM) data in combination of a 1:100,000 topographic map of the research area
Net primary production raster map for vegetation covers	2006	1 km × 1 km, global	Numerical Terradynamic Simulation Group (2013)
Net primary production of urban area	2006	Shanghai city district	Lin et al. (2008)
Net primary production of the Dianchi Lake	2009-2010	the Dianchi Lake	Sun et al. (2011); Chinnaiah and Madhu (2010)
Farming, animal husbandry and fishery yields	2010	Districts and counties, Kunming	Kunming Statistical Yearbook 2011, Statistics Bureau of Kunming
Road network	2007	China	National Administration of Surveying, Mapping and Geoinformation of China
Urban fabric	2010	1:100,000, Kunming prefecture-level city	Kunming Land and Resource Bureau
Population density	2010	Districts and counties, Kunming	Kunming Statistical Yearbook 2011, Statistics Bureau of Kunming

Do You Have 5 Minutes To Spare? –The Challenges Of Stakeholder Processes In Ecosystem Services Studies

Koschke, L.[1], Van der Meulen, S.[2], Frank, S.[1,5], Schneidergruber, A.[3], Kruse, M.[4], Fürst, C.[5], Neubert, E. [1], Ohnesorge, B.[6], Schröder, C.[7], Müller, F.[4], Bastian, O.[8]

1 Dresden University of Technology, Department of Soil Science and Site Ecology, Pienner Str.19, 01737 Tharandt Lars.Koschke@tu-dresden.de

2 Deltares, Ecosystem Services research group, Postbus 85467, 3508AL Utrecht, Suzanne.vandermeulen@deltares.nl

3 University of Vienna, Department of Botany and Biodiversity Research, Rennweg 14, 1030 Vienna anna.schneidergruber@univie.ac.at

4 Kiel University, Institute for Natural Resource Conservation, Olshausenstr. 75, 24118 Kiel mkruse@ecology.uni-kiel.de, fmueller@ecology.uni-kiel.de

5 University of Bonn, Center for Development Research, Department of Ecology and Natural Resources Management, Walter-Flex-Str. 3, 53113 Bonn, cfuerst@uni-bonn.de, Susanne.Frank@uni-bonn.de

6 Ecosystem Services Research Group, Berlin-Brandenburg Academy of Sciences. Jaegerstrasse 22/23, 10117 Berlin b.ohnesorge@posteo.de

7 Eberswalde University for Sustainable Development, Department of Landscape Management and Nature Conservation, Schicklerstr. 5, 16225 Eberswalde, Claudia.Schroeder@hnee.de

8 Leibniz Institute of Ecological Urban and Regional Development, Weberplatz 1, 01217 Dresden, o.bastian@ioer.de

Abstract

Operationalization of the ecosystem services (ES) concept for improved natural resource management and decision support cannot, thus far, be rated as satisfactory. Participation of stakeholders is still a major methodical and conceptual challenge for implementing ES. Therefore, we conducted an online survey and a literature analysis to identify benefits and challenges of the application of ES in participatory processes. The results show that the purpose of stakeholder engagement is very diverse as a result of varying objectives, spatial scales and institutional levels of analysis. The complexity, terminology and (lacking) coherent classification of ES are pivotal aspects that should be accounted for in the design of studies to improve stakeholder participation. Although limitations of time and financial resources are bigger challenges than ES related ones, tailoring communication strategies and information for different stakeholder groups are of major importance for the success of ES studies. Results support the view that the potential benefits of applying ES, e.g., consensus finding, and development of integrated solutions, cannot be realized consistently across the different spatial scales and decision-making levels. Focusing on stakeholder processes represents a means to increase the relevance, reliability and impact of study results and to move participation in ES research from theory to reality.

Keywords:
decision-making, spatial planning, natural resource management, operationalization, participative processes, questionnaire

*Corresponding author. Email: Lars.Koschke@tu-dresden.de

1 Introduction

In the last decade the concept of ecosystem services (ES) (e.g., de Groot et al. 2010; Haines-Young and Potschin 2010; MEA 2005) has gained increasing attention and recognition in environmental sciences. In line with recent mainstreaming activities integrating the ES concept into policy making (in Europe, e.g., Convention on Biological Diversity, 2012 EU Commissions Blueprint to safeguard the future of European Waters by 2015, EU Common Agricultural Policy) (Maes et al. 2012), the concept has been increasingly tested as a framework for an integrated environmental impact assessment to be applied in natural resource management, landscape planning, land use policy, and biodiversity conservation (e.g., Cowling et al. 2008; Fürst et al. 2011, 2012, 2013b; Grêt-Regamey et al. 2008; Termorshuizen and Opdam 2009; von Haaren and Albert 2011). Thus, after a phase where much conceptual work has been carried out (e.g., de Groot et al. 2010; Haines-Young and Potschin 2010; Kandziora et al. 2013; MEA 2005; TEEB 2010) and several methodological approaches on quantifying, modeling and mapping ES have been published (e.g., Burkhard et al. 2012; Haines-Young et al. 2012; Hermann et al. 2014; Tallis and Polasky 2009; Villa et al. 2009), the next step is application of the ES concept in practice, i.e. implementation of procedures for ES assessment into regional and local processes in planning and management (van der Meulen et al. 2012).

As ES put into focus the needs of humans and the benefits they obtain from ecosystems, it is a stakeholder-driven concept. Thus, participation is an integral part of ES research (Menzel and Teng 2009; Müller et al. 2011). However, usually only scarce information on participatory processes is provided, which often renders the process itself nontransparent. There is a lack of practical experience described in scientific literature. Also difficulties, failures or mistakes are not mentioned or discussed. Besides, an overview and analysis of major challenges within participatory processes appear to be currently lacking in literature.

ES have been considered to be useful in communicating the benefits people obtain from ecosystems, to decide jointly on the best use and allocation of natural resources, and to facilitate deployment of a common discussion basis among different stakeholder groups (Grêt-Regamey et al. 2012; Wainger et al. 2010). However, there are drawbacks and obstacles that can impact its application and lead to discussions on the utility of the concept for decision-making in practice, for instance improvement of land management (e.g., Baker et al. 2013; Ghazoul 2007; Menzel and Teng 2009; Sagoff 2010; Wainger et al. 2010).

Participation in ES-based assessments – which is the focus of this paper – is also impacted by the ability to classify and quantify ES properly (Fisher et al. 2009). Other issues in group processes include stakeholder selection, the present personalities and the groups they represent, the questions asked and the form of communication. All have an impact on the resulting consensus (Malone et al. 2010; Mullen 1991).

Participation of (local) stakeholders is of major importance for the success in ES-based assessments, management and spatial planning (e.g., Chettri et al. 2007; Frank et al. 2012; Grêt-Regamey et al. 2012; König et al. 2010; Saxena et al. 2001; Sheppard and Meitner 2005; Turpie et al. 2008) and there are fundamental conceptual and operational issues still to be solved. Consequently, the proper inclusion of stakeholders has proven to be a challenge in many of the projects the authors of this contribution have been involved in. Therefore, stakeholders need much more attention in ES studies (Menzel and Teng, 2009).

The objective of this study is to investigate the challenges and the benefits of the application of ES in stakeholder participation processes in detail in order to derive general recommendations for improving participation in studies related to ES assessments in planning and management. In contrast to existing studies that investigated the challenges in integrating the ES concept in landscape or regional planning (e.g., Koschke et al., 2013; Albert et al. in press, we want to bridge the topic to stakeholder participation by revealing how scientists tackle the multiple demands related to stakeholder processes in the context of ES.

The **key questions** to be addressed are:

(1) Which are major challenges related to the ES concept in the practice of participatory processes?

(2) Does the ES concept indeed support better and more informed decision-making in stakeholder processes or are potential demerits and practical drawbacks actual hindrances to achieve this?

(3) Are there scale-dependent differences in terms of ES related challenges and benefits?

(4) Can general recommendations be derived on how to deal with challenges, potential demerits and practical drawbacks?

2 Methods

2.1 Development of an online-questionnaire

To get an overview on several important issues related to ES and participation, we first did a literature analysis and held discussions within the IALE-D working group "Ecosystem Services" (Figure 1). In order to collect data, we designed an online survey in which we questioned researchers about their practical experiences with stakeholder participation processes in their ES-based studies and projects. Since we aimed at investigating stakeholder processes from the perspective of scientists, we addressed only scientists who had conducted research related to ES including stakeholder participation. The survey was advertised in the newsletter of the Global Land Project (GLP, e-News No. 53; January 2013), the Ecosystem Services Partnership (ESP, Update No. 2013.1, January 2013), and the IALE-D/EU newsletter/mailing list. The questionnaire was also announced and linked to the two web-pages: http://www.eli-web.com/ and www.iale.de. 19 researchers were invited individually by email subsequent to a science-direct search.

In order to capture the diversity of study approaches and to analyze in a coherent manner the participatory processes thereof, the first part of the questionnaire dealt with general questions about the frame conditions (e.g., location, adopted scales, study aim) of the projects/studies. The second part was dedicated to the stakeholder participation, such as the reasons for their involvement, the task(s) stakeholders were asked to do, and the ES terminologies and communication techniques that have been applied. The third part belonged to evaluation of the stakeholder process and was aimed at uncovering benefits and challenges in general, and was related to the ES approach (the whole questionnaire can be found in the Appendix).

The survey was hosted by www.soscisurvey.de. Most of the questions in the survey were multiple choice questions in order to facilitate their completion by participants. For some of the questions multiple answers were allowed. Answer categories were formed in discussions among the authors of this

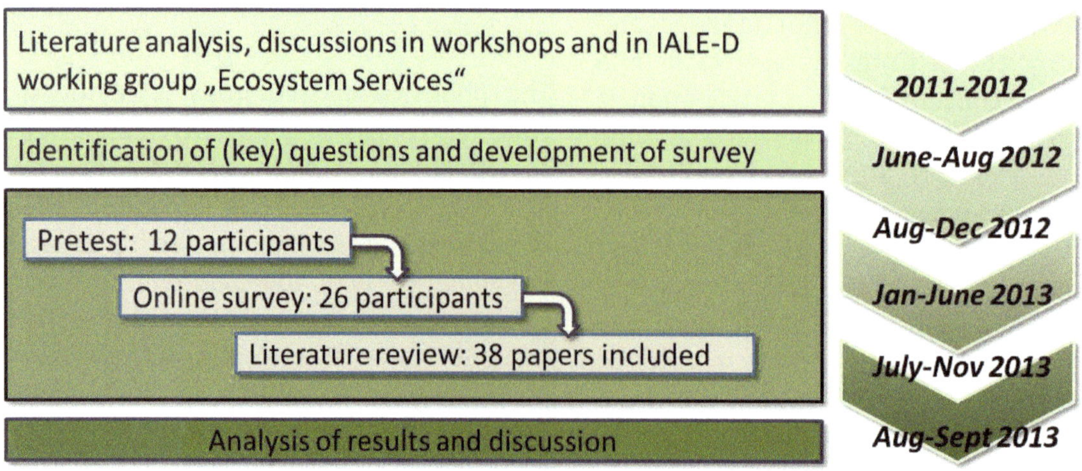

Figure 1: Overview of working steps

paper. There were also some open questions asking the interviewees to give free text answers. Subsequent to a pretest that was conducted with 12 participants, some flaws and some ambiguity were able to be eliminated before the survey was launched. The survey was online from January till June 2013. The survey results could be downloaded in *.csv format and were processed with Microsoft Excel to analyze the answers of the participants.

The results of the anonymous questionnaire are based on the answers of 38 interviewees mainly based in Europe (see results section). 26 participants answered all questions, while 12 datasets originate from the pre-test, which included fewer questions. Here, we provide only the most representative results with a focus on selected aspects.

2.2 Literature review

In addition to the questionnaire, a literature review was carried out. For the review of the studies a search on ISI Web of Knowledge was conducted to analyze peer-reviewed journal articles between the years 2000 to 2013 using the terms "ecosystem services", "stakeholder" and "participation" in their title, abstract, or keywords. The search results were examined for information on participation in ES studies. 38 papers were relevant and could be used to discuss and reflect survey findings. We classified the information provided in papers according to answer categories of the survey if possible in order to be able to compare selected results.

3 Results

3.1 General information of the online survey

We found very diverse goals and approaches of the analyzed studies. This documents the broad range and fragmented character of ES research. The relatively low number (n=38) of filled surveys was another reason that made analysis, i.e. structuring studies according to the addressed scale/decision-making level and comparing results, difficult. While the field the respondents were working in was not assessed, we could derive information on the institution they were working for in 21 cases (17 respondents answered the questionnaire anonymously). 13 were employed at a university, five at an applied research institute, two in a consultancy firm, and one at a public authority/government institution.

25 different countries were mentioned as study regions in the questionnaire. 76% of the studies were located in Europe and especially in Germany (13 out of 37) as the country of origin of survey respondents. 8% were situated in South America, 5% in both North America and Asia, and 3% in Africa and Australia respectively. Studies referred to by survey participants focused at a regional/landscape (74 %) or a local scale (14 %). Only two studies operated in a national or Pan-European area (3 %).

The ES concept was applied in projects thematically ranging from climate change, policies/decision-making, conservation and biodiversity protection to ecosystem and landscape management, energy crops and spatial planning. The distribution between scientifically-focused (46%) and application-oriented (41%) studies was almost balanced. 39% of the studies were dedicated to spatial and conservation planning and 26% to the testing/development of assessment approaches. The most frequent reason for using the ES approach was the need for integrated solutions (35%).

The references for the theoretical background of the ecosystem services concept mostly mentioned for ES were MEA (2005), de Groot et al. (2002, 2010), Haines-Young and Potschin (2010) or TEEB (2010), Burkhard et al. (2012), and Boyd and Banzhaf (2007). Among the 16 respondents who applied other references with lists and definitions of ES or combined references, ten other references were mentioned: Staub et al. (2011), Bolund and Hunhammar (1999), Bryan et al. (2010), Beaumont et al. (2007), Bastian et al. (2012), Defra et al., (2011; the complete reference could not be identified), Link (2010), Rutgers et al. (2008), Ranganathan (2008), and Fagerholm et al. (2012). Ten respondents (26%) used two or three and one even used seven different ES references. Five of the respondents explicitly stated that they tailored their approach by combining various references to better account for the aim of the study and to be more pragmatic in the use of ES with stakeholders (e.g., in case stakeholders were not experienced with ES).

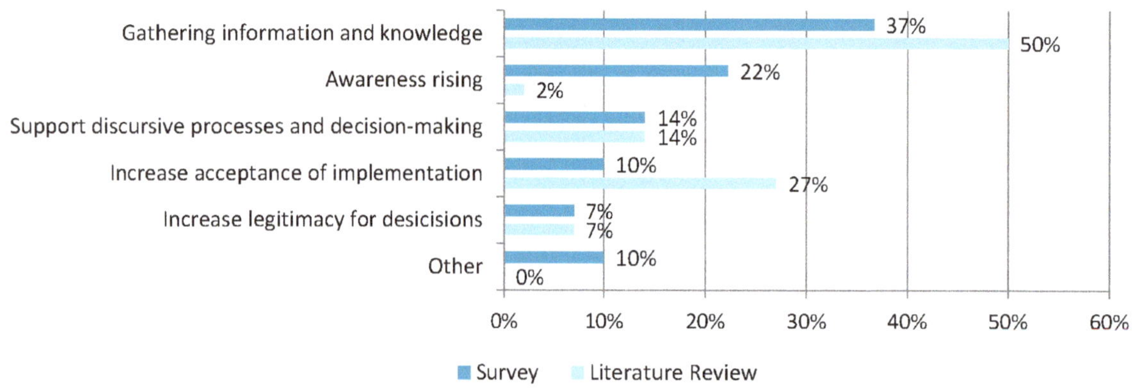

Figure 2: Reasons for stakeholder involvement (Survey: 90 answers, multiple answers allowed; Review: 44 mentions)

One respondent answered, *"ES were identified by stakeholders themselves [and] no a priori typology was used"*. One respondent stated that *"neither of these [references] were used directly, stakeholders modified them again to suit the purposes of the project and reflect the most up to date research on ES."*

3.2 Findings related to stakeholder processes

The main purposes of stakeholder involvement were gathering of information and knowledge with 37% and 50% of all mentions in the survey and literature review respectively (Figure 2).

Asking the respondents *"What type of information has been asked from stakeholders?"*, they stated that trade-off analyses (17 mentions), estimation of trends of supply (16), and the identification of relevant ES (14) were the most demanded information items from stakeholders. Less than every fifth study asked stakeholders for information on ecosystem service demands and only 13% of the studies asked for monetary valuation. Asked whether the information gathering was successful, 23% of respondents stated that it was not successful (i.e. unsatisfying or failed). For instance, reasons for unsuccessful trade-off analyses, which were mentioned in three of 17 cases, were problems *"to discriminate between similar or related ES"*. This caused stakeholders to be reluctant or unable to weight ES against each other in trade-off analyses (Figure 3). Further, the estimation of trends on demand and supply, quantification of impact on ES, and monetary valuation were not successful in some cases.

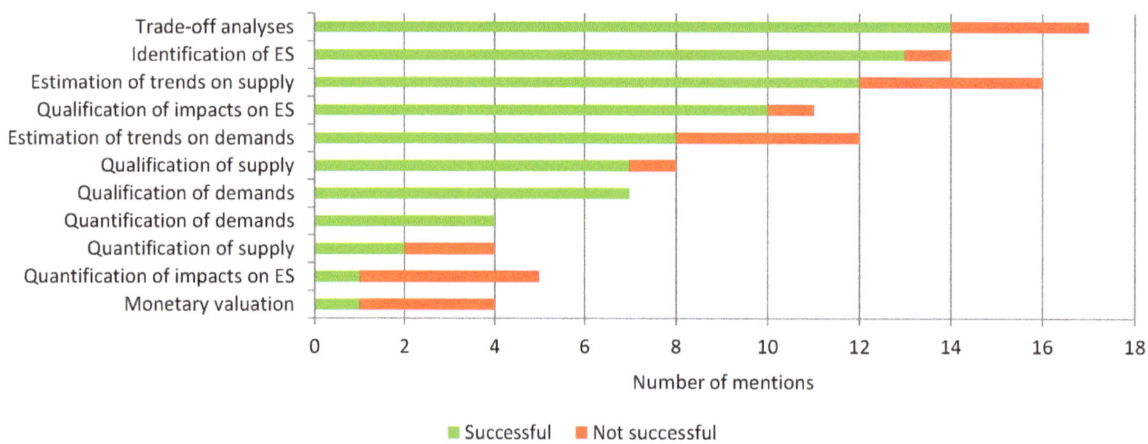

Figure 3: Type of information that has been asked from stakeholders (102 mentions, multiple answers allowed)

The techniques broadly applied to address stakeholders have been workshops (32% of overall 73 mentions), questionnaires (23%), interviews (22%) and round tables (18%). In 42% of studies only one technique was used, whereas the rest of the studies used one or more technique.

We further asked whether the respondents did explicitly use ES terminology and definitions during work with stakeholders. The ES approach was explicitly explained and the ES terms were applied in 53 % of the studies. 24 % of the respondents used the ES terms without any (extended) explanation of the ES concept (Figure 4, left). Among those who explained the ES concept and respective terms, almost all (90 %) were successful according to their own judgment. That is, the concept was mostly accepted and understood after explanation (see Table A1 for extended comments). This might be due to the fact that stakeholders were already familiar with the concept (as some of the respondents stated). However, understanding of the concept varied between stakeholder groups and educational level. To not additionally burden stakeholders with a new scientific concept, 24% did not provide an overview of the concept and used other terms. Either information was translated from stakeholders to ES, or ES terms were translated to fit stakeholders' perceptions/educational background (see Table A2).

65% of the respondents did include a social scientist or a mediator to facilitate communication. From the 17 survey participants who included a mediator, eight stated that he/she could significantly enhance the stakeholder process.

The distribution across spatial scales was rather unbalanced. Most of the studies addressed the landscape/regional scale. At local scales, ES was less often fully explained and more often 'translated' (Figure 4, right). The ad-hoc applicability of ES and the motivation to fully, i.e, holistically apply the concept seems to decline with decreasing scale level.

There have been quite contrasting opinions on ES communication. As the complexity of the concept can discourage stakeholders and the process can become an *"unsatisfactory experience"*, one respondent advocated ignoring the ES concept when dealing with (local) stakeholders. Another example provided evidence that the concept can be well understood and *"quite intuitively"* applied by stakeholders if a certain level of education and familiarity with concepts in general can be expected. One respondent mentioned that spatially available (statistical) data are not explicitly linked to ES and therefore it was challenging to apply the concept. Another respondent argued that the term 'landscape' (services) in comparison to 'ecosystem' (services) may be more attractive for stakeholders from non-ecological contexts or disciplines, and thus more suitable for being used in landscape planning projects (see also, e.g., Grunewald and Bastian 2010; Termorshuizen and Opdam 2009).

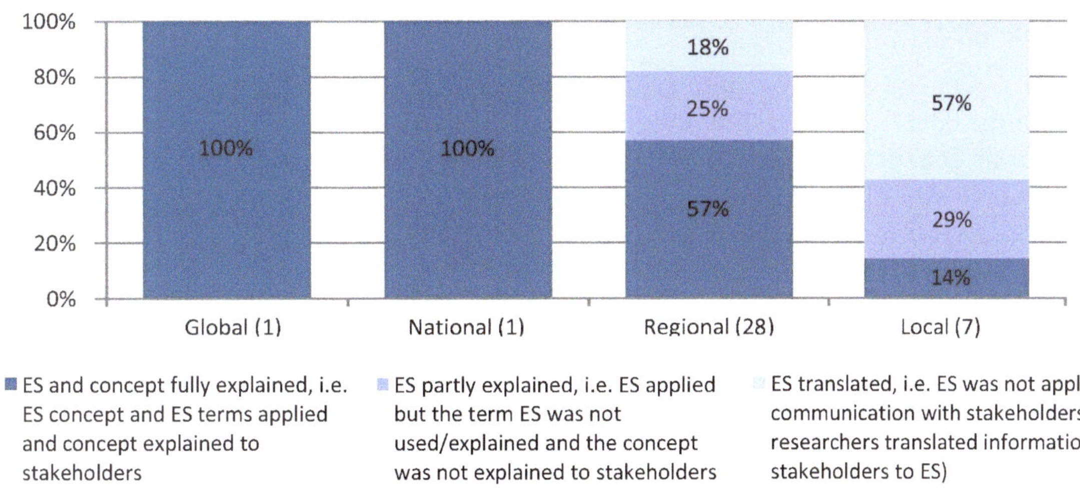

Figure 4: Overall comprehensiveness of communicating ES (left, 38 answers) and across the different scale levels (right, 37 answers). Number of considered answers in brackets

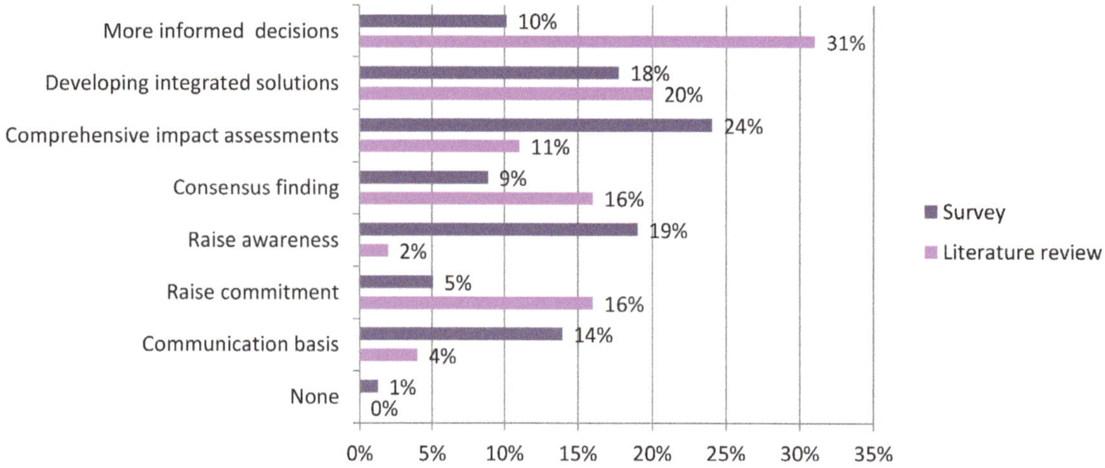

Figure 5: Benefits of using the ES approach in participatory processes as perceived by survey respondents (percentage of positive agreements, 79 answers) and based on mentions in literature (45 mentions)

3.3 Evaluation of the added value and applicability of the ES concept in stakeholder processes

Taking into account the survey results and the literature review, the greatest cumulative benefit of applying ES appeared to be related to achieving more informed decisions, developing integrated solutions and taking a comprehensive perspective for impact assessments. Further, ES were advantageous for awareness-raising, improved consensus finding and increased commitment (Figure 5). Only one respondent from the survey did not see a benefit.

In the survey, almost 70% of the respondents agreed that ES actually allow finding more informed decisions and 60% stated that benefits depend on the targeted decision-making level, (not displayed).

80% see clear potentials for improving stakeholder processes using ES. General improvements proposed by respondents (and efforts of the ES community) should aim at enhancing general publicity, acceptance, and familiarity of the concept by continuing to apply and spread it through various dissemination channels and through education. Specific improvements refer mainly to simplification of the ES terms. Many respondents would also opt to improve and simplify explanations for stakeholders, including a better preparation of the communication. Having more time available appears to be a major asset to deal with the challenges. Also methodical improvements concerning assessment, modeling, and visualization have been raised (see Table A3).

3.4 Challenges when using the ES concept in participatory processes

The most important ES related challenge for the survey respondents was dealing with the lack of data on ES (33% of all mentions). 12% of all answers stressed problems with the relevancy of assessed ES for stakeholders, which might result again from lacking suitable data on ES supply (e.g., different or opposed relevance of ES in a case study area for stakeholders). A failed analysis of stakeholders needs or an overall *"less tangible use"* of the ES concept for stakeholders might have been other reasons for this outcome. Further, the integration of the approach into existing concepts was demanding (30%). It was challenging for respondents to create a common understanding and *"communication basis between all participants"*, to deal with *"different notions of what belongs to a service"*, and to *"communicate the concept without creating confusion with respect to stakeholder's existing representations"*.

Challenges related to using ES identified in the survey were well supported by literature. Integration into existing concepts (30%), lack of data (20%), and lacking relevancy of the analyzed ES (10%) were found to be problems often mentioned in literature. The complexity of ES was found to be the most important challenge (40%) in the literature. In the respective multiple choice question in the survey, complexity was not offered as an answer (Figure 6).

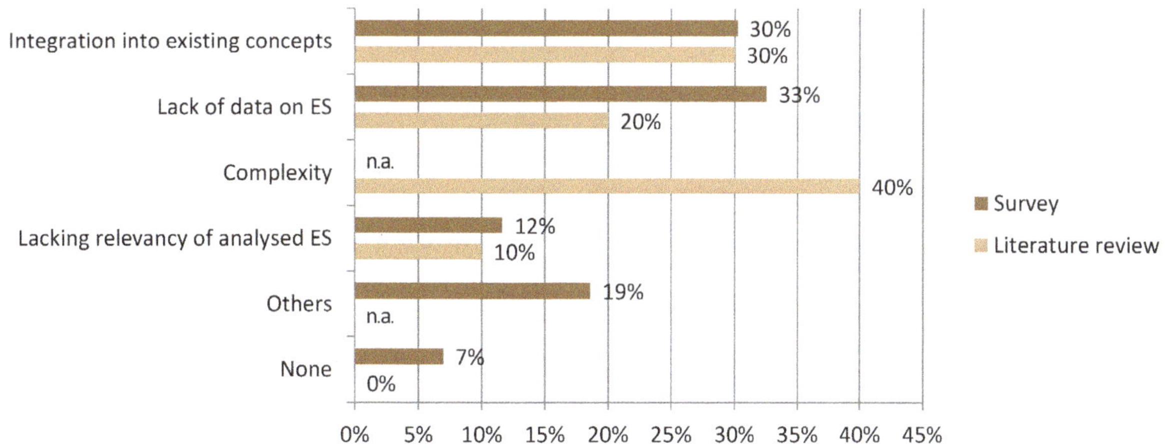

Figure 6: Challenges related to using the ES concept in participatory processes (survey: 67 mentions; literature review: 10 mentions)

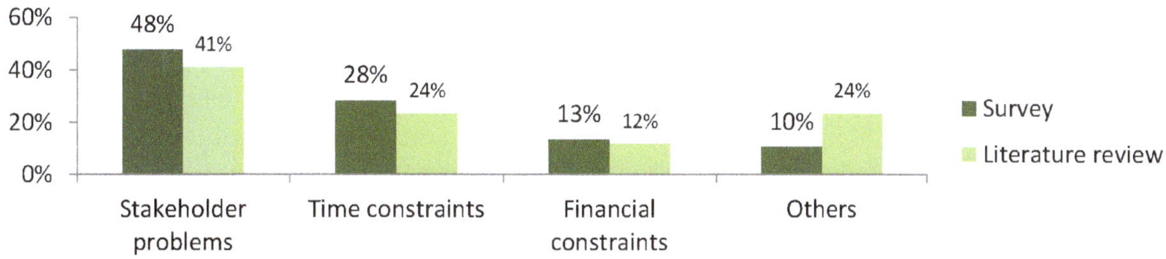

Figure 7: Overall challenges of stakeholder participation (survey: 67 mentions; literature review: 17 mentions)

Overall challenges were also quite similar in both the survey and in the review. Quite commonly, time constraints and financial restrictions were mentioned. Problems related to the actual inclusion of stakeholders however, were dominant in the survey as well as in the reviewed publications (Figure 7). Survey results allowed a more detailed differentiation in terms of stakeholder related issues. Hence, the identification of (motivated) stakeholders (26%), the (balanced) composition of stakeholder groups (12%) and gathering of all relevant stakeholders (9%) were challenging aspects related to stakeholder processes.

In comparison, 71% of the respondents agreed that stakeholder processes are hampered more by general challenges such as limitations of time and money than by ES related ones (e.g., complexity, integration of ES into existing concept etc.). Only around 19% perceived ES related challenges as a bigger obstacle.

In terms of the ES concept, major reasons for poor motivation of stakeholders, which was found in 17% of studies, were again the complexity (32%), the lack of tradition (and thus the competition with existing concepts and regulations; 18%) and a lack of understanding of the added value of ES (14%). In the comments, it was obvious that researchers were often not able to explain clearly the benefits of assessments for stakeholders and the importance of ES. Time restrictions of stakeholders (31%), communication problems (26%), and lacking identification with project aims (14%) were additional threats to stakeholders' motivation and commitments and thus for a successful participatory process.

4 Discussion

4.1 Challenges in participatory ES studies

Participation processes in ES studies are subjected to the main problems that have been identified in sociological literature on group processes (Jones and Stenseke 2011; Malone et al. 2010). Although the survey results suggest that general challenges, such as time constraints, lack of money and problems related to the (balanced) selection and involvement of stakeholders are more important, the use of the ES concept can pose additional threats to the success of stakeholder processes. This is due to its complexity, a lack of data, and terms unknown to stakeholders. For instance, Albert et al. (in press) reported that the ES concept was unknown by the majority of landscape and regional planners they questioned in their study. Further, the difficulty to integrate ES in accepted existing concepts and regulatory planning processes plays an important role (Albert et al. in press), as well as the inconsistent classification of ES. Merits and demerits of using ES might differ and depend on the specific decision-making context. Potential demerits of applying the ES concept in participation processes we found in our study were also often mentioned in ES literature (see Table 1). These demerits need to be considered since they can overcome merits.

Challenges related to the availability of data are not specific for ES studies of course, but the need to translate existing data to fit ES hampers the applicability of the concept. 60% of the survey participants believe that the benefits of the ES approach depend on the targeted decision-making level. At overarching, strategic decision-making levels where respective stakeholders are included, the need to translate ES terms is less pronounced in comparison to regional/ landscape and local scales where stakeholders ask for very precise and spatially explicit information. Employing the ES approach in natural resource management, Wainger et al. (2010: 1) said that *"the devil, truly, is in the details"* especially at local scales.

Table 1: Summary of potential merits and demerits of applying the ES concept in participatory processes in management and planning practice, based on ES literature.

Potential Merits (Benefits)	Potential Demerits (Drawbacks)
ES allow integration of disciplines, interests and sectors, bridge natural sciences and human well-being and can thus serve as a common communication basis between stakeholder groups in order to develop integrated solutions (Seppelt et al. 2011, de Groot et al. 2010; Barbier 2007; Carpenter et al. 2006, Tallis et al. 2009, Grêt-Regamey et al. 2012, Van der Meulen et al. 2013).	ES can hinder communication instead of facilitating it, which can aggravate conflicts (Menzel and Teng 2009).
	In the event there is no demand for ES, the concept might not serve as a useful management strategy (Grêt-Regamey et al. 2012).
ES explicitly involve beneficiaries, i.e. public demand for services (Grêt-Regamey et al. 2012).	Problems* come up when communicating the ES concept to the relevant actors (Meinke et al. 2006; Opdam et al. 2009, Koschke et al. 2012).
ES help to make more-informed management decisions and realize synergies between ES (Hauck et al. 2013).	There is a lack of clarity and definition standards (Broekx et al. 2013; Fisher et al. 2009).
ES provide a framework for a way of thinking that broadens the scope (Brauman et al. 2014)	There is a lack of methods for assessment and monetization (de Groot et al. 2010; Kroll et al. 2012).

* Problem might be: incomprehension of the purpose for applying ES, lack of clarity and definition standards or background for definitions

This might explain that within the survey two studies at the local/ farm level did not communicate ES. Respondents might not have been able to identify an added-value from the use of ES in their case studies. These respondents did not explicitly explain the term 'ecosystem services' but used either 'environmental services' or only single 'services' as a specific topic, i.e. soil erosion protection. In these cases it was more difficult to explain to practitioners (e.g., farmers) the added-value of ES and the different terminologies. Integrating ES into existing planning and decision-making approaches, especially at the local level, appears to be too demanding and not easily accepted. This, however, may also be related to the fact that the reported cases were research projects carried out mostly at universities and applied research institutes which hence dealt more with developing concepts and methods. Application of the ES concept might have been less demand-driven in these cases and there might be a weak point in connecting ES to ongoing processes in the case study areas.

Although common references such as MEA (2005) and de Groot et al. (2002, 2010) have been widely used, the range of other references clearly documents that there is no ecosystem services classification system that is totally consistent, can be commonly applied, and that there is a distinct need for refinements in practical applications. Some researchers working at the local scale tried to facilitate communication by not elaborating on ES. Experiences of survey participants have shown that, when discussing ES with expert stakeholders, one runs the risk of getting lost in definition and classification issues (e.g., which ones are final services, and which functions support which ES). These issues may make it more difficult for researchers to communicate with stakeholders.

4.2 Evaluating the impact of ES in stakeholder processes

Evaluating the applicability and the utility of ES, the concerned spatial scale and the related decision-making level are major drivers for the selection of stakeholders. Although based on a very low number of large scale cases, it seems that the ES concept is at the moment more easily understood or better suited to the stakeholders who work on a larger scale than at the local scale. This result is based on the decision of the researchers to not explain the ES concept or use alternative wording in the small scale cases. At higher levels, an adaption might be necessary in fewer cases, as the ES concept tends to be common to involved stakeholders (García-Llorente et al. 2011).

The survey and the literature review led to different results in terms of the role of ES to find more informed decisions. This benefit was mentioned in 10% of the answers from survey respondents while it was mentioned in 31% of the reviewed literature. In contrast, survey respondents put more focus on the ability to run comprehensive impact assessments (Figure 8, 24% of all mentions in comparison to 11% in the literature review). However, both answer categories are related, as they refer to the holistic, integrated character of ES. Asked if they agree or disagree with this statement, the majority of respondents stated that *"ES support finding better (more informed) solutions"* for landscape planning and management in practice. Yet, convincing examples could not be identified. It remains unclear whether the potential benefits of higher quality, better acceptance, and higher legitimacy in decision-making can be broadly realized in practice-oriented contexts.

Given the potential and perceived demerits, only 32% of the survey respondents evaluated the overall impact of ES on stakeholder processes as (rather) positive. Ambiguous findings were reported also from Albert et al. (in press) who asked regional and landscape planners about the usefulness of ES in planning processes. Interestingly, planners with no previous knowledge of ES believed it could be useful to integrate ES information in planning processes. Conversely, planners who already knew the ES concept had doubts and were rather pessimistic with respect to the benefits of integrating ES information.

4.3 How to handle challenges, potential demerits and practical drawbacks

Implementing participatory processes, the challenge starts with the selection of key stakeholders. A sound stakeholder analysis is the prerequisite for identifying values, interests, and needs of different and heterogeneous stakeholder groups (e.g.,

Grimble and Wellard 1997; Hein et al. 2006; Kasemir et al. 2003). The evaluation of the stakeholder process showed that a proper communication is of major importance. Although ES gained ground also in public discussion, many stakeholders might not quite understand what is meant by 'ecosystem' or relate it with natural processes and conservation instead of with human habitat and cultural patterns (Albert et al. 2012a). The advisable degree to which information on the conceptual background of ES needs to be conveyed to stakeholders varies.

The use (and extended communication) of the ES concept in the actual process might not always be necessary and advisable. Partly a certain reluctance of practitioners and (regional) planners to adopt the concept was reported by other scientists (e.g. Albert et al. in press). Some actors choose not to make the concept explicit when communicating with stakeholders, but merely mention specific services without labeling them ES, or by using other terms such as 'functions of the natural system' (Brauman et al. 2014; Everard 2009; van der Meulen and Brils 2011.) Thus, it is meaningful to adapt or translate the ES terminology and the amount of provided information to the needs of stakeholders in a way that the connections of the ES with stakeholders' well-being become more obvious and the information is useful and understandable for stakeholders with varying levels of expertise (cf. Granek et al. 2010; Nordström et al. 2010). Also technical support by interactive tools generally helps to trigger the interest of stakeholders for planning and can facilitate communication (Arciniegas et al. 2011; Berkers et al. 2006; Eikelboom et al. 2013; Frank et al. 2014).

Some of the caveats such as the problem of perceiving ES as a green concept, the unfamiliarity with terms and their connotations (Albert et al. 2012a) might diminish with time as the concept will be better known, acknowledged, and understood as an overarching paradigm. Other challenges however will persist (e.g., inconsistency of the concept, classification problems, and unit of data in official accountings) and might continue to impact stakeholder processes negatively.

Since ES studies do often include spatial data treatment and collection and application of participatory methods from social sciences, sound expertise from both fields should be present within the research team (Hauck et al. 2013). It can be recommended to include a social scientist or a mediator to increase the likeliness of a successful process.

5 Conclusions

In order to collect data on participation processes related to the application of the ES concept, we conducted an online questionnaire addressing researchers. In a subsequent step, a review of peer-reviewed journal articles was carried out to increase the information basis and to compare with findings of the questionnaire.

Referring to the **key question (1)** ("What are the major challenges related to the ES concept in the practice of participatory processes?"), results of the online survey support the view that the ES concept poses additional challenges to stakeholder processes (see also chapter 4.1). Most of the survey participants stated that the concept was understood to a great extent, if properly explained. However, often the survey participants did not, or did not completely, explain ES. Hence, there are particular requirements for a proper and well-prepared communication of the ES concept (terminology, definitions) and for keeping up the motivation of the stakeholders (relevance of the information provided). Although there are a number of potential drawbacks directly related to the ES concept, general challenges such as limitations of time and money and stakeholder choice appear to be more relevant.

Ambiguous results were found in terms of whether merits of applying ES outweighed demerits which has been addressed in **key question (2)**. Although several drawbacks were pinpointed, the overall impact of applying the ES approach in stakeholder processes within scientific studies was predominantly evaluated as being positive. The commonly conveyed benefits of the ES approach as a common basis for discussion, for consensus building, for the identification of integrated solutions or more informed decisions, for example, could mostly be realized in practice according to researchers who conducted the studies.

However, merits and demerits of using ES is also related to scale which was addressed in **key question (3)** ("Are there scale-dependent differences in terms of ES related challenges and benefits?"). Survey participants generally agreed that benefits and challenges of applying the ES concept depend on the targeted scale/decision-making level. There were results (for example, the fact that researchers did not explain the concept in local cases) and comments from survey respondents (for example, in terms of acceptance/unfamiliarity of the concept, communicating the added-value) indicating that the application of the ES concept at local scales is more demanding with respect to stakeholder integration. However, sound confirmation could not be achieved due to the lack of data on the scale-dependent success of stakeholder processes. How exactly the study scale impacts the benefits of applying ES, is a question that should be further addressed by the research community.

As to **key question (4)** ("Can general recommendations be derived regarding how to deal with challenges, potential demerits and practical drawbacks?"): Participatory processes need to be closely related to demands by stakeholders concerning language, information types, and results that are produced. Thus, for a successful application in practice, it is definitely advisable to explicitly bring the ES concept to local context, to provide many examples and to focus on communicating the added-value of ES. It is safe advice to develop an individual communication strategy for each of the involved stakeholder groups with respect to the used terms and the amount of information provided on the ES concept (cf. UNEP-WCMC 2011). Communication of knowledge generated in assessments has to be adapted also to the decision-making level. Planning for larger scales allows more general information while the focus on smaller scale requires more detailed knowledge. If (cross-scale) stakeholder groups are involved, an important question is how to bridge knowledge gaps and different decision criteria between them. Still, stakeholders do not often talk the ES language and it is difficult to adapt research language to their needs. If it cannot be ensured that information is understood and perceived as relevant, it will be difficult to keep the process and results significant for stakeholders and ensure continued motivation and commitment. This problem may not always be coherently accounted for within studies.

One should be aware, that for changing awareness of ES, participatory processes (i.e. reasoning together, exchanging arguments) and actual study results are important. Thus, during assessments, the coherency and consistency of the ES framework might be of lesser significance, even more so if work is carried out at local scale. An important consensus of discussions within the group of authors was that ES might be well suited as a common conceptual/theoretical and analytical framework rather than as a blueprint to conduct practice oriented projects and investigations.

A major concern of ES research is still to be able to provide 'actionable' knowledge, which is needed to overcome the apparent problem that the ES framework is a scientific concept rather than a concept that supports decision makers (Fürst et al., 2013; RR21, 2013). Dealing with stakeholders, we cannot present the ES approach as a new management approach in just one case study, but we do have to assess how the concept can support ongoing processes. Information allowing an evaluation of stakeholder processes appears to be largely missing currently in most studies/projects. Thus, the structured collection of stakeholder feedback on participatory processes should be a focal aspect in the future to better assess the success of ES studies. At times it seems that we need to take participation and the role and (information) needs of stakeholders more seriously instead of just *"muddling through stakeholder processes"* (Jennifer Hauck, personal communication, 07.05.2013), as it seems frequently to be the case in current ES projects. Together with efforts to develop standardized methods for ES assessment, we advocate a stronger focus on stakeholder processes as a key element for implementing ES into planning and management practice. This would help to develop new ideas for how to deal with particular challenges in ES, and it represents a means to increase the relevance and impact of study results in real life decision-making.

Acknowledgements

We would like to thank the many people who have contributed directly or indirectly to this paper through discussion in workshops and conferences. Above all, we would like to thank the survey participants for their patience, time and effort in filling in the questionnaire.

References

Albert, C., Hauck, J., Buhr, N. & C. von Haaren in press. What ecosystem services information do users want? Investigating interests and requirements among landscape and regional planners in Germany. Landscape Ecology, DOI 10.1007/s10980-014-9990-5

Albert, C., Galler C. & C. von Haaren 2012a. Integration von Ökosystemdienstleistungen in die Landschaftsplanung. Klimawandel – Was tun! IALE-D-Annual conference 2012, 24. 26. Oktober 2012, Eberswalde. http://www.iale.de/fileadmin/user_upload/PDFs/IALE-Jahrestagung_2012/IALE_Tagungsband_2012_Eberswalde.pdf (Date: 04.03.2014).

Albert, C., Von Haaren, C. & C. Galler 2012b. Ökosystemdienstleistungen - Alter Wein in neuen Schläuchen oder Impuls für die Landschaftsplanung. Naturschutz und Landschaftsplanung, 44 (5): 142-148.

Arciniegas, G., Janssen, R. & N. Omtzigt 2011. Map-based multicriteria analysis to support interactive land use allocation. International Journal of Geographical Information Science 25, 1931-1947.

Baker, J., Sheate, W. R., Phillips, P. & R. Eales 2013. Ecosystem services in environmental assessment - Help or hindrance? Environmental Impact Assessment Review 40, 3-13.

Barbier, E. 2007. Valuing ecosystem services. Economic Policy 22 (49), 177-229.

Bastian, O., Haase D. & K. Grunewald 2012. Ecosystem properties, potentials and services - The EPPS conceptual framework and an urban application example. Ecological Indicators 21, 7-16.

Beaumont, N.J., M.C. Austen, J. Atkins, D. Burdon, S. Degraer, T.P. Dentinho, S. Derous, P. Holm, T. Horton, E. van Ierland, A. H. Marboe, D.J. Starkey, M. Townsend & T. Zarzycki 2007. Identification, definition and quantification of goods and services provided by marine biodiversity: Implications for the ecosystem approach. Marine Pollution Bulletin 54: 253-265.

Bolund, P. & S. Hunhammar 1999. Ecosystem services in urban areas. Ecological Economics 29: 293 30.

Brauman, K.A., van der Meulen E.S. & J. Brils 2014. Ecosystem services and river basin management. In: Brils J., Brack W., Müller D., Negrel P., Vermaat J. (Eds.) Risk-informed management of european river basins, The Handbook of Environmental Chemistry Volume 29, 2014, pp 265-294. Springer Verlag 2014.

Bryan, B. A., Raymond, C. M., Crossman, N. D. & D.H. Macdonald 2010. Targeting the management of ecosystem services based on social values: Where, what, and how? Landscape and Urban Planning 97, 111-122.

Broekx, S., Liekens, I., Peelaerts, W., De Nocker, L., Landuyt, D., Staes, J., Meire, P., Schaafsma, M., Van Reeth, W., Van den Kerckhove, O. & T. Cerulus 2013. A web application to support the quantification and valuation of ecosystem services. Environmental Impact Assessment Review 40, 65-74.

Burkhard, B.; Kroll, F.; Nedkov, S. & F. Müller 2012. Mapping supply, demand and budgets of ecosystem services. Ecological Indicators 21, 17-29.

Chettri, N., Sharma E., Shakya B. & B. Bajracharya 2007. Developing forested conservation corridors in the Kangchenjunga landscape, eastern

Himalaya. Mountain Research and Development 27(3), 211-214.

de Groot, R.S., M.A. Wilson & R.M.J. Boumans 2002. A typology for the classification, description, and valuation of ecosystem functions, goods and services. Ecological Economics 41 (3), 393-408.

de Groot, R. S., Alkemade, R., Braat, L., Hein, L. & L. Willemen 2010. Challenges in integrating the concept of ecosystem services and values in landscape planning, management and decision making. Ecological Complexity 7, 260-272.

Carpenter, S.R., DeFries, R., Dietz, T., Mooney, H.A., Polasky, S., Reid, W.V. & R.J. Scholes 2006. Millennium ecosystem assessment: Research needs. Science 314 (5797), 257-258.

Cowling, R.M., Egoh, B., Knight, A.T., Reyers, B., Rouget, M., Roux D.J., Welz, A. & A. Wilhelm-Rechman 2008. An operational model for mainstreaming ecosystem services for implementation. PNAS 28, 9483-9488.

Eikelboom, T., Janssen, R., 2013. Interactive spatial tools for the design of regional adaptation strategies. Journal of Environmental Management 127, Supplement, S6-S14.

Everard, M. 2009. Ecosystem services case studies, better regulation science programme, Environment Agency. Productcode: SCHO0409BPVM-E-P.

Fagerholm, N., Käyhkö, N., Ndumbaro, F. & M. Khamis 2012. Community stakeholders' knowledge in landscape assessments - Mapping indicators for landscape services. Ecological Indicators 18, 421-433.

Fisher, B., Turner, R. K. & P. Morling 2009. Defining and classifying ecosystem services for decision making. Ecological Economics 68, 643-653.

Frank, S., Fürst C., Koschke L. & F. Makeschin 2012. A contribution towards a transfer of the ecosystem service concept to landscape planning using landscape metrics. Ecological Indicators 21, 30-38.

Frank, S., Fürst, C., Witt, A., Koschke, L. & F. Makeschin 2014. Making use of the ecosystem services concept in regional planning—trade-offs from reducing water erosion. Landscape Ecology: 1-15.

Fürst, C., Lorz, C. & F. Makeschin 2011. Integrating land management aspects into an assessment of the impact of land cover changes on ecosystem services. International Journal of Biodiversity Science, Ecosystem Services & Management, 1-14.

Fürst, C., Pietzsch, K., Frank, S., Witt, A., Koschke, L. & F. Makeschin 2012. How to better consider sectoral planning information in regional planning - example afforestation and conversion. Journal of Environmental Planning and Management 2012, 1-29.

Fürst, C., Helming, K., Lorz, C., Müller, F. & P.H. Verburg 2013a. Integrated land use and regional resource management - A cross-disciplinary dialogue on future perspectives for a sustainable development of regional resources. Journal of Environmental Management 127, Supplement, S1-S190.

Fürst, C., Frank, S., Witt, A., Koschke, L. & F. Makeschin 2013b. Assessment of the effects of forest land use strategies on the provision of ecosystem services at regional scale. Journal of Environmental Management 127, 96-116.

García-Llorente, M., Martin-Lopez, B., Nunes, P.A.L.D., Gonzalez, J.A., Alcorlo, P. & C. Montes 2011. Analyzing the Social Factors That influence willingness to pay for invasive alien species management under two different strategies: Eradication and prevention. Environmental Management 48(3), 418-435.

Ghazoul, J. 2007. Challenges to the uptake of the ecosystem service rationale for conservation. Conservation Biology 21, 1651-1652.

Granek, E. F., Polasky, S., Kappel, C. V., Reed, D. J., Stoms, D. M., Koch, E. W., Kennedy, C.J., Cramer, L. A., Hacker, S. D., Barbier, E. B., Aswani, S., Ruckelshaus, M., Perillo, G. M. E., Silliman, B. R., Muthiga, N., Bael, D. & E. Wolanski 2010. Ecosystem services as a common language for coastal ecosystem-based management. Conservation Biology 24(1), 207-216.

Grêt-Regamey, A., Walz, A. & P. Bebi 2008. Valuing ecosystem services for sustainable landscape planning in Alpine regions. Mountain Research and Development 28, 156-165.

Gret-Regamey, A., Brunner, S. H. & F. Kienast 2012. Mountain ecosystem services: Who cares? Mountain Research and Development 32, S23-S34.

Grimble, R. & K. Wellard 1997. Stakeholder methodologies in natural resource management: a review of principles, contexts, experiences and opportunities. Agricultural Systems 55, 173-193.

Grunewald, K. & O. Bastian 2010. Ökosystemdienstleistungen analysieren - begrifflicher und konzeptioneller Rahmen aus landschaftsökologischer Sicht. Geo-Öko 31, 50-82.

Haines-Young, R. & M. Potschin 2010. The links between biodiversity, ecosystem services and human well-being. In: Raffaelli, D.G & C.L.J. Frid (Eds) Ecosystem Ecology: A new synthesis. Cambridge University Press, British Ecological Society, pp. 110-139.

Haines-Young, R., Potschin, M. & F. Kienast 2012. Indicators of ecosystem service potential at European scales: mapping marginal changes and trade-offs. Ecological Indicartos 21, 39-53.

Hauck, J., Schweppe-Kraft, B., Albert, C., Görg, C., Jax, K., Jensen, R., Fürst, C., Maes, J., Ring, I., Hönigová, I., Burkhard, B., Mehring, M., Grunewald, C., Schwarzer, M., Meurer, J., Sommerhäuser, M., Priess, J., Schmidt, J. & A. Grêt-Regamey 2013. The promise of the ecosystem services concept for planning and decision-making. GAIA - Ecological Perspectives for Science and Society 22, 232-236.

Hein, L., van Koppen, K., de Groot, R. S. & E.C. van Ierland 2006. Spatial scales, stakeholders and the valuation of ecosystem services. Ecological Economics 57, 209-228.

Hermann, A., Kuttner, M., Hainz-Renetzeder C., Konkoly-Gyuro E., Tiraszi A., Brandenburg C., Allex B., Ziener K. & T. Wrbka 2014. Assessment framework for landscape services in European cultural landscapes: an Austrian Hungarian case study. Ecological Indicators 37(A), 229-240.

Jones, M. & Stenseke, M. (eds.) 2011. The European Landscape Convention - Challenges of participation, Springer, Berlin – Heidelberg – New York.

Kasemir, B., Jäger, J., Jaeger, C.C. & M.T. Gardner 2003. Public participation in sustainability science, A Handbook. Cambridge University Press, Cambridge, UK.

Kandziora, M., Burkhard, B. & F. Müller 2013. Interactions of ecosystem properties, ecosystem integrity and ecosystem service indicators - A theoretical matrix exercise. Ecological Indicators 28, 54-78.

König, H.J., Schuler J., Suarma U., McNeill D., Imbernon J., Damayanti F., Dalimunthe S.A., Uthes S., Sartohadi J., Helming K. & J. Morris 2010. Assessing the impact of land use policy on urban-rural sustainability using the FoPIA approach in Yogyakarta, Indonesia. Sustainability 2, 1991-2009.

Koschke, L., Fürst, C., Frank, S. & F. Makeschin 2012. A multi-criteria approach for an integrated land-cover-based assessment of ecosystem services provision to support landscape planning. Ecological Indicators 21, 54-66.

Kroll, F., Müller, F., Haase, D. & N. Fohrer 2012. Rural-urban gradient analysis of ecosystem services supply and demand dynamics. Land Use Policy 29, 521-535.

Link, J. 2010. Ecosystem-Based Fisheries Management: Confronting tradeoffs. Cambridge University Press, pp. 224.

Maes, J., Egoh B., Willemen L., Liquete C., Vihervaara P., Schägner J.P., Grizzetti B., Drakou E.G., Notte A.L., Zulian G., Bouraoui F., Luisa Paracchini M., Braat L. & G. Bidoglio 2012. Mapping ecosystem services for policy support and decision making in the European Union. Ecosystem Services 1, 31-39.

Malone, E.L., Dooley, J.J. & J.A. Bradbury 2010. Moving from misinformation derived from public attitude surveys on carbon dioxide capture and storage towards realistic stakeholder involvement. International Journal of Greenhouse Gas Control 4, 419-425.

MEA (Millennium Ecosystem Assessment) 2005. Ecosystems and human well-being: Synthesis. Island Press/World Resources Institute, Washington, DC.

Meinke, H., Nelson, R., Kokic, P., Stone, R., Selvaraju, R. & W. Baethgen 2006. Action-able climate knowledge: from analysis to synthesis. Climate Research 33, 101-110.

Menzel, S. & Teng, J. 2009. Ecosystem services as a stakeholder-driven concept for conservation science. Conserv. Biol. 24, 907-909.

Müller, F., de Groot, R. & L. Willemen 2011. Ecosystem Services at the landscape scale: The need for integrative approaches. Landscape Online 23, 1-11.

Mullen, B. 1991. Group composition, salience, and cognitive representations: the phenomenology of being in a group. Journal of Experimental Social Psychology, 27, 297–323.

Nordström, E.-M., Eriksson, L. O. & K. Öhman 2010. Integrating multiple criteria decision analysis in participatory forest planning: Experience from a case study in northern Sweden. Forest Policy and Economics 12, 562-574.

Opdam, P., Luque, S. & K. Jones 2009. Changing landscapes to accommodate for climate change impacts: a call for landscape ecology. Landscape Ecology 24, 715-721.

Ranganathan, J. ,Bennett, K., Raudsepp-Hearne, C., Lucas, N., Irwin, F., Zurek, M., Ash, N. & P. West 2008. Ecosystem services: A guide for decision makers. World Resources Institute, Washington, DC.

Rutgers, M., Mulder, C. & A.J. Schouten (Eds.), 2008. Soil ecosystem profiling in the Netherlands with ten references for biological soil quality. RIVM Report 607604009, Bilthoven, The Netherlands.

RR21, 2013. http://theiufroblog.wordpress.com/2013/10/21/the-ecosystem-service-concept-an-integrative-framework-for-land-use-planning/ (Date: 27.10.2013).

Sagoff, M. 2010. The quantification and valuation of ecosystem services. Ecological Economics 70, 497-502.

Saxena, K.G., Rao K.S., Sen K.K., Maikuri R.K. & R.L. Semwal 2001. Integrated natural resource management: Approaches and lessons learned from the Himalaya. Ecology and Society 5(2), 14-24.

Seppelt, R., Dormann, C. F., Eppink, F. V., Lautenbach, S. & S. Schmidt 2011. A quantitative review of ecosystem service studies: Approaches, shortcomings and the road ahead. Journal of

Applied Ecology 48, 630-636.

Staub, C. & W. Ott 2011. Indikatoren für Ökosystemleistungen. Systematik, Methodik und Umsetzungsempfehlungen für eine wohlfahrtsbezogene Umweltberichterstattung. Bundesamt für Umwelt, Bern.

Sheppard, S. R. J. & M. Meitner 2005. Using multi-criteria analysis and visualisation for sustainable forest management planning with stakeholder groups. Forest Ecology and Management 207, 171-187.

Tallis, H., Goldman, R., Uhl, M. & B. Brosi 2009. Integrating conservation and development in the field: implementing ecosystem service projects. Frontiers in Ecology and the Environment 7 (1), 12-20.

Tallis, H. & S. Polasky 2009. Mapping and valuing ecosystem services as an approach

for conservation and natural-resource management. Annals of the New York Academy of Sciences 1162, 265-83.

TEEB 2010. The economics of ecosystems and biodiversity: Mainstreaming the economics of nature: a synthesis of the approach, conclusions and recommendations of TEEB.

Termorshuizen, J.W. & P. Opdam 2009. Landscape services as a bridge between landscape ecology and sustainable development. Landscape Ecology 24, 1037-1052.

Turpie, J.K., Marais C. & J.N. Blignaut 2008. The working for water programme: Evolution of a payments for ecosystem services mechanism that addresses both poverty and ecosystem service delivery in South Africa. Ecological Economics 65(4), 788-798.

Wainger, L. A., King, D. M., Mack, R. N., Price, E. W. & T. Maslin 2010. Can the concept of ecosystem services be practically applied to improve natural resource management decisions? Ecological Economics 69, 978-987.

Van der Meulen, S. & J.M. Brils 2011. Do ecosystem services provide a common language to facilitate participation in water management? Session report, Resilience conference, Tempe, Arizona, USA. Available at: http://kennisonline.deltares.nl/2/m/search/products.html?q=resilience&qtype=1 (Date: 27.10.2013).

Van der Meulen, S., Brils, J.M. & I. Borowski 2012. Lessons learned from stakeholder interviews on ecosystem services with a special focus on scale issues. Presentation at the RegioResources21 Conference, Dresden, Germany.

Van der Meulen, S., Brils, J.M., I. Borowski-Maaser & U. Sauer 2013. Payment for ecosystem services (PES) in support of river restoration. Water Governance 4, 40-45.

Villa, F., Ceroni, M., Bagstad, K., Johnson, G. & S. Krivovet 2009. ARIES (ARtificial Intelligence for Ecosystem Services): A new tool for ecosystem services assessment, planning, and valuation. 11th International BIOECON Conference on economic instruments to enhance the conservation and sustainable use of biodiversity, September 21-22, Venice, Italy. http://www.ucl.ac.uk/bioecon/11th_2009/Villa.pdf (Date: 27.10.2013).

Von Haaren, C. & C. Albert 2011. Integrating ecosystem services and environmental planning: Limitations and synergies. International Journal of Biodiversity Science, Ecosystem Services & Management 7 (3), 150-167.

UNEP-WCMC 2011. Developing ecosystem service indicators: Experiences and lessons learned from sub-global assessments and other initiatives. Secretariat of the Convention on Biological Diversity, Montréal, Canada. Technical Series No. 58, 118 pages, http://www.naturalcapitalproject.org/pubs/cbd-ts-58-en.pdf (Date: 04.03.2014).

Appendix

Table A1: Extended optional answers to the question whether the concept was accepted and understood by stakeholders (question 15.1 in the questionnaire)

(Rather) Yes	Partly, [...] it was difficult for them which ES might be more important over other ES!
	No noticeable obstacles
	No, the term and concept was already known by all involved stakeholders
	The scientists were familiar with it
	If explained properly, the concept can definitely be made understandable. Some terms (e.g. biodiversity) are used differently in different contexts.
	I did not have major obstacles, but I worked with people who are fairly familiar with these terms.
	[...] some concepts are difficult to understand for the community.
	Stakeholders basically understood the approach but due to lacking time not every single detail could be clarified
	Definition was mostly accepted.
	We had to clearly explain the concept, but afterwards it was in our experience quite intuitive for stakeholders. I have to state that all our stakeholders were quite well educated, e.g. forest engineers, city planners that have most probably been in contact with research and "concepts" in general.
	The concept has been completely understood.
(Rather) No	Broad unfamiliarity of ES concept to many non-academic audiences. Researcher's attempt to 'reclassify' relationships that were implicit in local discourse connotations of monetary fungibility in 'services' terminology was particularly problematic within the study area. The 'services' terminology can be counter-productive because it seems to be associated with commodification.
	ES is for most stakeholders quite difficult to understand. Farmers were the most easy, but only interested in production, government was more difficult at the time.
	At the beginning it was difficult to communicate the complex concept in an understandable, easy way. Additionally, the term "ecosystem services" is complicated in German.

Table A2: Reasons for not (fully) explaining the ES concept to stakeholders (question 15.2)

[...] it seemed to be too complicated
We communicated the concept but embedded it in broader discussions about uses and values of different land-cover and land-use options
To be sure that stakeholders understand the questions
Communication with stakeholders is more easy, if other terms are used
We did not want to ask too much of the stakeholders and did not want to discourage them
[...] to keep the process simple and to make it easier for the stakeholder to understand [the aim]
To help general understanding by not-technical stakeholders
[...] we know from other experiences that the concept is very hard to understand, even by environmental experts and that it raises discussion about the definition. The specific ES are generally well understood by interviewees, with a few exceptions.
Communication with stakeholders was easier when the term ES was not mentioned, since in most times it was directly linked to monetary values.
The method applied for eliciting preferences and evaluating alternative solutions was multi-criteria decision analysis (MCDA) and within this framework terms like "values, goals, criteria, objectives, attributes" are commonly used.

Farmers are mostly not familiar with the ES-concept. Thus a term was used that has a stronger connection to environmental measures on farms, pretending that by implementing these measures real ES are provided.
Use of existing data and preprocessed spatial data of administrative authorities which are not explicitly linked to ES concept!
It is not necessary to fully communicate the ES concept, to make it workable
'Landscapes' may be more attractive to non-ecological scientific disciplines in contrary to the term 'ecosystem' and may be associated with peoples local environment. The term landscape services might be more suitable for implementing the concept into landscape planning projects. In contrast, 'ecosystem' may be related with natural processes and conservation instead of with human habitat and cultural patterns.

Table A3: Selected statements on things that can be improved in stakeholder processes based on experiences of survey respondents (question 22)

[Better] preparation of the questionnaire
Provide more scientific background information to stakeholders and/or scientists, provide a specific challenge or problem to be tackled by the process
More previous networking
Adapt methodology better to situation and stakeholders
[Address] issues of representativeness [better]
Consider the difference in demand and tradition of participant
Increase stakeholder involvement upstream of research design
Explain in a better way the concept of ES
Better explanation of the ES concept. Facilitate methods for ES weighting. Keep stakeholder contact as simple as possible. Provide only essential information. Focus more on social-science related communication techniques and methods. Try to make assessment (ES and indicators) more relevant to stakeholders.
[More] time [is necessary]
Involve other stakeholder groups, other methods to promote discussion for consensus between different stakeholders and other topics of consults (e.g. identification of ESs)
Introduction through open-ended discussion of environmental/landscape benefits [to increase familiarity with the concept]
In this case: different groups of stakeholders need to be involved and more people needed for quantification
Identifying ES in a common language for all stakeholders OR produce a two step process 1. identify a common language for environmental quality aspects and 2. produce a framework for translating environmental quality into ES.
Financial incentives might enhance participation in conservation measures
Include experts on financial aspects and decision makers/managers of support programs
Better methods for WTP/ Choice Analysis
Starting earlier, trying to keep explanations as easy as possible, more direct contact (if time allows)
structure of workshops have to be well (and maybe strictly) organized, in order to be really productive
Better preparation concerning the choice of participants, reduce choice to local people with scientific background to ensure a better understanding of the concept (question of feasibility!?)
(1) Other ways to deal with supportive services, (2) learn how to deal with recognition, valuing and addressing beneficiaries for ES that are relevant at large scales, (3) use larger study area.
To facilitate the visualization of service provision (maps and diagrams are sometimes too abstract (too scientific) for stakeholders. It is very important to speak in 'their' language and to communicate them the benefits they will receive from the ES assessment

Table A4: Questionnaire

Variables Listing

Questionnaire - internal data

Apart from your questions, you will find in the data record additional variables, as far as you have not deactivated this option when downloading the data record.

CASE Serial number of the subject
REF Reference, if such has been provided in the link to the questionnaire
LASTPAGE Page number of the questionnaire that has been edited and sent last
QUESTNNR ID of the questionnaire that has been edited
MODE Information if the questionnaire has been started by pretest or by a project member
STARTED Time the interviewee opened the questionnaire
FINISHED Information if the questionnaire has been completed up to the last page
TIME_001... Time that an interviewee has spend on a questionnaire page

Please note that you can not read the internal variables of the questionnaire with the function value(). For interview number and reference there are the PHP functions PHP function caseNumber() and PHP function reference()

Details about additional variables can be find in the manual: Additional Variables in the Data Set

Section A1: General Information

[A101] Text Input
Q01 - Country
"Where is the study region located?"

A101_01 Country
Text input

[A102] Selection
Q02 - Scale level
"What is the analysis or decision making level addressed within your study?"

A102 Q02 - Scale level
1 = Global or Pan-European or similar
2 = National
3 = Regional/ Landscape
4 = Local
5 = Other:
-9 = Not answered

A102_05 Other
Text input

[A103] Selection
Q03 - Organizational scale
"What was the targeted organizational scale?"

A103 Q03 - Organizational scale
1 = Regional planning district
2 = Municipality
3 = Farm level/Company level/ Neighbourhood level (in urban cases)
4 = Field block level/ Stand level/Neighbourhood level
5 = Landscape
6 = Water management unit (catchment)
8 = Ecosystem
7 = Other:
-9 = Not answered

A103_07 Other
Text input

[A104] Text Input
Q04 - Project aim
"What was the aim of the research project?"

A104_01 [01]
Text input

[A107] Multiple Choice
Q05 - Case study aim?
"What was the aim of the case study/regional project?"

A107_01 Spatial or strategic planning support
A107_02 Conservation planning support
A107_03 Nature protection
A107_04 Adapted management practices
A107_05 Test/development of assessment approach
A107_06 Implementation of measures
A107_07 Other
1 = Not checked
2 = Checked

A107_07a Other (free text)
Text input

[A105] Text Input
Q06 - Why ES?
"Why was the ES approach applied?"

A105_01 [01]
Text input

[A106] Text Input
Q07 - Who asked for it?
"Who asked for the application of the ES approach (e.g. the project holder, stakeholders, research group, etc.)?"

A106_01 [01]
Text input

[A109] Multiple Choice
Q08 - Which ES ref.?
"Which reference for definition and classification of ES or which other concept did you apply?"

A109_01 MA (e.g. MA, 2005. Ecosystems and human well-being: Synthesis. A Report of the Millenium Ecosystem Assessement. Island Press, Washington)
A109_02 de Groot et al. (e.g. de Groot, R. S., Wilson, M. A., Boumans, R. M. J., 2002. A typology for the classification, description and valuation of ecosystem functions, goods and services. Ecological Economics. 41, 393-408// de Groot, R. S., Alkemade, R., Braat, L., Hein, L., Willemen, L., 2010. Challenges in integrating the concept of ecosystem services and values in landscape planning, management and decision making. Ecological Complexity. 7, 260-272)
A109_03 Burkhard et al. (Burkhard, B., Kroll, F., Müller, F., Windhorst, W., 2009. Landscapes' Capacities to Provide Ecosystem Services – a Concept for Land-Cover Based Assessments. Landscape Online. 15, 1-22)
A109_04 Haines-Young and Potschin (Haines-Young, R., Potschin, M., The links between biodiversity, ecosystem services and human well-being. In: D. Raffaelli, C. Frid, Eds.), Ecosystem ecology: a new synthesis. BES ecological reviews series, CUP, Cambridge, 2009) or TEEB (e.g. TEEB, The Economics of Ecosystems and Biodiversity: Mainstreaming the Economics of Nature: A synthesis of the approach, conclusions and recommendations of TEEB. 2010)
A109_05 Boyd and Banzhaf (Boyd, J., Banzhaf, S., 2007. What are ecosystem services? The need for standardized environmental accounting units. Ecological Economics. 63, 616-626)
A109_06 Other
A109_07 Other
A109_08 Other
A109_09 Other
1 = Not checked
2 = Checked

A109_06a Other (free text)
A109_07a Other (free text)
A109_08a Other (free text)
A109_09a Other (free text)
Text input

[A108] Selection
Q09 - Focus?
"Has your study been rather ... "

A108 Q09 - Focus?
1 = scientifically-focused, or
2 = application-/implementation-oriented ... ?

3 = Other/Comments:
-9 = Not answered

A108_03 Other/Comments
Text input

[A110] Free Mentions
Q10 - Which ES?
"Which specific ES did you consider (e.g. food and fodder provision, landscape aesthetics , soil erosion regu..."

A110 Number of mentions
A110x01 Mention 1
A110x02 Mention 2
A110x03 Mention 3
A110x04 Mention 4
A110x05 Mention 5
A110x06 Mention 6
A110x07 Mention 7
A110x08 Mention 8
A110x09 Mention 9
A110x10 Mention 10
A110x11 Mention 11
A110x12 Mention 12
A110x13 Mention 13
A110x14 Mention 14
A110x15 Mention 15
A110x16 Mention 16
A110x17 Mention 17
A110x18 Mention 18
A110x19 Mention 19
A110x20 Mention 20
A110x21 Mention 21
A110x22 Mention 22
A110x23 Mention 23
A110x24 Mention 24
A110x25 Mention 25
Text input

Section A2: Stakeholder Participation

[A201] Multiple Choice
Q11 - Why SH involved?
"Why did you involve stakeholders anyway? "

A201_01 Awareness rising/ Provision of information
A201_02 Support discursive processes and decision making (e.g. for a specific problem)
A201_03 Increase legitimacy for decisions
A201_08 Increase acceptance of implementation and compliance with measures
A201_04 Gathering information and knowledge (e.g. on relevant ES, preferences, quantification etc.)
A201_05 Other
A201_06 Other
A201_07 Other
1 = Not checked
2 = Checked
A201_05a Other (free text)
A201_06a Other (free text)
A201_07a Other (free text)
Text input

[A202] Scale (fully labeled)
Q12 - Infotype?
"What type of information has been asked from stakeholders? Was the information collection successful? "

A202_15 None
A202_02 Identification/definition/selection of ES
A202_03 ES prioritization/weighting/trade-off analyses
A202_04 Qualification of supply (stocks, flows)
A202_05 Qualification of demands (stocks, flows)
A202_06 Qualification of impact on ES as result of change
A202_07 Quantification of impact on ES as result of change
A202_08 Quantification of supply (stocks, flows)
A202_09 Quantification of demands (stocks, flows)
A202_10 Monetary valuation
A202_11 Estimation of past trends on ES supply
A202_12 Estimation of future trends on ES supply/ scenario building
A202_13 Estimation of past trends on ES demand
A202_14 Estimation of future trends on ES demand/scenario building
1 = Fully successful +++
2 = Rather successful ++
3 = Partly successful/ Don't know +/-
4 = Rather not successful --
5 = Not successful ---
-1 = Not assessed
-9 = Not answered

[A204] Text Input
Q12b - Infotype?
"Please enter the information you asked from stakeholders which have not been captured in Question 12 - or ch..."

A204_05 [05]
Text input
A204_05a [05]: No additional information/ comments
1 = Not checked
2 = Checked

[A205] Multiple Choice
Q13 - SH groups?
"Please indicate which stakeholder groups you involved."

A205_01 Representatives of municipalities or administrative authorities/
A205_02 Local politicians/
A205_03 Policymakers/
A205_04 Water managers/
A205_05 Investors/
A205_06 Stakeholders from recreation and tourism/
A205_07 Project managers of municipalities/
A205_08 Landscape and regional planners/managers/
A205_09 Managers of protected areas or nature conservationists/
A205_10 Land owners/ land users (farmers, foresters) or their representatives/
A205_11 Citizens, interested people/
A205_12 Stakeholders from education and research/ Scientists
A205_13 Other (Stakeholder group/number)
A205_14 Other (Stakeholder group/number)
A205_15 Other (Stakeholder group/number)
1 = Not checked
2 = Checked
A205_01a Representatives of municipalities or administrative authorities/ (free text)
A205_02a Local politicians/ (free text)
A205_03a Policymakers/ (free text)
A205_04a Water managers/ (free text)

A205_05a Investors/ (free text)
A205_06a Stakeholders from recreation and tourism/ (free text)
A205_07a Project managers of municipalities/ (free text)
A205_08a Landscape and regional planners/managers/ (free text)
A205_09a Managers of protected areas or nature conservationists/ (free text)
A205_10a Land owners/ land users (farmers, foresters) or their representatives/ (free text)
A205_11a Citizens, interested people/ (free text)
A205_12a Stakeholders from education and research/ Scientists (free text)
A205_13a Other (Stakeholder group/number) (free text)
A205_14a Other (Stakeholder group/number) (free text)
A205_15a Other (Stakeholder group/number) (free text)
Text input

[A206] Selection
Q14b - Work load for SH?
"Please estimate the work load dedicated to stakeholder participation with respect to other study work packages?"

A206 Q14b - Work load for SH?
1 = Most of the time was dedicated to intensive stakeholder participation (approx. >75% of total work load related to it)
2 = Stakeholder participation was the central component within our study (approx. 50 to75% of total work load related to it)
3 = Stakeholder participation was an important component within our study (approx. 25 to 49% of total work load related to it)
4 = Stakeholder participation was a necessary component within our study (approx. 10 to 24% of total work load related to it)
5 = Stakeholder participation was of secondary/marginal importance (approx. less than 10% of total work load related to it)
-9 = Not answered

[A207] Selection
Q15 - ES Terminology?
"Did you explicitly use ES terminology and definitions during work with stakeholders?"

A207 Q15 - ES Terminology?
1 = (a) ES applied AND the term ES used AND concept explained to stakeholders
2 = (b) ES applied BUT the term ES was NOT used and concept was NOT explained to stakeholders
3 = (c) ES was not applied in the communication with stakeholders but researchers translated information from the stakeholders to ES If you used another term instead of ES, what was the term that was communicated with stakeholders (e.g. landschape services): [Please use the text field to provide further information]
-9 = Not answered
A207_03 (c) ES was not applied in the communication with stakeholders but researchers translated information from the stakeholders to ES If you used another term instead of ES, what was the term that was communicated with stakeholders (e.g. landschape services): [Please use the text field to provide further information]
Text input

[A208] Selection
Q15.1
"If you answered with (a) under 17., was the concept accepted and understood? "

A208 Q15.1
1 = Yes; Please describe shortly if there have been any obstacles, challenges or drawbacks?
2 = No; Please describe shortly which obstacles, challenges or drawbacks you faced?
3 = Not applicable
-9 = Not answered
A208_01 Yes; Please describe shortly if there have been any obstacles, challenges or drawbacks?
A208_02 No; Please describe shortly which obstacles, challenges or drawbacks you faced?
Text input

[A209] Selection
Q15.2
"If you answered with (b) under 17., why did you not fully communicate the ES concept?"

A209 Q15.2
1 =
2 = Not applicable
-9 = Not answered
A209_01 [No Description] 01
Text input

[A210] Selection
Q15.3
" If you answered with (c) under 17., what was your motivation to not apply ES terminology or using a concept..."

A210 Q15.3
1 =
2 = Not applicable
-9 = Not answered
A210_01 [No Description] 01
Text input

[A211] Multiple Choice
Q16 - Techniques/Methods
"Which techniques/methods did you use to involve stakeholders?"

A211_01 Workshop
A211_02 Questionnaire
A211_03 Focus group discussion/ Round table
A211_04 Interview
A211_05 Other
A211_06 Other
1 = Not checked
2 = Checked
A211_05a Other (free text)
A211_06a Other (free text)
Text input

[A214] Selection
Q16.1 - Technical support
"Did you use technical support such as planning/ software tools, touch tables etc.?"

A214 Q16.1 - Technical support
1 = Yes. Please specify kind of tool and purpose of its application:
2 = No.
-9 = Not answered
A214_01 Yes. Please specify kind of tool and purpose of its application
Text input

[A213] Selection
Q16.2 - Technical support
"Did these techniques enhance the process?"

A213 Q16.2 - Technical support
1 = Yes.
2 = No.
3 = Don't know.
4 = None/ Not applicable
-9 = Not answered

[A215] Selection
Q16.3
"Did you involve a social scientist or a mediator/communication facilitator to accompany the process? "

A215 Q16.3
1 = Yes.
2 = No.
-9 = Not answered

[A216] Selection
Q16.4
"Could he/she contribute to enhance the process and communication?"

A216 Q16.4
1 = Yes.
2 = No.
3 = Don't know.
4 = None/ Not applicable
-9 = Not answered

[A217] Multiple Choice

Q17 - Presentation of results
"How did you present results of your study to stakeholders/ participants?"

A217_01 Maps
A217_02 Charts
A217_03 Diagrams
A217_04 Texts
A217_05 Other
A217_06 Other
A217_07 Other
1 = Not checked
2 = Checked
A217_05a Other (free text)
A217_06a Other (free text)
A217_07a Other (free text)
Text input

[A218] Scale-like Selection
Q14a - Importance of SH consultation?
"Please evaluate the importance of stakeholder consultation within your study."

A218 Q14a - Importance of SH consultation?
1 = Stakeholder consultation was of major importance
2 = Stakeholder consultation was important
3 = Stakeholder consultation was of medium/ average importance
4 = Stakeholder consultation was less important
5 = Stakeholder consultation was not very important
-9 = Not answered

Section A3: Evaluation of the process of stakeholder integration

[A301] Multiple Choice
Q18 - Benefits of ES?
"In the process of stakeholder participation, which benefits/opportunities related to the ES approach did you..."

A301_06 None
A301_01 ES useful as common communication basis
A301_05 ES useful for developing integrated/interdisciplinary solutions (e.g. linking the biophysical and socioeconomic)
A301_10 ES allow a comprehensive impact assessments (to identify synergies/trade-offs)
A301_08 ES helps consensus finding between conflicting interests
A301_02 ES useful to raise awareness of ecological principles/ to increase societal relevance of conservation efforts
A301_09 ES helps to raise commitment
A301_07 ES helps to make more informed decisions
A301_04 Other
A301_03 Other
1 = Not checked
2 = Checked
A301_04a Other (free text)
A301_03a Other (free text)
Text input

[A302] Multiple Choice
Q19 - Challenges of ES?
"In the process of stakeholder participation, which have been overall challenges/drawbacks you faced related ..."

A302_09 None
A302_01 Methodical difficulties to assess and value ES/ Lack of data on ES
A302_05 Lacking relevancy of analysed ES with study targets or problem level
A302_02 Integration into existing planning instruments and concepts
A302_07 Other
A302_08 Other
A302_10 Other
A302_11 Other
1 = Not checked
2 = Checked
A302_07a Other (free text)
A302_08a Other (free text)
Text input

[A314] Scale (fully labeled)
Q18.1 - Benefits of ES?
"Given your personal experiences, please indicate your (dis)agreement with the following statements!"

A314_01 ES useful as common communication basis
A314_02 ES useful for developing integrated/interdisciplinary solutions
A314_07 ES allow a comprehensive impact assessments (to identify synergies/trade-offs)
A314_03 ES useful to raise awareness of ecological principles/ to increase societal relevance of conservation efforts
A314_04 ES helps to make more informed decisions
A314_05 ES supports the implementation of conservation/ sustainable management measures
A314_06 Benefits of applying the ES concept in planning practice strongly depend on the targeted decision making level
1 = strongly disagree
2 = disagree
3 = neither agree nor disagree
4 = agree
5 = strongly agree
-9 = Not answered

[A305] Multiple Choice
Q19.1 - Challenges/not ES?
"Are there other – not directly ES related – overall challenges that impacted stakeholder participation?"

A305_06 No
A305_01 Financial constraints
A305_05 Time constraints
A305_02 Identify motivated stakeholders
A305_09 Assemble all relevant stakeholders
A305_03 Achieve balanced composition of stakeholders
A305_04 Other
A305_07 Other
A305_08 Other
1 = Not checked
2 = Checked
A305_04a Other (free text)
A305_07a Other (free text)
A305_08a Other (free text)
Text input

[A318] Two-sided Slider
Q19.2 - Which issue more important?
"Please click on the scale and drag the emerging button to indicate which issue posed the bigger challenge on..."

A318_01 Stakeholder process was impacted more by ES related challenges/drawbacks/Stakeholder process was impacted more by other (not ES related) challenges/drawbacks
1 = Stakeholder process was impacted more by ES related challenges/drawbacks
11 = Stakeholder process was impacted more by other (not ES related) challenges/drawbacks
-9 = Not answered

[A304] Selection
Q21 - Motivation of SH?
"How did you perceive the willingness/motivation of stakeholders to participate?"

A304 Q21 - Motivation of SH?
1 = (a) Very good
5 = (b) Rather good
2 = (c) Rather poor
3 = (d) Poor
4 = (e) Depending on stakeholder group (if several were included)

6 = (f) Don't know
-9 = Not answered

[A309] Scale (fully labeled)
Q23 - SH Process vs. Results
"How would you evaluate RELEVANCY of participation process (e.g. discussion, consensus finding, dialogue betw..."

A309_01 Participation process
A309_02 Study results
1 = Low Relevancy
2 = Rather Low Relevancy
3 = Medium Relevancy
4 = Rather High Relevancy
5 = High Relevancy
-1 = Don’t know
-9 = Not answered

[A315] Multiple Choice
Q21.3 - Reasons?
"If you answered with (c), (d), or (e) under 29., what might have negatively impacted motivation or acceptanc..."

A315_01 Complexity
A315_02 Lacking tradition
A315_03 Lacking practical relevance, e.g. trough scale level of analysis
A315_04 Lacking practical relevance, e.g. trough level of detail/vagueness of analysed ES
A315_08 Lack of understanding what ES are
A315_09 Lack of understanding of the added value of addressing ES
A315_06 Other
A315_07 Other
1 = Not checked
2 = Checked
A315_06a Other (free text)
A315_07a Other (free text)
Text input

[A307] Free Mentions
Q21.2 - Reasons - not ES?

A307 Number of mentions
A307x01 Mention 1
A307x02 Mention 2
A307x03 Mention 3
A307x04 Mention 4
A307x05 Mention 5
A307x06 Mention 6
A307x07 Mention 7
A307x08 Mention 8
A307x09 Mention 9
A307x10 Mention 10
Text input

[A306] Free Mentions
Q21.1 - Reasons?
"If you answered with (a) or (b) under 29., what might have positively impacted motivation or acceptance of s..."

A306 Number of mentions
A306x01 Mention 1
A306x02 Mention 2
A306x03 Mention 3
A306x04 Mention 4
A306x05 Mention 5
A306x06 Mention 6
A306x07 Mention 7
A306x08 Mention 8
A306x09 Mention 9
A306x10 Mention 10
Text input

[A316] Multiple Choice
Q21.4 - Reasons - not ES?

A316_01 Communication techniques inappropriate
A316_02 Communication techniques too elaborate/detailed/complex questionnaire
A316_03 Communication techniques not elaborated enough
A316_05 External factors, e.g. time restrictions of stakeholders
A316_06 External factors, e.g. financial restrictions of stakeholders
A316_07 External factors, e.g. lacking identification with/appreciation of project/study objective
A316_08 External factors, e.g. lacking relevancy of project/study objective
A316_09 Other
A316_10 Other
1 = Not checked
2 = Checked
A316_09a Other (free text)
A316_10a Other (free text)
Text input

[A311] Scale (fully labeled)
Q25 - Neg. Impact of ES?
"Did the ES approach support the participation process or did it pose a negative impact on the process?"

A311_01 Impact of ES approach
1 = ++ Overall positive impact
2 = + Rather positive impact
3 = +/- Balanced (equal positive and negative impact)
4 = - Rather negative impact
5 = - - Overall negative impact
-1 = Don't know/ Not applicable
-9 = Not answered

[A308] Multiple Choice
Q22 - Improvements?
"Based on your experiences, can you name topics you would put (more) emphasis on to improve participation pro..."

A308_02 No, there is no potential for improvements.
A308_03 Yes, the following issues can be improved
1 = Not checked
2 = Checked
A308_03a Yes, the following issues can be improved (free text)
Text input

[A310] Scale (fully labeled)
Q24 - Process or resutls?
"From your point of view, what was more important for supporting decision making in the scope of your study:"

A310_01 (1) participation process, or (2) actual study results?
1 = (1) was much more important
2 = (1) was more important
3 = (1) and (2) were equally important
4 = (2) was more important
5 = (2) was much more important
-1 = don’t
know
-9 = Not answered

[A312] Text Input
Q26 - Reference?
"If you like - and if available - you can now provide a published reference study to which you referred."

A312_01 [01]

Text input

[A317] Text Input
Q27 - Name
"If you are available for additional inquiry from our side if need be, please enter your data. Otherwise clic..."

A317_01 First Name
A317_02 Last Name
A317_03 E-Mail
A317_04 Comments (e.g. on your general experiences with ES and stakeholder participation)
Text input

Permissions

All chapters in this book were first published in LO, by International Association for Landscape Ecology (IALE); hereby published with permission under the Creative Commons Attribution License or equivalent. Every chapter published in this book has been scrutinized by our experts. Their significance has been extensively debated. The topics covered herein carry significant findings which will fuel the growth of the discipline. They may even be implemented as practical applications or may be referred to as a beginning point for another development.

The contributors of this book come from diverse backgrounds, making this book a truly international effort. This book will bring forth new frontiers with its revolutionizing research information and detailed analysis of the nascent developments around the world.

We would like to thank all the contributing authors for lending their expertise to make the book truly unique. They have played a crucial role in the development of this book. Without their invaluable contributions this book wouldn't have been possible. They have made vital efforts to compile up to date information on the varied aspects of this subject to make this book a valuable addition to the collection of many professionals and students.

This book was conceptualized with the vision of imparting up-to-date information and advanced data in this field. To ensure the same, a matchless editorial board was set up. Every individual on the board went through rigorous rounds of assessment to prove their worth. After which they invested a large part of their time researching and compiling the most relevant data for our readers.

The editorial board has been involved in producing this book since its inception. They have spent rigorous hours researching and exploring the diverse topics which have resulted in the successful publishing of this book. They have passed on their knowledge of decades through this book. To expedite this challenging task, the publisher supported the team at every step. A small team of assistant editors was also appointed to further simplify the editing procedure and attain best results for the readers.

Apart from the editorial board, the designing team has also invested a significant amount of their time in understanding the subject and creating the most relevant covers. They scrutinized every image to scout for the most suitable representation of the subject and create an appropriate cover for the book.

The publishing team has been an ardent support to the editorial, designing and production team. Their endless efforts to recruit the best for this project, has resulted in the accomplishment of this book. They are a veteran in the field of academics and their pool of knowledge is as vast as their experience in printing. Their expertise and guidance has proved useful at every step. Their uncompromising quality standards have made this book an exceptional effort. Their encouragement from time to time has been an inspiration for everyone.

The publisher and the editorial board hope that this book will prove to be a valuable piece of knowledge for researchers, students, practitioners and scholars across the globe.

List of Contributors

Matthias Rosenberg
Leibniz Institute of Ecological Urban and Regional Development, Weberplatz 1, 01217 Dresden, Germany

Ralf-Uwe Syrbe
Leibniz Institute of Ecological Urban and Regional Development, Weberplatz 1, 01217 Dresden, Germany

Juliane Vowinckel
Leibniz Institute of Ecological Urban and Regional Development, Weberplatz 1, 01217 Dresden, Germany

Ulrich Walz
Leibniz Institute of Ecological Urban and Regional Development, Weberplatz 1, 01217 Dresden, Germany

Marion Kandziora
Kiel University, Institute for Natural Resource Conservation, Department of Ecosystem Management, Olshausenstr. 40; 24098 Kiel

Katja Dörnhöfer
Kiel University, Department of Geography, Remote Sensing & Environmental Modelling, Olshausenstr. 40; 24098 Kiel

Natascha Oppelt
Kiel University, Department of Geography, Remote Sensing & Environmental Modelling, Olshausenstr. 40; 24098 Kiel

Felix Müller
Kiel University, Institute for Natural Resource Conservation, Department of Ecosystem Management, Olshausenstr. 40; 24098 Kiel

Markus Leibenath
Leibniz Institute of Ecological Urban and Regional Development, Weberplatz 1; 01217 Dresden; Germany

Antje Otto
University of Potsdam, Institute of Earth and Environmental Science, Karl-Liebknecht-Str. 24-25, 14476 Potsdam-Golm

Benjamin Burkhard
Kiel University, Institute for Natural Resource Conservation, Department of Ecosystem Management, Olshausenenstr. 40 ; 24098 Kiel
Leibniz Centre for Agricultural Landscape Research ZALF, Eberswalder Straße 84, 15374 Müncheberg; Germany

Marion Kandziora
Kiel University, Institute for Natural Resource Conservation, Department of Ecosystem Management, Olshausenenstr. 40; 24098 Kiel

Ying Hou
Kiel University, Institute for Natural Resource Conservation, Department of Ecosystem Management, Olshausenenstr. 40; 24098 Kiel

Felix Müller
Kiel University, Institute for Natural Resource Conservation, Department of Ecosystem Management, Olshausenenstr. 40; 24098 Kiel

Franka Papendiek
Leibniz Centre for Agricultural Landscape Research, Eberswalder Straße 84, D-15374 Müncheberg

Hans-Peter Ende
Leibniz Centre for Agricultural Landscape Research, Eberswalder Straße 84, D-15374 Müncheberg

Uta Steinhardt
Department of Landscape Management and Nature Conservation, School of Sustainable Development Eberswalde (FH), Friedrich-Ebert-Straße 28, D-16225 Eberswalde

Hubert Wiggering
Leibniz Centre for Agricultural Landscape Research, Eberswalder Straße 84, D-15374 Müncheberg
Institute of Earth and Environmental Science, University of Potsdam, Karl-Liebknecht-Straße 24/25, D-14476 Potsdam

Friedrich-Karl Holtmeier
Institute of Landscape Ecology, University of Münster

Olaf Bastian
State Capital of Dresden, Environmental Authority, P.O. Box 12 00 20, D-01001 Dresden, Germany

Christian Stein
Leibniz Institute of Ecological Urban and Regional Development, Weberplatz 1, D-01217 Dresden, Germany

Gerd Lupp
Chair for Strategic Landscape Planning and Management, TU Munich, Emil-Ramann-Str. 6, 85354 Freising, Germany

Jan Behrens
Technische Universität Dresden, Chair of Tourism Economics and Management, D-01062 Dresden, Germany

Christina Renner
Alaunstr. 90, D-01099 Dresden, Germany

Karsten Grunewald
Leibniz Institute of Ecological Urban and Regional Development, Weberplatz 1, D-01217 Dresden, Germany

Ying Hou
Christian Albrechts University of Kiel, Institute for Natural Resource Conservation, Olshausenstr. 75, D 24118 Kiel, Germany
Beijing Normal University, College of Resource Science and Technology, Beijing, China 100875, Beijing, China

Felix Müller
Christian Albrechts University of Kiel, Institute for Natural Resource Conservation, Olshausenstr. 75, D 24118 Kiel, Germany

Bo Li
Beijing Normal University, College of Resource Science and Technology, Beijing, China 100875, Beijing, China

Franziska Kroll
Regierungspräsidium Darmstadt, Abteilung Arbeitsschutz und Umwelt Frankfurt, Gutleutstraße 114, D 60327 Frankfurt am Main, Germany

Sam Riffell
Department of Wildlife, Fisheries & Aquaculture, Mississippi State, MS, 39762 USA

Brian Keas
Department of Biological & Allied Health Sciences, Ohio Northern University, Ada, OH, 45810 USA

Thomas Burton
Zoology Department, Michigan State University, East Lansing, MI, 48824 USA

Holger Jäckle
University of Freiburg, Faculty of Biology, Department of Geobotany, Schänzlestraße 1; D - 79104 Freiburg; Germany

Michael Rudner
University of Freiburg, Faculty of Biology, Department of Geobotany, Schänzlestraße 1; D - 79104 Freiburg; Germany

Ulrich Deil
University of Freiburg, Faculty of Biology, Department of Geobotany, Schänzlestraße 1; D - 79104 Freiburg; Germany

L. Koschke
Dresden University of Technology, Department of Soil Science and Site Ecology, Pienner Str.19, 01737 Tharandt Lars

S. Van der Meulen
Deltares, Ecosystem Services research group, Postbus 85467, 3508AL Utrecht

S. Frank
Dresden University of Technology, Department of Soil Science and Site Ecology, Pienner Str.19, 01737 Tharandt Lars
University of Bonn, Center for Development Research, Department of Ecology and Natural Resources Management, Walter-Flex-Str. 3

A.Schneidergruber
University of Vienna, Department of Botany and Biodiversity Research, Rennweg 14, 1030 Vienna

M.Kruse
Kiel University, Institute for Natural Resource Conservation, Olshausenstr. 75, 24118 Kiel

C.Fürst
University of Bonn, Center for Development Research, Department of Ecology and Natural Resources Management, Walter-Flex-Str. 3

E.Neubert
Dresden University of Technology, Department of Soil Science and Site Ecology, Pienner Str.19, 01737 Tharandt Lars

B.Ohnesorge
Ecosystem Services Research Group, Berlin-Brandenburg Academy of Sciences. Jaegerstrasse 22/23, 10117 Berlin

C.Schröder
Eberswalde University for Sustainable Development, Department of Landscape Management and Nature Conservation, Schicklerstr. 5, 16225 Eberswalde

F. Müller
Kiel University, Institute for Natural Resource Conservation, Olshausenstr. 75, 24118 Kiel

O.Bastian
Leibniz Institute of Ecological Urban and Regional Development, Weberplatz 1, 01217 Dresden,

Erich Szerencsits
Agroscope Reckenholz-Tänikon Research Station ART, Agricultural Landscapes and Biodiversity, Reckenholzstrasse 191; CH-8046 Zürich